Environmental Solutions

Conceived, Solicited, and Edited by

Franklin J. Agardy
and
Nelson Leonard Nemerow

Environmental Solutions

Editors

Franklin J. Agardy
Nelson Leonard Nemerow

AMSTERDAM • BOSTON • HEIDEIBERG • LONDON
NEW YORK • OXFORD • PARIS • SAN DIEGO
SAN FRANCISCO • SINGAPORE • SYDNEY • TOKYO

Elsevier Academic Press
30 Corporate Drive, Suite 400, Burlington, MA 01803, USA
525 B Street, Suite 1900, San Diego, California 92101-4495, USA
84 Theobald's Road, London WC1X 8RR, UK

This book is printed on acid-free paper. ♾

Library of Congress Cataloging-in-Publication Data

Application submitted.

British Library Cataloguing in Publication Data
A catalogue record for this book is available from the British Library

ISBN-13: 978-0-12-088441-4
ISBN-10: 0-12-088441-0

For all information on all Elsevier Academic Press publications visit our Web site at www.books.elsevier.com

Printed in the United States of America

05 06 07 08 09 10 9 8 7 6 5 4 3 2 1

To all of you with the wisdom and courage to replace studying and reporting with assessing and solving environmental dilemmas

Contents

Introduction

During the 20th Century, we wrote and published countless books on the environment—beginning with "Theories and Practices of Industrial Waste Treatment" in 1963. Most, if not all, of these books centered on the theoretical aspects of environmental components. By the end of the century environmentalists worldwide possessed all the basic ingredients to enable them to ameliorate a rapidly deteriorating resource quality condition. Some real progress was made in improving environmental quality, but primarily theories were expounded and scientific papers were generated. And environmental quality did not improve sufficiently enough to overcome critical and dangerous situations. Toxic and hazardous wastes were identified and then discharged indiscriminately and rather secretly into the environment.

The time has arrived for us to write and propose to practicing professional environmental engineers potential solutions to vexing problems, instead of theories of their origin, characteristics, and potential treatment. You, as the reader, can select a particular area of interest and peruse the chapters' potential solutions; then select the appropriate one(s) which apply most directly to your situation. Economic comparisons of solutions make your selection even simpler.

For example, let us presume that you have a complex organic chemical waste that is potentially damaging to the environment. And, further, that you want to examine chapter 8 to pinpoint chemical solutions. Suppose you find in this chapter that you can safely discharge this waste after either chemically oxidizing it with chlorine or substituting a non-polluting chemical with the same production qualities. Then you may also compare these two solutions to ascertain the least costly one. You may also wish to compare potential solutions in other chapters which would also alleviate the problem such as the use of political solutions (Chapter 1) or forensic solutions (Chapter 2). You may then be in a better position to decide which solutions would serve you best on an overall basis. You may even continue to compare your selected solution with the idea of utilizing an environmentally-balanced industrial complex as described in chapter 11, or send the waste to a nearby industrial plant as described in chapter 12. These latter two solutions may be feasible and even less costly than your other selected solution.

In any event, you are encouraged to consider all solutions proposed in chapters which may contain potential answers to your problem. The authors of each chapter were aware of the possibility of your need to make such decisions when they wrote their chapters.

In Chapter 1, Bob Perciasepe recognizes from his many years of dealing with governmental agencies that politics enter into most environmental solutions. For example, when environmental contamination crosses established governmental boundaries, political pressures and influences will be brought to bear on any solution selected by the professional person. Also when money has to be spent to abate a pollution situation, the particular political party in control must weigh the expenditure against its political future. The challenge for the professional is to understand the political situation and to work with it to affect a suitable environmental solution. Political solutions to environmental problems are usually the result of the actions of thoughtful people in "contriving" a means of abating the degradation of any area of our environment. Political solutions are often considered "contrived" not for the benefit of humankind's environment, but for the good of a segment of the people and usually at the expense of most of the people. And behind it all lurks the ugly head of the money gained by this segment of the people. However, it is our intent to present *political solutions* in its favorable light—one of benefiting the majority of mankind. And it may not necessarily result in a net cost to this society to attain a "political solution in a favorable light."

In Chapter 2, Franklin Agardy gives the reader an opportunity to use forensic solutions to solve environmental dilemmas. The threat and use of legal remedies often effect more abatements than physical, chemical, or biological treatments. Unfortunately forensic solutions have been used mainly 'after the fact' of environmental contamination rather than for prevention. However, if threats of legal action can be brought to bear "a prioi" to avoid contamination, practical and economical solutions will result. In any event, the environmentalist should consider the use of this method for preventing environmental deterioration. When you compare other conventional solutions to forensic ones, the environmentalist can make decisions based upon economics.

Dr. James Mihelcic, a Professor of Civil and Environmental Engineering at Michigan Technological University, describes his graduate school and engineering service program with the United States Peace Corps as an excellent example of solving environmental problems. He also describes the education of pre-college teachers to develop the State of Michigan's middle school curriculum related to energy and pollution prevention. His chapter

(Chapter 3) gives guidance to using education to aid in the solution to environmental problems of all kinds.

In Part 2 we have segregated the scientific and technical solutions from the politics and policies. In Chapter 4, Ernest Lowe describes the combined crisis in resource supply and environmental pollution as a major economic development opportunity for both developing and developed countries. One objective of this chapter is to provide the reader with economically feasible strategies of reducing the depletion of resources and abating contamination of the environment. This author considers economic frameworks for encouraging cleaner, more efficient production, such as the Recycling Economy laws in Japan and Germany and the Circular Economy initiative in China. The Chinese initiative, for instance, integrates cleaner production and industrial ecology approaches with the goal of gaining a 7 to 10 factor improvement in resource utilization.

Even a Recycling or Circular Economy is only a partial economic solution. The deterioration of ecological systems—locally, nationally, and globally—demands economic solutions that will enable their restoration. When looking at these systems in terms of human welfare, ecological economists use the term "natural capital." Natural capital is the source of all natural resources and the sinks that absorb the by-products of human activities. The accounts of natural capital must be recharged through major investments in restoration of forests, grasslands, deserts, farms, watersheds, oceans, and atmospheric balance. The balance between human system unrecovered outputs and the sinks of natural capital must also be restored. The "restoration economy" must also recover the sunk investment in human habitat and infrastructure so as to extend its life. Here too the investments will create massive opportunity for venture development and job creation.

Dr. Tewari, in Chapter 5, offers you physical, or engineering, solutions. These are specific "concrete" types of systems designed to remove some or all of contaminants from the environment. He tries to point out which processes are most suited to eliminate certain of the multi-varied pollutants. Without being armed with these methods the reader will not be able to compare physical solutions to alternate techniques presented in the other chapters.

On the other hand, Dr. Oerther uses microorganisms of all types to solve existing as well as potential contaminant problems. Since most organic matter can serve as food sources for specific bacteria, the use of this technique to prevent environmental contamination is prevalent today. Even inorganic contaminants such as phospates and carbonates can be utilized by some flora—such as algae—to remove them from the environment. In Chapter 6 you will discover many of these biological processes which you may find as useful solutions to your situations.

In Chapter 7, Drs. Veziroglu, Sherif, and Barbir offer hydrogen as an alternate fuel source for moving vehicles and power plants. Products of the combustion of hydrogen not only give off energy, but also are completely non-polluting to the environment into which they are released. They present production cost comparisons of hydrogen with other forms of fuels. Also, the authors give the reader several methods of producing hydrogen fuel along with their relative utility. It is only a question of time when this fuel will replace conventional limited resources of oil and gas.

Dr. Patrick Sullivan describes in Chapter 8 the various chemical processes currently in use to ameliorate existing contaminants. In addition, he offers the use of chemical substitutions in certain situations as alternatives to toxic or hazardous chemical contaminants. An important contribution to this chapter is the use of chemical methods of detoxifying wastes prevalent and persistent in landfills and underground soils.

Dr. John Wilcox tackles solutions to contaminants by electrical and thermal means in Chapter 9. Of special interest are high temperature treatments such as closed system organic matter destruction at 600+degrees centigrade. Electrical systems are also offered as treatments for microorganism reduction/removal. Although relatively expensive capital and operating costs are usually involved, systems proposed in this chapter are often economically feasible for smaller volume, highly concentrated wastes. Alternative energy sources such as wind, solar, geothermal, and ocean wave are also offered by your editor, Dr. Nemerow as potential solutions in proper situations.

In Chapter 10, Drs. Kilbourne and Falk cover the various medical solutions available to avoid environmental diseases. Not only are the prevention of conventional and historic environmental diseases such as dysentery, typhoid and poliomyelitis discussed, but also solutions and systems are offered for the more recent and dangerous ones. Needless to say the prevention of and solution for biological methods of mass destruction are vitally important in today's world living. Illustrative of these is the transmission of arsenic powders through the postal systems, and prevention, detection, and medical solutions for the diseases resulting from them.

Your editor becomes an author in Chapter 11 and presents for industrial collaboration the manufacturing of products in environmentally-balanced industrial complexes. In these complexes, a group of industries located together reuses wastes from ancillary plants and produces useful products without any adverse environmental impacts. I illustrate, for the first time, with a real case scenario how the benefits of such a complex will result in lower production costs for the industries and, at the same time, environmental preservation.

Many potential planned and designed industrial complexes are suggested to stimulate the reader to plan for environmental solutions. In the following chapter complexes which are created from initially unplanned groupings of industries are described.

In Chapter 12, Erkman and Ramaswamy show how industrial ecology utilization can lead to solutions to the environmental contamination dilemma. Many industries have already been built and operate in so-called "industrial parks" or "industrial estates," or at least nearby one another. With proper cooperation these industries might avail themselves of the residual wastes from the other plant(s) to aid in manufacturing their products. Thus, a savings in money for raw materials results as well as a substitute for waste treatment.

These authors have already published several books related to industrial ecology and qualify as leaders in this area of environmental solutions.

In Chapter 13 we enter the international area of the book. Dr. Salah El Haggar proposes specific solutions especially applicable to rural, developing country environmental problems. These problems are unique to low-population areas of nations which are in the process of developing their regulatory systems. Problems of economics and lack of scientific and technical know-how are of uppermost importance in these country situations.

Dr. Balkau, having a "world" of experience with the United Nations, examines the possibilities open to the world's countries for solving environmental problems. Economic aid and technical assistance—so needed as mentioned in Chapter 13—are available through various international agencies listed in Chapter 14. Often the political support of an agency such as the World Health Organization is all that is needed to provide the necessary impetus for a country environmental pollution problem. The United Nations Industrial Development Organization (UNIDO) may also promote industrial collaboration such as proposed in Chapter 11.

The entire book of 14 chapters is summarized by our author and editor Dr. Agardy, in Chapter 15, so that you, the reader, are afforded a consolidated view of what you have read. You should obtain the feeling of how the many facets of environmental solutions are related and fit together to strive toward a state of zero pollution.

Nelson L. Nemerow

Biographies

Franklin J. Agardy, Ph.D. is President of Forensic Management Associates, a company focused on environmental litigation support and expert witness services. Dr. Agardy received a B.S. in civil engineering from the City College of New York in 1955, an M.S. in sanitary engineering from the University of California at Berkeley in 1958, and a Ph.D. degree in sanitary engineering from the University of California at Berkeley in 1963. He taught civil and sanitary engineering at San Jose State University and left the faculty as a tenured full professor in 1971. He spent 19 years with URS Corporation, retiring in 1988 as President of the corporation. After retirement he took the position of President/CEO/Chairman of In-Process Technology, Inc. in Sunnyvale, California, and at the same time formed Forensic Management Associates. Dr. Agardy is a Director of Komex Corporation, a Director of EGG Corporation, and holds an advisory seat on the Board of The Environmental Company. Dr. Agardy is a former member of the Dean's Advisory Council, School of Engineering, University of California, Berkeley, and currently is a guest lecturer at the university. Dr. Agardy has published over 50 articles and reports, has authored or co-authored, and co-edited four textbooks, and is a life member of a number of professional societies. During his career he has consulted to numerous federal, state, and local agencies on subjects ranging from environmental matters, nuclear weapons countermeasures, and international business development.

Fritz Balkau, Ph.D. is Head of the United Nations Environmental Programme's Production and Consumption Branch in the Division of Technology, Industry and Economics.

Dr. Balkau graduated as a research chemist from Monash University in Australia in 1973. After some years teaching at Victoria University, he worked for the Environment Protection Authority in various functions concerned with environmental planning, waste and pollution management, chemicals, and environmental policy. He spent one year with the Chemicals Division of the Organisation for Economic Co-operation and Development before joining UNEP's Industry and Environment Office in Paris in 1987.

The Production and Consumption Branch of UNEP promotes and facilitates the worldwide implementation of cleaner and safer production approaches and more systematic industrial pollution management in key industry sectors, including resource industries such as mining, oil and gas. It also leads activities in UNEP to promote more sustainable patterns of consumption in civil society and industry. The unit has an active program of information exchange, environmental education, and training support on environmental control systems and tools to help governments and industry to adopt more systematic approaches to environmental management (see http://www.uneptie.org/pc/home.htm).

Frano Barbir, Ph.D. is currently a Professor-in-Residence at the Connecticut Global Fuel Cell Center at the University of Connecticut. Prior to joining UConn in 2003, Dr. Barbir served as Director of Fuel Cell Technology and Chief Scientist at Proton Energy Systems in Wallingford, CT, and Vice President of Technology and Chief Scientist at Energy Partners, West Palm Beach. In these positions, he assembled and led research teams covering all areas of Proton Exchange Membrane fuel cell technology including electrochemistry, materials, heat transfer and fluid mechanics. He led teams that developed novel fuel cell stacks (from 1 to 5 kW) with emphasis on low cost manufacturing, and applied them in a variety of working fuel cell systems, including five fuel cell powered vehicles. He is co-inventor on several fuel cell-related patents, and has authored or co-authored more than 100 publications, mostly on fuel cells and hydrogen energy, that appear in scientific journals, books, encyclopedias, and conference proceedings. He is currently an Associate Editor of the *International Journal of Hydrogen Energy*, serving as Editor-in-Charge of that publication's special issues on fuel cells. He is also a member of the editorial boards of the *Journal of New Materials for Electrochemical Systems, EGE (Croatian Journal of Energy, Environment and Economics)*, and the *Fuel Cell Virtual Journal.*

Salah M. El Haggar, Ph.D., P.E. is the Professor of Energy and Environment at the American University in Cairo, Mechanical Engineering Department, Cairo, Egypt. Dr. El Haggar received a B.Sc. (1972) and an M.Sc. (1976) from Ain Shams University, Cairo, and a Ph.D. in Mechanical Engineering from Washington State University in 1983.

Dr. El Haggar has more than 30 years experience in energy and environmental consulting and university teaching. He has been a visiting professor at Washington State University and at University of Idaho. Dr. El Haggar has more than 18 academic honors, grants, and awards. He received the Outstanding Undergraduate Teacher Award at The American University in Cairo in 1995, as well as a number of outstanding AUC trustees awards. In addition, Dr. El Haggar has 105 scientific publications in environmental and energy fields, 29 invited presentations, 42 technical reports, and 10 books.

Dr. El-Haggar's environmental consulting experience includes more than 40 environmental/industrial audits for major industrial identities, 20 compliance action plans, and nine environmental impact assessments. In addition, he has extensive consulting experience in environmental engineering, environmental auditing, environmental impact assessment, environmental management systems, cleaner production, industrial ecology, energy management, hazardous and non-hazardous waste management, recycling, pollution prevention and waste minimization, zero pollution, biogas/solar/wind technology, community/desert development, solid and industrial waste, and environmental assessment for the local government and private industries. Dr. El-Haggar is a member/board member of 14 national and international societies in the area of mechanical engineering, environmental engineering, and community development.

Suren Erkman, Ph.D. has an academic background in philosophy and biology, and holds a Ph.D. in Environmental Sciences from the University of Technology of Troyes (France). In 1994, after working for a number of years as a science and business journalist for various media, he created an organization headquartered in Geneva (Switzerland), the Institute for Communication and Analysis of Science and Technology (ICAST). ICAST's mission is to provide independent information in a readily accessible form on scientific, technological and environmental issues, to companies, governments, international organizations, academic institutions and NGOs. In 1995, he launched an international network, Industrial Ecology Praxis, devoted to the dissemination and implementation of ideas relating to industrial ecology. He also teaches industrial ecology in various universities in Switzerland and abroad. Suren Erkman is a member of the Managing Board of the *Journal of Industrial Ecology* (MIT Press) and an elected member of the Council of the International Society for Industrial Ecology (ISIE).

Henry Falk, M.D., M.P.H. currently heads two organizations at the center of the Department of Health and Human Services' work in environmental health. He serves as Director for both the National Center for Environmental Health (NCEH) and the Agency for Toxic Substances and Disease Registry (ATSDR). In 2003, these two entities consolidated to form NCEH/ATSDR.

Dr. Falk arrived at the Centers for Disease Control and Prevention (CDC) in 1972. He is also a 30-year veteran of the U.S. Public Health Service Commissioned Corps. This service culminated with his being named rear admiral and an appointment as assistant U.S. Surgeon General.

At NCEH, Dr. Falk heads the Center's national effort to prevent or control environment-related diseases, illness, and deaths. He served NCEH for 14 years as Director of the Division of Environmental Hazards and Health Effects. At ATSDR, which was created by the 1980 Superfund legislation, Dr. Falk leads the federal agency whose mission is to protect public health from hazardous releases of toxic substances.

Dr. Falk earned his medical degree from the Albert Einstein College of Medicine in 1968. He received a master's degree from the Harvard School of Public Health in 1976, and he is board-certified in pediatrics and in public health and general preventive medicine. Throughout his career at the CDC, Dr. Falk has lent knowledge and leadership to myriad public health projects around the United States and the world. His work includes contributions to the federal responses to Three-Mile Island, Mount St. Helens, Hurricanes Hugo and Andrew, and the September 11th attacks.

Dr. Falk has also authored or coauthored more than 100 publications in a variety of subjects, including vinyl chloride-induced liver cancer, prevention of lead poisoning, and the health effects of environmental hazards.

During his career, Dr. Falk and his distinguished work have been recognized many times. His honors include the Vernon Houk Award for Leadership in Preventing Childhood Lead Poisoning and the Homer C. Calver Award from the American Public Health Association. He has also received the CDC's William C. Watson Jr. Medal of Excellence, as well as the Distinguished Service Award from the U.S. Public Health Service.

David R. Hokanson, Ph.D. is the Operations Manager of the Sustainable Futures Institute at Michigan Technological University. Dr. Hokanson is also an Adjunct Assistant Professor at Michigan Technological University, Department of Civil and Environmental Engineering. He worked for several years as a Research Engineer affiliated with the National Center for Clean Industrial and Treatment Technologies (CenCITT) at Michigan Technological University, a U.S. Environmental Protection Agency Center of Excellence and a partnership between Michigan Technological University (lead institution), the University of Minnesota-Twin Cities, and the University of Wisconsin-Madison. Through CenCITT and the Sustainable Futures Institute, Dr. Hokanson has experience working on several federally funded projects in the area of pollution prevention, water treatment, and sustainability, including educational and outreach applications at many levels. Dr. Hokanson was also instrumental in the development of the textbook *Water Treatment: Principles and Design*, 2nd ed. (2005).

Edwin M. Kilbourne, MD is Chief Medical Officer Division of Environmental Hazards & Health Effects, National Center for Environmental Health (NCEH), Centers for Disease Control & Prevention (CDC) in Atlanta, Georgia. In this capacity he deals with emerging issues in medical toxicology, including prevention and management of biological/chemical terrorism. He also promotes CDC's involvement with poison control centers in the service of public health.

Ernest Lowe is the Director of Indigo Development. He has played a central role in creating the concept of eco-industrial parks (EIP) as a model for local sustainable development. Current work also includes whole systems approaches to sustainable farming and building design. He is advisor to the Policy Research Center for Environment and Economy, State Environmental Protection Bureau, on EIPs and the Circular Economy.

In 2001–2002 Mr. Lowe served as industrial ecologist working on a year-long Asian Development Bank project promoting cleaner production in industrial estates in Sri Lanka. He was strategic planner for the Industrial Estates Authority of Thailand in 1999 in a project seeking to turn the authority's 28 estates into eco-industrial estates.

Mr. Lowe guided the Dalian Development Zone in an eco-planning process for a 430-km^2 economic and technological development zone in Northeast China and the planning of an eco-industrial park there. This process explicitly linked the eco-industrial park with fulfillment of the zone's obligation to model the Circular Economy mandated by Liaoning Province.

Mr. Lowe has served as a strategic planning consultant for eco-industrial park developments in Northern California, Puerto Rico, Texas, Maryland, West Virginia, Johannesburg, South Africa, Thailand, the Philippines, China, and Vietnam. He has led seminars and workshops on eco-industrial development in five Asian countries.

Mr. Lowe is author of the *Eco-Industrial Handbook for Asian Developing Countries* (2001), prepared for the Environment Department, Asian Development Bank (revised and updated from an earlier publication produced for the U.S. EPA.) This is available in a Chinese edition as *Industrial Ecology and Eco-Industrial Parks* (Beijing: Chemical Industries Press; see http://www.Indigodev.com/ADBHBdownloads.html). He is also co-author of *Discovering Industrial Ecology: An Executive Briefing and Sourcebook* (1997), prepared for the U.S. EPA Futures Group under a cooperative agreement with Pacific National Northwest Laboratory.

James R. Mihelcic, Ph.D. is a Professor of Civil and Environmental Engineering at Michigan Technological University and co-directs the Sustainable Futures Institute. He also directs the Master's International program in civil and environmental engineering that allows students to combine graduate school with engineering service in the U.S. Peace Corps. Additional information on both efforts can be obtained at www.cee.mtu.edu/peacecorps and www.sfi.mtu.ed.

Dr. Mihelcic has worked with several precollege teachers to develop a middle school curriculum related to energy and pollution prevention. He is currently an elected member of the Board of Directors of the Association of Environmental Engineering and Science Professors (AEESP) and is a past recipient of the AEESP-Wiley Interscience Award for Outstanding Contributions to Environmental Engineering & Science Education. He is also the lead author of the textbook *Fundamentals of Environmental Engineering* (1999), which has been translated into Spanish as *Fundamentos de Ingenieria Ambiental* (2001). Dr. Mihelcic has studied environmental policy as an AAAS-U.S. Environmental Protection Agency Environmental Fellow and is currently writing a book titled *Sustainable Development Engineering: Engineering Appropriate Solutions for the Developing World.*

Nelson L. Nemerow, Ph.D. is a retired professor of environmental engineering after over 50 years of teaching at various universities across the country. He is currently a consulting environmental engineer. He has published over 210 technical papers and 23 textbooks, primarily on the subjects of industrial waste treatment and stream pollution. His first textbook was published in 1963 on industrial waste treatment and his last book in 2004 on environmental engineering with Joseph Salvato and Franklin Agardy. He is also an editor of this book and a contributing author to Chapter 11. His major current field of special interest is the utilization of environmentally balanced industrial complexes to attain zero pollution from industrial plants.

Dr. Nemerow has taught in several universities providing his knowledge to many graduate students who are now actively engaged in a variety of important areas of the environment. As a consultant to industry and advisor to the United Nations Industrial Development Organization in Vienna and the World Health Organization, Dr. Nemerow has worked in many developing countries solving complex environmental problems. His research findings, largely on various industrial wastes, have been used to solve many significant waste problems.

Dr. Nemerow has been married to the same woman, a dedicated teacher of children and a talented artist, since he returned from WWII service with the Merchant Marines in 1948. He and his wife have 3 children and 4 grandchildren and reside in Encinitas, California. Dr. Nemerow and his children have given their working lives to the scientific betterment of humanity.

Daniel B. Oerther, Ph.D. earned a B.A. in biological sciences and a B.S. in environmental engineering at Northwestern University in 1995, an M.S. in environmental engineering, and a Ph.D. in environmental engineering, both from the University of Illinois, Urbana-Champaign, in 1998 and 2001, respectively. Since the autumn of 2000, Professor Oerther has been teaching and researching environmental biotechnology at the University of Cincinnati. In the classroom, Dr. Oerther employs problem-based-learning and encourages students to work in design teams. He offers undergraduate courses on Applications of Computers in Civil and Environmental Engineering as well as Introduction to Environmental Engineering. At the graduate level, Dr. Oerther offers courses in the Chemistry and Microbiology of Environmental Systems as well as Molecular Biology Methods for Environmental Engineers. His research program, funded by various sources, including the National Science Foundation and the U.S. Environmental Protection Agency, focuses on the identification and control of pathogenic microorganisms in drinking water and biofilm systems, as well as the encouragement of biocatalytic microorganisms in sewage treatment and bioremediation systems. As a licensed Professional Engineer in the State of Ohio, Professor Oerther performs consulting for local industry in Ohio as well as national clients including the City of New York. Among his awards and honors, Dr. Oerther is the recipient of the Engelbrecht Fellowship (1997) from the University of Illinois, the Distinguished Thesis Award (1998) and the Outstanding Educator Award (2004) from the Association of Environmental Engineering and Science Professors, the Young Engineer of the Year for the State of Ohio (2003) from the National Society of Professional Engineers, and an early CAREER Award (2003) from the National Science Foundation. Dr. Oerther's service to the environmental engineering community includes chairing the Research Symposium Committee for the Water Environment Federation Technical Exposition and Conference, serving as an associate editor for *Water Environment Research*, serving as treasurer of the Cincinnati chapter of Sigma Xi, and serving as President of the Cincinnati chapter of the National Society of Professional Engineers.

Bob Perciasepe is Chief Operating Officer for the National Audubon Society and responsible for overall operational management of the organization and public policy initiatives both at the national level and in coordination with Audubon state and chapter offices. He has 30 years of extensive experience in environmental and natural resources management, legislative and governmental affairs, and creative problem solving.

Mr. Perciasepe received a B.S. in natural resources from Cornell University and an M.S in planning from the Maxwell School of Syracuse University.

Mr. Perciasepe has been confirmed twice by the U.S. Senate, both times as an Assistant Administrator for the EPA (October 1993 and October 1998). First serving as Assistant Administrator for Water, and then Air and Radiation, he managed all aspects of air pollution control and water pollution control, and drinking water protection programs for the United States and coordinated with 50 state programs.

From 1990 to 1993, Mr. Perciasepe was Maryland's Secretary of Environment and directed all aspects of pollution control and environmental protection in Maryland. He served as the first Chairman of the Northeast Ozone Transport Commission and on the Board of the Chesapeake Bay

Trust. He also worked for Baltimore City as the Assistant Director of Planning.

Ramesh Ramaswamy is the convener of the Resource Optimization Initiative (ROI), an international effort to promote Industrial Ecology as a platform in the economic and environmental planning processes of developing countries. He consults internationally, as well as in India, and specializes in the application of industrial ecology in developing countries.

Mr. Ramaswamy has a basic degree in science and an MBA from the Indian Institute of Management, Ahmedabad. He also has a diploma (Advanced Training in Management) from RVB, Holland (now called the Maastricht School of Management). After 16 years of experience in senior management positions in organized business, in 1989 he started his own consulting company, Technology Exchange Network (TEN), in Bangalore, India. He has been associated with industrial ecology since the early 1990s and has directed many field studies in India.

S. A. Sherif, Ph.D. is Professor of Mechanical and Aerospace Engineering and is the Assistant Director of the Industrial Assessment Center and the Founding Director of the Wayne K. and Lyla L. Masur HVAC Laboratory at the University of Florida. He served on the faculties of the University of Florida (1991–present), University of Miami (1987–1991), and Northern Illinois University (1984–1987).

Dr. Sherif holds a Ph.D. degree from Iowa State University (1985), a B.Sc. degree (Honors, 1975) and an M.Sc. degree (1978) from Alexandria University, all in Mechanical Engineering. He is a Fellow of ASME, a Fellow of ASHRAE, an Associate Fellow of AIAA, and a member of Commission B1 on Thermodynamics and Transfer Processes of the International Institute of Refrigeration. He has edited or co-edited 25 bound volumes, and published 14 book chapters, 100 referenced journal papers, and 200 conference papers and technical reports. He is the recipient of the E.K. Campbell Award of Merit from ASHRAE in 1997 for "outstanding service and achievement in teaching" and a TIP Teaching Excellence Award from the University of Florida in 1998. In 2000, he was selected by the University of Florida's College of Arts and Sciences as an Anderson/CLAS Scholar Faculty Honoree for "Excellence in Teaching." In December 2002, Dr. Sherif was selected to be the sole recipient of the 2001 Kuwait Prize in Applied Sciences. He is also the 2003 recipient of the ASHRAE Distinguished Service Award.

Patrick Sullivan, Ph.D. is a Partner with Forensic Management Associates, a company focused on environmental litigation support and expert witness services. Dr. Sullivan received his B.S. in geochemistry from the University of California at Riverside in 1974 and a Ph.D. degree in soil chemistry from the University of California at Riverside in 1978. He was a senior environmental analyst at the Jet Propulsion Laboratory, California Institute of

Technology. He later taught Environmental and Soil Sciences at Ball State University and left the faculty as a tenured associate professor in 1985. He was Manager of Environmental Chemistry at Western Research Institute at the University of Wyoming for three years. Since 1988, he has been an environmental forensics expert. Dr. Sullivan has been involved with environmental research, teaching, and environmental consulting for approximately 30 years. During this period, he has managed or directed an extensive number of research and laboratory studies, field investigations and legal cases dealing with (1) the environmental chemistry of soil, water, solid and hazardous wastes, (2) waste and wastewater treatment, (3) soil and groundwater geochemical modeling, and (4) the development of analytical test methods and QA/QC. Dr. Sullivan has published 35 articles and reports and has authored two textbooks. He also holds a patent on the application of laser Raman Spectroscopy for *in situ* continuous monitoring of organic compounds in groundwater.

Ram Tewari, Ph.D. is the Director of Solid Waste Operations Department (SWOD) for Broward County, Florida. He is a graduate of the University of Miami and has a Ph.D. in environmental engineering.

T. Nejat Veziroglu, Ph.D. graduated from the Imperial College of Science and Technology, University of London, with degrees in mechanical engineering (A.C.G.I., B.Sc.), advanced studies in engineering (D.I.C.), and heat transfer (Ph.D.). In 1962, after serving several governmental agencies and working in the private sector, he joined the faculty of University of Miami Engineering. In 1965, he became the Director of Graduate Studies for Mechanical Engineering, and two years later initiated the first Ph.D. Program in the School of Engineering and Architecture. He served as Chairman of the Department of Mechanical Engineering (1971–1975) and was the Associate Dean for Research of the School of Engineering and Architecture (1975–1979). He has published some 300 scientific papers and reports, edited more than 170 volumes of books and proceedings, and is the Editor-in-Chief of the monthly scientific journals *International Journal of Hydrogen Energy* and *International Journal of Energy-Environment-Economics*. He has been an invited lecturer and/or consultant on energy research and education to the former Soviet Union, People's Republic of China, India, Pakistan, Japan, Canada, Germany, England, France, Italy, Switzerland, Turkey, Egypt, Saudi Arabia, Kuwait, the Philippines, Brazil, Argentina, Venezuela, and Colombia. He organized the first major conference on hydrogen energy: The Hydrogen Economy Miami Energy (THEME) Conference, Miami Beach, Florida, in March 1974. This conference considered hydrogen as a replacement for fossil fuels, as a clean, renewable fuel, and aroused wide interest throughout the world. Subsequently, several conferences and symposia have been organized by Dr. Veziroglu on energy- and environment-related subjects. Dr. Veziroglu is a member of 18 scientific organizations, has been elected to the grade of Fellow in the British Institution of Mechanical Engineers, the American Society of Mechanical Engineers, and the American Association for the Advancement of Science, and is the Founding President of the International Association for Hydrogen Energy. In recognition of his many research contributions, Dr. Veziroglu has been the recipient of several international awards. He was presented the Turkish Presidential Science Award in 1974, made an Honorary Professor at Xian Jiaotong University of China in 1981, awarded the I.V. Kurchatov Medal by the Kurchatov Institute of Atomic Energy of U.S.S.R. in 1982, and the Energy for Mankind Award by the Global Energy Society in 1986, and was a Noble Prize nominee in Economics for the year 2000.

John B. Wilcox, Ph.D. is Senior Scientist with Conestoga-Rovers & Associates. He has a doctorate in chemistry and over 20 years experience in the environmental field focusing on providing environmental permitting, compliance and engineering services for a wide range of manufacturing, service, and waste disposal industries. Dr. Wilcox is also a recognized expert in the field of thermal treatment technology, with special expertise in medical waste management and disposal. He has prepared thermal treatment-based remedial designs and managed air quality issues for numerous RCRA facilities and CERCLA sites. In addition to environmental consulting, Dr. Wilcox's direct industry experience includes applied research and development of thermal treatment systems and semiconductor fabrication equipment. He has authored numerous papers and patents in his fields.

Part I

Politics and Policies of Pollution

CHAPTER 1

The Political Environment

Bob Perciasepe

Introduction

Virtually all environmental solutions are implemented in a political context. While that statement can send chills down the back of many environmental experts and practitioners, the reality is that most decisions must further a set of goals or expectations derived from a political process. How the "solution" advances those goals and expectations and how the community or society in general perceive it, can result in some tough sledding or smooth sailing.

Understanding this reality and embracing the fact that politics are the tools that govern how a democratic society works out its conflicting interests, can improve the ability of practitioners to successfully solve environmental problems. Politics, to a large degree, created the modern environmental legal and regulatory framework within which most solutions reside. Politics can also be a significant barrier to new approaches and significant "paradigm shifts" as the broader community and their elected and appointed representatives work to sort out the conflicting values. Likewise the politics of necessity can be the lubricant for innovation.

Recognizing that there is often a political aspect to the solutions we seek to have implemented can actually be an empowering observation, because it means that solution advocates can be part of the discussion.

By now you recognize that the term "political" here is not referring to the partisan politics of elections but the practical politics of society dealing with multiple objectives. Should economic considerations be more important than clean-up goals? Should our objective be to protect the general population or some sensitive sub-population like children? Should a stream restoration end before it has been returned to some pristine condition? These are examples of policy or political considerations that are "value driven." If society can instruct the process, in a unified way, of where it wants to end up on any of those continuums, the technical work can be done to devise the solution. If on the other hand, a solution is put forward that is ahead of that normalizing process, it can be rejected before it has the chance to be explained and debated. Allocating public resources, mitigating involuntary risks, and preserving other societal demands are all subject to political debate as to what is acceptable.

These "policy continuums" are present in most environmental debates and solution designs. They overlay the technical and scientific aspects of our work. Where the community at large "frames" the solution and, how the solution fits into that frame is very important. In most policy debates, there are preconceived perspectives from different audiences in advance of any detailed or specific discussion. As a simplistic example, it would be easy to see an environmental group being skeptical of a solution being offered by an industry representative. Likewise, one could visualize an industry environmental practitioner expecting a proposal from a local environmental organization to be unworkable before listening to the concerns. While the real world experiences never follow this kind of simple calculus, these perceptions are real and understanding they exist can be very helpful in developing fact-based discussions about solutions.

We will discuss the nature of these *policy continuums* with some examples to better understand how they can influence environmental solutions.

The debates over policy can also happen in many different *forums*. There is no "official" place where you can go to have your political discussion about the environment. Some issues will be debated and discussed at forums as diverse as the United Nations (UN) to a local town council. Some are discussed in official meetings with agency experts and decision makers, while others may take place on the floors of the Congress in Washington, D.C.

Issues related to global warming or climate change are good examples of where debates can occur at the international level under the sponsorship of organizations like the UN or at the local town or city level. Trying to develop a global framework and greenhouse gas target will likely be debated at the appropriate international forums. Statements in favor of one approach or the other could come from local resolutions sent to the larger international forums as well as local contributions toward solutions (a town energy ordinance or a state-level renewable fuel portfolio are examples of more local policies). Here you have a large-scale environmental problem that can have solutions or actions at the international and local levels.

The debates over these different aspects of an environmental issue take place in many different *forums*. It is important to understand what forum you are in and what *format* the decisions are being made or codified. The basic formats include: government budgets, where funding can be allocated for a solution; laws passed at local, state, and national levels; regulations or guidelines that provide the road maps for implementing laws; and agreements between parties from interstate agreements to international treaties.

Part of the dynamics of politics includes understanding how policy implications can change over time. We like to think there is a solution for any environmental problem, and that is generally true. What changes over time can be society's values concerning how much risk is acceptable, our understanding of technology and our ability to continually innovate less costly solutions or our understanding of the environmental problem, and the nature of the impact. All of these factors play a role and influence the many different *interest groups* that engage in the political debate. Understanding the various interests in a debate is another essential element of understanding how environmental solutions can be influenced by politics. The risk acceptable to one group may not be acceptable to another; one group's perception or confidence in an innovative solution will differ from another's. These are all-important pieces of the political discussion and include public interest groups, groups with economic interests, groups substantially value driven, and groups directly impacted by actions or solutions.

We can summarize the elements of understanding and interacting with the politics of environmental solutions into four overarching categories:

The Political Forums
Local, state, national, international, and agency decision-making processes
The Format of the Debate
Budgets, laws, regulations or guidance, and agreements
The Interest Groups
Public, economic, values oriented, and those directly impacted
Policy/Political Continuums
Use of public resources, health risk, economic, environmental restoration, and science

These four elements are discussed in detail in the following sections of this chapter. It is important to remember that this chapter is an overview of politics and environmental solutions and as such should give the reader a basic understanding and appreciation of how it affects their work.

Political Forums

A strong argument can be made that the ultimate forum for politics is public opinion. That thought needs to be kept in mind throughout this discussion and many environmental solutions will have their day in that court. However, there will always be an institutional and geographic setting where public opinion is formed during the process of reconciling conflicting public and private interests. The location and design of a landfill or solid waste transfer station will be of keen interest at the local level; it may be of moderate interest at the state level and may have no public interest at the national level. While national standards for landfills can have strong interest at a national level by the interested parties, they may have very spotty interest at local levels.

Where public interest and therefore opinion can be influenced, is a fact that needs to be understood when working on environmental issues and solutions. We talk about this in tandem with the next sections on formats and interest groups, but the initial key to understanding how politics may impact the decisions around environmental solutions is to understand the forum.

So how do we go about the business of determining the political setting for the work we are doing? Like many things in the policy world there are no hard and fast rules, but there are certain helpful attributes.

The easiest attribute to examine is the geographic area of influence. The more site specific an environmental problem is the more local the solution will be. As you read through the later chapters of the book you will be going into great detail on physical, biological, chemical, and even economic solutions. Each of these can be very site specific. They can also have more far-reaching effects if they are first-time solutions or a new approach with limited prior exposure. In these instances a very local solution with all of the local impacts and interests at play can also gain state and even national attention. Thus, the site-specific nature of a solution or issue can be expanded geographically if it may be replicated and applied at other sites or locations. The key here is that the geographic effect plays a role in determining the political forum.

Working the other way, from large-scale application to local impact, can easily occur as well. Economic incentives and market-driven solutions can certainly have these attributes.

To be effective, economic solutions or incentives usually are applied at a state or national level. That will stimulate the political debate and interest groups at that level, while the application of the solutions might create concern at the local level, initially due to some uncertainty related to these approaches. For example, a pollutant trading effort will require a solution framework over a large geographic area; however, the exact impact on a local facility cannot be immediately known. Here an aggregation of local issues can become a national issue.

Understanding the geographic extent of the issue will then lead the practitioner to be able to identify the levels of government that can be expected to have an interest. The interest can be direct (e.g., issuing of permits, funding, or other action) or it can be indirect (e.g., representing the interests of local constituencies). A state-wide renewable energy standard for utilities can have wide support in the state capital, but some levels of concern locally if there is a sense that the requirement will create incentives for local impacts. Communities on ridgelines (or with views of ridgelines) could be concerned about the aesthetics of wind-driven turbines in their community, and you can expect the local governments and representatives of those communities to want a say in the outcome. Understanding the geographic nature of the issue or solution will provide a roadmap to the likely political forums. The political forums are the governmental entities with an interest and the media that routinely covers those areas. When working solutions, know your forums.

Format of the Debate

Environmental solutions by definition are designed to solve an environmental problem. The problem is one that you, as a professional or an entrepreneur, have been asked to address or that you simply see a better way to make progress on the issue. Under all circumstances the solution will need either approval or acceptance to be implemented. In many cases approval will come in the form of meeting existing regulatory requirements. In other cases it may require a more widespread economic or legal acceptance. In all cases public budgets, regulations, laws, or agreements will be involved.

In many respects what we are discussing here is the answer to: Who has to approve this? and What actions are necessary to implement the solution? Let's look at some examples.

Permits, agency approvals, allocations of budgeted funds, and site requirements are all typical examples of actions necessary to implement a solution already approved by law. There may be debates and proofs involved to determine if the proposed solution meets the regulation or existing law, and there can be many detailed technical discussions, reports, and analyses to make these determinations. The format here is "meeting the existing laws and the regulations designed to implement them." There might be a specific performance standard or numerical discharge or clean-up limit to meet. Here the goal is to demonstrate the ability of the solution to meet those established objectives. It is very important to note that any existing regulations or guidelines will likely have been determined by a past political process. Those political processes could be reignited if the solution proposed has never been tried before or if there is some degree of uncertainty around performance. Even if you are working on solutions within existing rules and regulations you can still have political implications.

Solutions or proposals not covered by existing law or regulation will require a more extended public process to enable implementation. Modifying or creating new regulations within existing law or modifying the law itself requires extensive administrative and legislative action. Executive branch-appointed and elected officials as well as legislative representatives will all become involved, and advocates for the proposal will need to begin discussing the ideas with them in advance of any action.

Knowing which elected officials will be involved, either because of the geography of their districts or the jurisdiction of certain committees and boards, is an important step. For executive branch administrative actions, knowing the agency personnel is essential. Getting information to all of these potential interested representatives will be crucial. Also critical will be understanding the time frames involved. Starting off with a difficult process with expectations of quick action can be a fatal mistake. That is not to say action can't happen quickly, but you should realize that action is often proportional to the political difficulty of the issue. The policy continuums discussed later can be helpful in determining the level of difficulty involved.

A key component of a strategy for working on a new regulation or law is determining who will propose it. Usually a key government official in the executive branch must be the proponent of a new regulation and an interested elected legislator, who sees this as a prudent political move, will offer to sponsor new law. This is an important element of implementing a solution that cannot be done under existing law or regulation. Without the appropriate sponsor, you will not get far.

It should be reiterated here that the step of getting regulations changed or modified and laws enacted to implement an environmental solution should be done only when existing laws and regulations just won't work or when they are not certain enough for the many interested parties in the solution.

In addition to working with existing or new laws and regulations, some environmental solutions can be implemented or may require some form of an agreement. Here solutions that go beyond the law or that are interstate or international in nature can benefit. A group of states could voluntarily agree to put into effect incentives for green house gas reductions. They could agree to put funding into some innovative technology ideas (such as natural gas or even hydrogen fueling stations). Interjurisdictional agreements are often used to try out new ideas or to implement solutions uniformly over a larger geographic area. Getting the appropriate levels of government to enter into these agreements is a political process with an expanded playing field. International agreements are an additional format discussed in more detail in Part III of this book.

Interest Groups

Up until now we have been reviewing, for the most part, the mechanical aspects of politics: Where will the impact be? Which part of government will be involved? Who will sponsor my piece of legislation? Now we will look at the forces that can come to bear on those forums and formats. Certainly, understanding where the issues and solutions you are working on will have an effect and the legal framework governing decisions are important. However, with that information in hand, moving to understand the interest groups involved or potentially involved will be the next critical step to engaging the politics.

Interest groups will be and are influenced by the mechanics of political policy making discussed above. A national public health interest group may not become engaged in a local sanitation issue, while a local open space advocacy group will likely pass on a national energy issue. While there is no doubt a strong relationship between the *forum and the format* and which interest groups become engaged or that you might convince to become engaged, the dynamics of public involvement are at the heart of politics.

From a broad perspective there are several categories of interest groups from an environmental policy and solutions standpoint. This is by no means a comprehensive summary of the many different interest groups that can evolve or engage around issues but gives a simple overview.

With environmental issues many general public interest groups can become involved. Groups engage in environmental issues across the board, from natural resources to technology to pollution. There are national level groups with some depth of expertise and local groups with little specific science or technical capacity.

The interests of these groups relate to good environmental management, and they position themselves as advocates for the environment. The restoration or preservation of natural resources and the reduction of health risks are higher priorities than the economic impact of the solution. In most cases it would be a mistake to assume that environmental groups have no interest in the economic impact, but it will vary greatly.

Often on environmental issues where solution cost can become high, there will be another set of interest groups that will engage on those issues. This could be local chambers of commerce or national trade associations representing affected businesses. There can also be economic interest groups advocating for a solution that will benefit certain classes of business. A local recycling business group could advocate for more recycling requirements and a national pollution control industry association could find itself advocating for more controls. Depending on circumstances, citizen groups can also be opposed to certain solutions viewed as too expensive. Like the environmental groups and their interest in economic impact, you would also make a mistake to assume that most groups concerned with the economics of environmental solutions as their primary point of interest do not have some concern about environmental issues.

There can be interest groups more driven by a particular issue or value. Energy impacts, open space, water pollution, air pollution, wilderness, birds and wildlife, sport fishing, or outdoor recreation—in virtually any area where environmental issues or solutions can be devised, there can be an interest group specifically focused on that area. In addition there are often many interest groups focused on any particular area and they may come at the issue from many different points of view.

Interest groups most directly impacted by any environmental solution are one of the most important types of groups. Here you can find a neighborhood association organized to prevent new development, or a local group created around cleaning up a lake or stream. A local group of businesses that could be displaced would be another example of a directly impacted group.

When working on environmental solutions it is essential to cultivate an understanding of who these different groups are and how they are interested in your project or proposal. It is recommended that a simple matrix be developed to lay out the forum, the format, and the interest groups as described in the preceding sections. All of these work together to paint a picture of where, how, and with whom the politics involved with your efforts will unfold. In every case there will be leaders to contact and work with; and the better the primary, secondary, and tertiary impacts of an environmental solution can be projected the more precise the identification of the forum or geography, the format or legal framework, and the interest groups involved will be. Identifying the leadership of the political jurisdiction, the legal processes, and the interest groups will give anyone working on environmental solutions the necessary information to begin to engage in the politics of their project or proposal. Get out and talk to people.

While this introductory chapter is not designed to cover the public relations aspect of the political arena, there are a few points that should be made. It is always better to get ahead of the public disclosure of information about your project. Once you start discussing with government officials and interest groups you should expect that information will find its way to the media. You should include the media as part of the process and as a significant interest in and of itself. The matrix analysis above will also help identify what media you need to begin discussions with.

Political Continuums

Having a good understanding of the setting and participants in the politics of any environmental solution is a significant undertaking. However, a key question remains to be addressed: What will the political debate be about and how can they be framed?

In order to get a firm sense of this it is very helpful to consider the concept of "policy continuums." You probably have heard the saying "you are either for it or against it"—that well describes the concept. "For it" is on one end of the continuum, and "against it" is at the other. While there will be some (and in some cases many) people and groups at one of those ends, the reality of much environmental work is that there is a scale or continuum between those two ends. Several of these continuums are relevant to most environmental policy debates and how those debates can be perceived and communicated. These are general categories and can be further broken down as needed or desired but they will give the reader a good sense of this concept.

The first example that demonstrates this idea of policy continuums is the use of public land. The public holds title to land generally for public purposes and those purposes can be as diverse as a football stadium or wilderness preservation. Since the land resource is a publicly owned asset, the public has an interest in how that land or asset is used. Remembering the discussions earlier, these lands can be owned by the federal, state, or local governments, thus the use issues can vary and the related interest groups can vary accordingly.

The use of land continuum is essentially straightforward: The land can be set aside as a preserve at one end of the scale, or at the other end, it can be intensely developed and used.

The debate over the Alaska's Arctic National Wildlife Refuge is linked to many other intense public debates but one of the essential components of the debate is the use of the land resource. Keep it wilderness or let it be developed. The national discussion has really taken place near one side of the continuum, to keep it wilderness or develop a small part of it. No one has discussed developing the entire refuge. However, the land that has been identified for development potential also happens to be one of the more sensitive areas of the refuge from a wildlife perspective. This debate will continue to play out for the next decade, one way or the other. Again, relating back to our earlier discussions, keeping an area as wilderness is in part a values debate versus a science debate. The interest groups attracted to the debate come to it with strong foundations in the value of wild public land. How the debate is framed sits on this continuum.

Health risk is another overarching continuum framing environmental policy debates. Here, in simple terms the scale has low risk at one end and higher risk at the other. The debate usually takes place along a "section" of the larger continuum of all the health risks we may be exposed to. Many sophisticated statistics-based tools have been developed to technically frame this debate. Risk assessment is one such tool that attempts to translate the facts on to some scale usually expressed in numbers of people exposed, or occurrence of a disease or mortality. This is usually normalized by a ratio to some number of overall population and time (e.g., 234 cases per year per 1,000,000-population). Here again, the values interest groups bring to the debate play a role on where they may be on this continuum. Especially embedded in this policy scale is the concept of voluntary risk versus involuntary risk. Often in communicating risk, comparisons are made to the risk of everyday activities like driving or playing some sport. The key is that during these discussions most people assign different values when it comes to risks they will assume themselves by choice versus those risks they may be exposed to that "someone else" decides for them. A key here when dealing with risk reduction solutions is the politics of how risks are debated in our democratic society will affect public acceptance.

Economic impact or cost is another significant element of political debate. In this arena, many technical tools have been developed to help frame the decisions and debates around environmental solutions. Cost-benefit ratios and cost-effectiveness analyses are two widely used methods in public debate. The essential continuum here runs on the scale of cost, lower cost in one direction and higher cost in the other. Like risk debates, the economic debate over a particular policy or solution will take place in one area of a larger continuum of all costs to the economy. Here the impacts that can be discussed include broader economic impact and direct economic impact to individuals. There is an interesting aspect to this part of many debates, that is, virtually everyone will want the solutions that cost less if they are all equally effective. This is one element of any political debate very hard to look at alone (hence the ratios on benefit and effectiveness).

Absent a set of solution options that are all equally effective to choose the least cost approach, the interests will begin to move to different locations on the continuum on the basis of their other attributes. Values-oriented groups will be willing to spend more; directly impacted groups will want a more costly solution if it reduces their impact; and those who will have to pay the costs will often move toward a lower cost spot on this scale.

There is another factor to consider in the economic cost debate related to the benefits and effectiveness ratios (which of course have there own continuums) and that is the weakness inherent in the calculation of benefits. Particularly in the environmental solutions arena, benefits can be very difficult to "monetize." What is the economic benefit of preserving a wilderness area? Many societal values come into play. In the discussion here, however, the key for a practitioner considering the politics of any situation concerning environmental policies and solutions, is that there may be a strenuous debate over benefits and where interests may stand on any of these continuums including the economic one.

Natural resource restoration debates provide another example of a class of policies and their associated politics residing on a continuum. The goals of local clean-up projects for streams or former waste sites, for example, may be expressed with the words "to restore." To someone thinking in terms of a continuum the immediate thought has to be how to define when this resource is restored. Any degraded resource (stream, lake, forest, Mississippi River, or local park) can be restored, but the debate will be around how far down the continuum to pristine or natural does the project go.

Most restoration projects eventually visualize a direction toward a more natural state or a state where natural processes and biological resources are improved. The scale along which this debate will occur is how far down that line are we able to or willing to go? Issues of cost and impact on other political issues, as well as technical capabilities, all come into play.

One way to construct a policy model for this continuum is to consider restoration as general movement toward a reference condition (another less degraded stream in the same region, for example) and accept that the solution may be composed of many smaller solutions that over time and when taken together produce that movement. Given the many components of the political debate and the many interests involved, an approach to manage the political and technical processes at the same time is called "adaptive management." Here recognition is included up front that the exact path is not precisely known, some things will be tried and perhaps discarded, we will learn while we "do," thus moving along the continuum. This model requires all interest groups to buy into the process.

This approach is also a way to deal with a level of uncertainty. There may be uncertainty in the full political forum on the definition of restoration and uncertainty about how to solve all the problems. Adaptive management in many ways accepts the uncertainty and strives to work with it and provides time and a vehicle to anticipate technological innovation. Another issue that faces this kind of environmental problem and the potential solutions associated with it is the ability to get agreement to move forward in the face of that uncertainty. An adaptive management—type approach can help overcome inertia and build the political support for moving forward. All the interests can continue to hold their key positions but agree to the direction of movement.

This issue of uncertainty is the final example of a continuum presented here to illustrate this approach of thinking along policy continuums. Science and its continuous development and incompleteness provide the basis for much of the uncertainty embedded in environmental issues and their solutions. The important factor here is the question: When along this continuum of scientific understanding does a decision maker have enough information to make a decision? In some cases, the law sets up a schedule for a decision and it must be made with the best information available at that time. In other cases, decisions, solutions, or even processes to develop solutions can be delayed by scientific uncertainty. In a public debate interests will line up in different locations on this scale. Those interest groups with a more precautionary view will be willing to act with less certainty. The consequence of inaction can push the political process along faster if it is large or wide spread. Some interest groups will tend to the end of the scale demanding more scientific certainty before any action is taken, usually because they will be more directly impacted by any decision. It is important to understand the basics of why different groups and interested will take different views on what action should be taken with the same level of science understanding.

Summary

This is not a comprehensive review of environmental politics. We hope that those reading this book to gain insight and advice on environmental solutions will also gain some appreciation of how the political process unfolds. As we said in the introduction and stands repeating here, politics are the tools that govern how a democratic society works out its conflicting interests. It is a reason to understand how politics can and will affect your work.

This chapter provides a framework to organize your analysis of the politics of most environmental issues and hence the solutions. By definition solutions are solving a problem; in most cases the reason there is a problem to solve is because a previous political process has helped society decide that the condition or situation is not acceptable. It also then stands to reason that the solution will need to stand the same political scrutiny.

Understanding the geographic forum, the legal format, the various interest groups, and where and why they stand on different places along a set of policy continuums, will enable anyone working on environmental solutions to begin to engage in the existing political process or to begin a new one.

Working with all the interests, developing common sets of understanding, and working to develop acceptance are necessary for long-term acceptance. Finding the leadership of the different interests, engaging them early, addressing the issues in a framework they will be responsive to and that deals with the issues that bring them to the table as an interested party are all necessary for success.

Review Questions

1. Do you think "politics" and their role in developing environmental solutions is a positive or negative factor? Why?
2. How does the geography associated with a particular environmental problem affect the politics of its solution?
3. Try to identify the interest groups that may be involved with an environmental solution you are working on or that you might work on?
4. What "policy continuums" would your effort be affected by? How would you determine the optimal location on the continuums?
5. Take an example of an environmental problem and identify the geographic forum, the legal format, the interest groups, and the three most important policy continuums the solution would affect?

CHAPTER 2

Forensic (Legal) Perspectives

Franklin J. Agardy

Preface

At the outset the reader should appreciate that this discussion of the role of forensics in the environmental process is predicated on the assumption that legal action has dictated the use of forensic experts in addressing environmental issues. So, despite the fact that much of this chapter is devoted to the history of litigation and law, it must be recognized and appreciated that the subject in question is seen through the eyes and experience of a forensic engineer/scientist and developed accordingly. Although I have chosen to begin "environmental history" by citing examples going back to the 13th century, with current environmental citations beginning with a court ruling in 1861, others who have addressed the development of environmental law (Sive 1995) begin their "history" with the passage of the National Environmental Policy Act (NEPA) in 1969. (One must credit David Sive, the author of the cited reference for including in his article the following observation: "In no other political and social movement has litigation played such an important and dominant role [as in the environmental movement].") Suffice it to say, regardless of where one proposes to begin, there is no doubt that litigation has played and continues to play a significant role in the environmental movement. Forensic experts follow in the footsteps of attorneys, furnishing the scientific and engineering underpinnings, without which the environmental movement would be severely limited. Forensic experts are highly prized participants in the snake pit of litigation.

In summary, where gaps exist or existed in environmental laws and regulations, litigation often has filled the voids. Of equal importance, litigation is often the only "forcing function" by which environmental regulations are met. The fact that laws exist does not, by itself, guarantee compliance. In the final analysis, litigation and the role of forensic experts has played and continues to play a major role in seeing to it that the best "environmental solutions" are used.

Introduction

Historically, before specific laws were passed addressing issues of environmental pollution, the courts were the vehicle for dealing with such problems. In fact it is not unreasonable to state that the beginning of the environmental movement, as such, started with citizens, communities, and industries bringing suit against their neighbors over the consequences of pollution. Although environmental insults were often addressed as "nuisances," the term "nuisance" had broad connotation, covering a variety of environmental problems often far more serious, despite being referred to as "nuisances." It is reasonable to conclude that in the early years, the courts were often the environmental regulators and the attorney's were the environmental advocates.

Early History

As far back as 1846, professionals in the manufactured gas industry advised their colleagues of the consequences of leaking tars, that is, they could enter the environment and work their way down to groundwater, thus damaging neighboring wells (Peckston 1841). In England and the United States, the courts were asked to address damage caused by the release of pollutants simply because appropriate and focused laws and regulations either did not exist, or if such laws were on the books they were either vague or unenforced. Recognizing that during the 19th century and certainly by the turn of the 20th century, pollution was most readily identifiable by sight, smell, taste, and fish kills, the parties most knowledgeable regarding "hidden pollution" (soluble chemicals) were the producers of pollution, such as the industries releasing them. Examples include the release of arsenic and cyanide, which are soluble chemicals and not easily identified by smell or taste. Both of these chemicals clearly could and did cause harm—and death when released to the environment. Other chemicals released during that period included residuals (from manufactured gas plants), such as spent lime, ammonia, benzene, toluol, toluene, paraffin, xylol, xylenes, phenol, creosote, naphtha, naphthalene, calcium sulphide, sulpho-cyanide, thiocyanate, mesitylene, and anthracene. Mining operations and smelters released such chemicals as sulfuric acid, arsenic, and cyanide, as well as heavy metals such as lead, cadmium, zinc, and copper. Pulp and paper mills discharged acids, sulfates and sulfides, dissolved and suspended solids, and in many cases discharged coffee-colored water (Nemerow and Agardy 1998). The textile industry was famous for discharging "waters of many colors." Lastly, many communities were plagued with the environmental damage associated with the release of untreated domestic wastes.

Pollution and Sources of Pollution

At this juncture, it is appropriate to develop a brief discussion of "pollution" and the sources of pollution. As with most developing communities and industries, the dependence on raw materials, a labor force, and modes of transportation were the cornerstones around which communities grew and industries were founded. Waterways such as rivers, lakes, and seaports initially were the most opportune locations. Water bodies, specifically surface waters such as rivers and lakes, as well as the atmosphere, served both as a resource and a waste depository. It wasn't long, however, before the discharge of wastes to the atmosphere and to the water resource created problems requiring a solution. Legal action often predated regulatory action as the most common approach to dealing with pollution problems. Legal action forced industry to begin treating their wastes or alternately to redirect residuals to their own property (a practice that would haunt industry in later years). Suffice it to say, virtually every operation, whether it be domestic or industrial, contributes residuals to the environment, and if not properly treated or disposed of, results in some manifestation of pollution or contamination of the surrounding environment—liquid wastes to bodies of water, solid wastes to the land, and gasses to the atmosphere—all are natures sinks.

The Litigation Environment

Environmental literature (including the large body of technical literature) includes many examples of just how the courts have dealt with issues of pollution and contamination (it should be noted that there are many, although often not dissimilar, definitions of what constitutes pollution and contamination). As will be shown, with the passage of time, and as courts increased their knowledge and understanding of the often complex issues of pollution, the ability to adequately address and rectify damage caused by pollution became much more focused. However, as the reader will see, the early litigation was often "right on the mark" despite lacking the knowledge or the level of sophistication common today.

Early Cases

One of the examples of early litigation includes the 1861 Pottstown lawsuit. This litigation, which resulted from an alleged contamination of Mr. Murphy's well by the neighboring gas works, is, in many ways, classic. The incident, as described, could well have taken place in 2005, with the same result. The Pottstown Gas Company commenced construction of the gas works in June 1856 and the plant was completed in September of that year. Mr. Murphy, an unhappy neighbor, owned a hotel adjacent to the newly established gas works and complained about the possibility of damages to his property even before the gas works began operation. Shortly thereafter (September 14, 1858) Murphy initiated an action against Pottstown Gas, laying the cause of action as a nuisance.

The gas company constructed a pit, in sandy soil, in which the main tank and gas holder were to be located. The soil was sandy and during the sinking of the pit for the tank, veins of water were discovered. Shortly after the tank was placed into the pit it was found to leak in several places. The site also contained an ammonia well into which the water from the gas washer was discharged. This well was lined with rough stone without cement, and with no designed outlet, the water simply soaked into the ground. The stage was now set for the inevitable consequence: pollution of Murphy's well.

The Court ruled on a number of significant issues, including the following:

(a) As to the issue of negligence, the Gas Company offered, "If the jury find that the defendants have not been guilty of negligence in the creation and in the carrying on of said works, the plaintiff cannot recover."

(b) The Gas Company also offered the following defense: "That the defendants were authorized by law to erect said works, and have the right to carry them on for the purpose of manufacturing gas for the public, and are not responsible in damages for the ordinary and usual smells that usually proceed from such works, nor are they liable to pay damages for injuring the plaintiff's water, unless done by their negligence."

(c) This issue of nuisance was defended as follows, "That in no sense can the gas-works be considered a nuisance, if conducted and carried on in the usual and customary way that similar works are conducted and carried on."

In response, the Court stated, "The question is not one of negligence or no negligence, but of nuisance or no nuisance. If the defendants have either so constructed, or carried on and conducted their works, or both, as to create an abiding nuisance to the particular injury of the plaintiff's property, they are liable in reasonable damages therefor, whether there was negligence or not [. . .]" The Court did say that, "A certain degree of offensive odour is unavoidably incident to the business, and must be endured by the public, or the business must stop."

As to the issue of Murphy's well, the Court ruled that, "A gas company is answerable for consequential damages, such as the corruption of the plaintiff's ground and well, by the fluids percolating from the works; and is not exempted, as a corporation authorized by statute to carry on the business of making gas and to purchase in fee simple the real estate necessary therefore."

The Court also stated that, "The law is not as contended for, that a gas company may pollute the air which man breathes, and the water which he uses, provided that it be done skillfully, and from no worse motives than selfishness. The offer to the public of a cheap, safe light, is no proper substitute for pure air and water."

Finally, in addressing the issue of Murphy's well, the Court went on to say, "Then, as to the corruption of the plaintiff's ground and well, by the fluids percolating from the defendant's works. This was disposed of in a similar way. But the defendants think that as a corporation, authorized by statute to carry on this business, and to purchase in fee simple such real estate as may be necessary for it, they are not answerable for such consequential damages as are complained of here. We cannot adopt this view. No such exemption is involved in the fact of incorporation, nor in the privilege of buying land. The principle they invoke applies only where an incorporation, clothed with a portion of the state's right of eminent domain, takes private property for public use on making proper compensation, and where such damages are not part of the compensation required."

The Pottstown case clearly dealt with issues that we still face today, and the Court responded in words that rang true then and ring true today.

Other state courts acted in similar fashion within this time frame. In 1861, the Columbus Gas Light and Coke Company was found to have "*adversely affected a well of water*" on the premises of the plaintiff by having deposited waste material from manufacturing operations. One year later (1862), The Ottawa Light and Coke Company contaminated a neighbor's well. The deposition of "*oily, tarry, resinous, gaseous and deleterious substances*" flowed from the gas works and permeated into the ground. In 1889, the Pensacola Gas Company was found to have "*rendered a neighbor's well foul and unwholesome*" because of onsite practices whereby impurities from the gas works percolated to the groundwater. In 1908, Ballantine & Sons brought suit against the Public Service Corporation of New Jersey because impurities from the adjacent gas works polluted the water in the Ballantine well. The water was "*rendered foul and unwholesome.*" However, this suit was not so easily settled and in 1914 this pollution issue was revisited. The pollution terminology now included the observation that "*tar and heavier than water sank down into and permeated surrounding soil and underground percolating waters,*" and ultimately carried into the neighbor's well.

Clearly, these issues of groundwater pollution as addressed by the courts, depended not only on the observation of contamination (by either, taste, odor, and/or testing) but on the ability of engineers to recognize and define the travel of pollutants through soil, into groundwater, and subsequently travel of contaminants within the groundwater regime.

Water Rights and Pollution

Issues of water rights and pollution have been addressed by many courts throughout the United States. An excellent paper published by the United States Geological Survey in 1905 (by Douglas W. Johnson) discusses numerous legal actions relevant to water rights related to underground waters. At the time of publication underground waters were classified as "underground waters of the first class" and "underground waters of the second class." The differentiation was based on the method of transmission underground: (a) in definite channels or (b) by general percolation. Generally, in the 1905 time frame, it was commonly accepted that one had the right to interfere with a neighbor's well by driving a well on his own property, irrespective of the negative impact on the neighbor. However, the courts did make exceptions when evidence warranted exceptions. Such was the case "where a corporation or similar body, having secured control of certain land, makes such use of that land as to damage the well of a neighboring landowner, the corporation being required to pay damages in such cases." A related case was cited where excavations by a railroad company resulted in damage to a neighbor's well. The ruling considered that the company did actually own the land but only acquired a special right in the land on the condition of paying all damages which the company might cause (Parker v. Boston & Maine R.R., 3 Cush.[Mass.], pp. 109, 114 [Douglas 1905]). When one considers groundwater or well water pollution, references are made to several cases, not unlike the Pottstown ruling. Douglas states, "The fact that a man has absolute right to the underground waters within his territory, and abstract those waters entirely, even to the point of draining his neighbor's land, does not give him the right to poison or foul those waters and allow them to pass into his neighbor's land in such condition. Such an act is illegal, and he who causes the damage is generally held liable even if he is not guilty of negligence, he whose filth it is being required to keep it on his own premises at his peril."

Douglas cites several other cases touching on the same subject. The first of these dealt with stored casks that leaked oil into the ground and subsequently polluted a neighbor's spring. The owners of the warehouse were held liable even though they had no knowledge of the leaking oil affecting the spring (Kinnard v. Standard Oil Co., 89 Ky., p. 468 [Douglas 1905]). In similar fashion, where sewage was permitted to flow into a neighbor's well, the party releasing the sewage was held liable for the damage: "The right to foul water is not the same as the right to get it. . . . The law of nuisance is not based exclusively on the rights of property" (Ballard v. Tomlinson, 29 Ch. Div., pp. 115, 126 [Douglas 1905]). The next cited case addresses both surface and ground water as follows: "The true cause of action . . . is not exactly that the defendant contaminated underground percolating water, but that he allowed his impure sewage to escape from his premises to the plaintiff's, and the circumstances that it reached there by underground percolation instead of by a surface stream is quite immaterial." (E.H. Bennett, 24 Am. Law Reg., pp. 638, 640 [Douglas 1905]). As with the Pottstown case, "when filthy water from a vault percolates through the ground and injures the cellar and well of a neighboring proprietor, the party who maintains the vault is liable for the damage" (Ball v. Nye, Mass., p. 582 [Douglas 1905]).

A Preemptive Strike

In the mid 1950s the Supreme Court of Michigan addressed a very interesting aspect of pollution of underground waters (*L. A. Darling Company, H.A. Douglas Manufacturing Company and Bronson Plating Company v. Water Resources Commission*). In 1950, the water resources commission ordered the installation of a suitable waste treatment and disposal system for two manufacturing plants. Up to this point it appears as a normal action. However, up to this point in time there had been no evidence of groundwater pollution. Because of the proximity of one of the city wells (located 840 ft. from infiltration ponds), the commission engineers felt that there was a very definite possibility that the city well would be contaminated. Because of the close proximity, the engineers felt that it would be too late to wait until the contamination reached the well. It was felt that once the well was contaminated it could never be corrected. Thus the call for preemptive action appeared appropriate. The appellants argued that tests had been made and that the groundwater was not flowing in the direction of the city well, and further, there was a solid stratum of clay that would prevent any seepage from the ponds. After hearing the testimony of witnesses for both parties, the court, in its written opinion, stated: "Once a fair hearing has been given by a State agency to which power to abate dangers to public health has been given, the proper findings made and other statutory requirements satisfied, *a court cannot intervene in the absence of a clear showing that the limits of due process have been overstepped or that the order was unreasonable, arbitrary, unlawful, capricious or confiscatory*" (italics added). An ounce of prevention is worth a pound of cure.

A Time of Confusion

A major difference in how litigation played a role in finding causes of pollution and the prevention of same, at least as the litigation applies to groundwater, has to do with the belief, at least as of 1905, that there were two classes of groundwater, referred to as first class and second class. As of 1905, no distinction was made between surface streams and underground waters of the first class. Further, there was no such clarity with respect to underground waters of the second class. Waters of the second class included those underground waters where there was a question as to the pattern of flow, whereas waters of the first class were described as being somewhat channelized such as in the case of surface streams. Today there is very little difference, if any, as regards litigation and laws affecting pollution of "waters of the state."

Historically (at least through 1940), confusion regarding legal action to restrain pollution was not uncommon, often resulting from "legal misunderstanding of ground-water hydrology and of legal proceedings that have been set up (and a very serious one) is illustrated in decisions which may be summarized as follows: Connecticut, New York, Michigan, and Kentucky" (Treadwell et al. 1940). Treadwell and his colleagues pointed out that when underground waters contaminated by wastes disposed in or on the ground with subsequent leaching to groundwater, relief from such contamination was not granted because of "the inability to prove either the intent to injure or negligence on the part of the discharger."

The Regulatory Environment: Water Pollution

The modern era of environmental laws and regulations in the United States can be considered with the inception of the Rivers and Harbors Appropriation Act of 1899. Congress passed this legislation which prohibited the discharge of refuse matter "of any kind or description whatever" into the navigable waters of the U.S. without a permit from the U.S. Army Corps of Engineers (Rivers and Harbors Appropriation Act, 1899).

Because of the difficulty in "policing" as well as much vocal resistance, the "Act" was rarely used after passage. However, a hundred years later, this Act has been employed (through very broad interpretation) as a regulatory tool. That not withstanding, pollution is as old as time and actions to address pollution were equally compelling.

Common Law

Olds (1952) defines the common law principle by referencing English common law and points out that the basic principle was that for every wrong there should be a remedy. He follows up by stating that courts rarely find such a simple solution, and, indeed, even equitable courts are loath to use the injunctive process except when obvious harm will result if no action is taken. In discussing contamination of underground waters, Olds cites three diverse aspects:

(a) Some courts have applied the English rule holding that a property owner has absolute right to make whatever use of his property he desires and is immune from the consequences such as damage caused to a neighbor, particularly if the injury occurs through the seepage of matter into subterranean waters. Simply stated, if you own the land you also own the water underneath and can do to it as you wish.

(b) A somewhat similar approach is based on the assumption that so long as negligence cannot be proven, there is no liability associated with the contamination of percolating waters.

Finally, Olds references the most favored interpretation, namely that:

(c) Each property owner is required to make only a reasonable use of his property and, if he makes an unreasonable use which results in injury to a neighboring property, he must respond in damages without the necessity of proving negligence. This is based on what is commonly referred to as correlative rights.

The Michigan Case

Olds (1952) discusses early legislation in the state of Michigan, referencing the Stream Control Commission. He points out that under the Stream Control Commission of 1929:

> It shall be unlawful for any person to discharge or permit to be discharged into any of the lakes, rivers, streams, or other waters of this state any waste or pollution of any kind that will tend to destroy fish life or be injurious to public health.

Olds also references Michigan law as it relates to underground waters:

> As of the effective date of these regulations any person who is introducing onto or into the ground any toxic, offensive, or objectionable wastes as herein defined which do or may pollute the underground water, shall, within ninety (90) days after such effective date or within such additional time as may be granted by the Commission, file a statement with the Commission, describing the existing method of disposal. Such statement shall include a chemical and physical description of the waste, its concentration, the volume of waste being disposed of and the method now in use for such disposal. Upon receipt of said statement the Commission will cause such disposal to be investigated. If the Commission finds that the underground waters are or may be polluted thereby it may disapprove the stated method. The Commission will establish such restrictions as it deems necessary to protect and conserve the underground water.

So it would appear that as of 1929, in Michigan at least, laws did exist so as to prevent and or curtail contamination of state waters, with specific attention directed at underground waters.

U.S. Laws: Water and Pollution

Because pollution of surface and ground waters was more closely related to local environments, states generally developed laws and regulations, as they saw fit, ahead of the federal government (Sullivan et al. 2001). Many states had established laws by the beginning of the 20th century. In fact, Massachusetts passed a "Special Act" in 1897 to deal with pollution of sources of water supply. Under the Act, the State Board of Health had the authority to examine all streams and ponds together with all springs, streams, and tributaries with regard to their purity with regulatory capability for the prevention of pollution. In similar fashion, New Jersey enacted legislation (*An Act to Secure the Purity of the Public Supplies in This State*) in 1899 in an attempt to protect waters of the state from being corrupted or impaired by discharges of "sewage, drainage, domestic of factory refuse, excremental or other pollution matter of any kind whatsoever." In 1905, the Pennsylvania legislature created the Department of Health and the Advisory Health Board (Stahl 1958) with powers "to make such reasonable rules and regulations, not contrary to law, as may be deemed by the Board necessary for the prevention of disease and for the protection of the lives and health of the people of the Commonwealth." Louisiana joined the parade in 1910 with the passage of legislation aimed at protecting rice planters by making it unlawful to pollute natural streams or drains from which water is taken for irrigation purposes. Wastes included oil, salt water, or other noxious or poisonous gases or substances which would render waters unfit for irrigation purposes or would destroy fish. In similar fashion, in 1911, Illinois formed a Rivers and Lakes Commission to see that the streams and lakes of the state are not polluted or defiled by the deposit or addition of any injurious substances. Not to be left out, in 1919 the Texas legislature enacted a statute giving the Texas Railroad Commission the authority to protect fresh water, whether above or below the surface, from discharges created by the gas and oil industry, "whether from drilling or plugging."

By the 1940s most states had recognized the serious nature and consequences of pollution and had passed laws attempting to directly regulate waste discharges to state waters. Many of these state laws were extrapolations and expansions of earlier state regulations such as those of Massachusetts, New Jersey, Illinois, Pennsylvania, and Texas.

State laws and regulations dealing with pollution control (Sullivan et al. 2001) include:

1897 Massachusetts Special Act
1899 New Jersey Act to Secure the Purity of Public Water Supplies of Potable State Waters
1910 Louisiana Act No. 183
1911 Illinois Rivers and Lakes Commission
1911 Illinois Sanitary Water Board
1925 Ohio House Bill No. 113
1929 Michigan Public Act No. 245
1937 Pennsylvania Clean Streams Law
1949 California Water Quality Control Act (the Dickey Act)
1957 Georgia Water Quality Control Act
1961 Texas State Water Pollution Control Act

Typically, the more recent laws expanded the definition of state waters, expressly including ground waters; expanded on the types and nature of pollutants, including a plethora of synthetic chemicals; and attempted to more closely "police" discharges through the permit process.

California's Dickey Act

Many environmentalists view the Water Quality Control Act of 1949 (commonly called the Dickey Act) in California as the first truly comprehensive legislation addressing the protection of water quality, including both surface and ground waters. The 1929 Michigan Act predated the Dickey Act by 20 years. With the benefit of a greater abundance of technical information, the Dickey Act was far more outspoken and detailed in addressing environmental issues. The Dickey Act had its origins in the formation of a Fact-Finding Committee on the Problems of Water Pollution, created in 1947. The committee report became the foundation of the Dickey Act (Pollution Control Legislation, California 1949). The Act created the California State Water Pollution Control Board; defined waters of the state to include both surface waters and ground waters; further defined pollution in terms of "impairment of water quality"; and defined other wastes (other than sewage) as "any and all liquid or solid waste substance, not sewage, from any producing, manufacturing or processing operation of whatever nature." The Act required reporting and filing (with the regional board of that region) of any proposed discharge (similar to the reporting requirements in the U.S. Rivers and Harbors Appropriation Act of 1899), and, after necessary hearings and review, receiving a discharge permit. The regional board had the authority to prescribe requirements as to what could be discharged (disposed of) or by describing the discharge in terms of the condition to be maintained in the receiving waters—or by a combination of both approaches (commonly referred to as discharge standards and/or receiving water standards). Of significance in the Act was the definition of the term "disposal area." Since the disposal area could address indirect discharges into the ground, the board had the authority to impose requirements for the indirect discharge of wastes, including industrial wastes.

The Clean Water Act and the NPDES Permit Program

Although the Clean Water Act was passed in 1972 (during the Nixon administration), implementation suffered primarily due to costs and "foot dragging." Once again the threat of legal action became the forcing function. As an example, Friends of the Earth (FOE) and others brought suit against Powell Duffryn Terminals (PD) for multiple violations of its NPDES permit for both conventional and toxic pollutant discharges (*Public Interest Research Group of New Jersey, Inc. v. PD*). In order to appreciate why FOE had to bring suit a little background discussion is necessary. In the mid 1970s, the EPA brought and subsequently settled an enforcement action in Federal District Court which required PD to construct a new treatment plant by no later than 1977. By the mid 1980s the new plant was still "on the drawing boards" and the EPA had neither enforced nor modified the consent decree. This is when FOE brought suit under the citizen suit provisions of the Clean Water Act (one must admire the wisdom of Congress in including within the Clean Water Act the citizen suit provisions—possibly anticipating that the government, in this case the EPA, would not necessarily strictly enforce compliance). Subsequently, the District Court ordered PD to construct the treatment plant, comply with its NPDES permit, and pay a fine of $3.1 million (which actually was $1 million less than the statuary maximum fine). The reason behind the reduced fine was the failure of the EPA to enforce its previous consent decree. As with most cases, PD appealed. The Circuit Court upheld all trial court rulings, but increased the fine to the statuary maximum of $4.1 million. These actions resulted in PD upgrading its wastewater and stormwater treatment at all of its East Coast facilities.

Another such action was brought by the America Canoe Association (ACA) against Murphy Farms (confined hog feeding operations in Rose Hill, North Carolina) (*ACA v. Murphy Farms, Inc.*). The hog feeding operations were discharging swine wastes to the waters of the United States without an National Pollutant Discharge Elimination System (NPDES) permit. Subsequently, the United States joined the ACA in the suit. Again, the court (4th Circuit Court) ruled in favor of the ACA/U.S. ruling that an NPDES permit was required. As a result of the action, all confined hog feeding operations in North Carolina (and effectively nationwide) became subject to NPDES permits, including significant best management practice (BMP) and record keeping requirements. Interestingly one would not normally think of the American Canoe Association as an "intervener" in environmental litigation. However, the Clean Water Act has received much scrutiny and has been and continues to be a powerful tool in the hands of "concerned citizens."

As recently as 1999, the Santa Monica BayKeeper, along with others, brought a citizens' suit against the City of Los Angeles for illegal sanitary sewer overflows (*Los Angeles Region v. City of Los Angeles*). In 2001, the citizens were joined in the suit by the United States as well as the State of California. The suit was settled in 2004, resulting in a commitment by Los Angeles to improve its operation and maintenance program, including sewer cleaning, root removal, grease control, and flow monitoring. The city also committed to over $1.5 billion in capacity enhancements, rehabilitation, and/or replacement of aging sewers. True to its word, the city began implementing portions of the final settlement during negotiations and reduced its sanitary sewer overflow rate by almost one-third between 1999 and 2004.

It should be obvious by now that these suits, initially brought by organizations under the citizen suit provisions, required the services of environmental professionals acting as forensic experts to evaluate the problem, present the evidence, and make the technical arguments in support the suits.

The Regulatory Environment: Air Pollution

Problems related to the release of vapors and gasses from the burning of fuel go back as far as recorded history. If one were to footnote the first "environmental insults" recognized and addressed, one is likely to focus on smoke and the obnoxious effects of smoke.

In 1306 a royal proclamation during the reign of Edward I prohibited the use of "sea coales" by artificers in their furnaces and set up a commission of inquiry to seek out violators and punish them "with great fines and ransomes." In Zwickau, Germany, 40 years later, metal plants were denied the use of coal as fuel through whose smoke "die Luft verpestet werde" [contaminates the air]. In the middle of the 16th century Queen Elizabeth, "greatly grieved and annoyed by the smoke of sea coales," issued a proclamation forbidding the burning of coal in London while Parliament was in session (Swain 1949). It is evident from these quotes that the issue of pollution was well recognized and aggressively addressed by those suffering the consequences—especially if those suffering were in power. One can only imagine what some of the punishments might have been if the polluter was unable to pay "great fines and ransomes." Therefore, it should come as no surprise that the first regulation on smoke appeared in England in 1343.

Legal Aspects of Air Pollution (Davenport and Morgis 1954)

Although the early efforts to curtail air pollution focused on smoke, every manner of contaminants has been identified with the release of vapors and gasses from industrial processes. Having said that, "smoke nuisance" was the driving force behind both litigation and regulation in the early years of air pollution control.

A brief history of regulations attempting to control smoke include the Public Health Act of 1891 (London, England) (Anonymous 1899), which attempted to classify sources of smoke and requirements to control releases. Sources included fireplaces and furnaces, which might be employed in mills, factories, dyehouses, breweries, bakehouses, gasworks, or "any manufacturing or trade process whatsoever." The law required that any release of black smoke in such quantity as to be a nuisance, shall be deemed to be a nuisance liable to be dealt with summarily.

The smoke issue was just as much a concern in the United States as it was in England. In the early years of manufacturing in the U.S., smelting operations were often the target of law suits over the huge amounts of pollutants spewing out of smelter stacks. Although there were numerous incidents of damage caused by smelter emissions (such as sulfur fumes, arsenic, and lead), because of the importance of the industry and the large number of people employed, the concept of the *greatest good for the greatest number* often held sway in legal determinations addressing pollution (Anonymous 1907). Often, there were two results from legal action: the first being damages assessed by the court; and the second being an injunction to restrain, which, in effect, could cause the closing of the facility. It was this second action that addressed the benefit/loss question. In most cases, the economic benefit outweighed the pollution issue—at least in the early years of litigation and laws. For example, in 1906, farmers in the Deer Lodge Valley area of Montana brought suit against the adjoining smelter in the Montana Federal Court claiming damage to their farm products from the release of sulfur fumes (Anonymous 1908). Even President Theodore Roosevelt was considering federal action to enjoin operation of the Washoe smelter at Anaconda because of damage to forest reserves in the area. However, the fact that some 100,000 persons in the Butte district depended on the economic viability of the local mining industry caused considerable soul searching. And any action relative to the smelting operation was effectively "put on hold" when the judge refused to issue a temporary injunction—the judge's decision was based on the principle of the greatest good for the greatest number. He further concluded that the smelting operation was of greater importance than agriculture and dissolved the farmers' injunction on the grounds that the smelting company had exhausted all means known to modern technical skill to control the enormous volume of fume released. Notwithstanding, in 1910, the United States brought suit against the Anaconda Copper Mining Company for an injunction to close the plant (Anonymous 1910).

Blast furnaces came under scrutiny in 1904 when the Pennsylvania Supreme Court reversed a lower court ruling and perpetually enjoined Jones & Laughlin from such operation of its blast furnaces, which had caused dust to be distributed upon adjoining property (Anonymous 1904). Needless to say, the legal action forced blast furnace operators to once again turn to technology improvements to get around the problem. Dust control became the answer (for that period in time) to controlling emissions, thus allowing operations to continue. Once again, the bringing of legal action forced operators to seek and employ advanced technical solutions to alleviate pollution problems.

In 1906 several smelters in the Salt Lake Valley were enjoined from operating. This resulted in drawn out litigation (Ebaugh 1910). The court decreed that in order for the smelters to continue in operation, the smelters had to treat ores, to a level equal to those which in their natural state contained less than 10% of sulphur. In an attempt to circumvent the court order, recognizing that it was a "commercial impossibility" to directly comply with the court, the smelter owners proposed to mix ores in a way that the average sulphur content was in compliance. The court was not swayed by the "dilution" argument and the petition was denied. Two other smelters were able to beat the court's requirements by employing modified technologies, but even in these two cases, *the injunction was suspended, but not removed.*

In 1907 the United States Supreme Court, acting at the request of the State of Georgia, granted permission for the state to issue a permanent injunction restraining the Tennessee Copper Company and the Ducktown Copper, Sulphur & Iron Company from operating their plants. The cause of action was based on fume damage that brought about the denuding adjacent wooded areas (Fulton 1915). Again, technology came to the rescue in that by constructing sulphuric acid plants the companies were able to treat the smelter smoke and in so doing, gain another commercial product. In Shasta County, California, the Mountain Copper Company was closed down as a result of legal action brought by the Shasta County Farmer's Protective Association (Fulton 1915). Similar actions were brought against the Mammoth Copper Company and the Balaklala Consolidated Copper Company. As a result of these actions, bag houses and diluting fans were installed at the Mammoth Copper Company, while an electrical precipitation process coupled with diluting fans were installed at the Balaklala Company. The use of these technologies allowed the operations to continue. Not to be left out, in the San Francisco Bay area, claims of damage to farms caused by releases from local smelting and acid plants led to frequent suites in local and state courts (Fulton 1915). Companies such as the Selby Smelting & Lead Company, the General Chemical Company, and the Mountain Copper Company were targets of law suits. Solutions to the pollution problem included such actions as curtailment of operations during the months of March through October (Selby Smelting & Lead), installation of bag houses, use of electrical precipitation, and diversion of high sulphur to adjacent plants for the manufacture of sulfuric acid.

By the 1940s most industrial communities in the United States had some form of legal regulation relative to the discharge of pollutants to the atmosphere (Hartman 1949). These regulations addressed such discharges as dense smoke, fly ash, metals, and noxious gases (e.g., sulfur dioxide). These regulations implemented by inspection and sampling (never enough inspectors), included the issuance of operating permits for existing equipment, permits to build or install new equipment and control of the kind and quantity of fuel used. A major drawback to this regulatory approach was that in most cases the final inspection consisted of "visual observation" rather than actual stack sampling. Thus, the litigation route was often the only way to actually address specific pollution releases. Law suits allowed engineers and scientists to actually examine and measure pollutant releases on a case by case basis. Again, the combination of regulations with litigation brought about the desired abatement of releases.

More often than not, states set up air districts to address air pollution issues. For example, the California Air Pollution Control Act of 1947 (Kennedy 1952) created countywide Air Pollution Control Districts, which had the authority to "permit" facilities. However, as current as this statewide regulation was, the standard for regulation was based on measurement of "visible smoke" using the Ringlemann chart as the standard. The regulations also addressed "invisible fumes" such as sulfur compounds. Standards were also established for solid products of combustion, particulate matter and dusts.

In summary, even today most industries meet their air pollution requirements by filing annual reports that show their mass load discharge (for each pollutant listed in their permit) has not exceeded the permitted amount(s). This author has had extensive experience with the introduction of advanced state of the art air pollution technology at a number of plants in the United States. When actual field measurements were carried out to determine the "real" amounts of chemicals released in stack discharges, the amounts released were often considerably higher than the estimated amounts reported under existing discharge permits. Nothing is more revealing as regards chemical releases than to design a piece of air pollution control equipment, based on estimated mass discharges, only to find that the equipment "blows up" when the actual chemical concentrations are introduced. One has to experience this event only once. From that point forward, estimated chemical discharges always become subject to "direct measurement certification."

Summary

There is little doubt that as people recognized the consequences of pollution they turned first to the courts and then to their legislators. Although in the early years most pollution events were classified or referred to as nuisances, this term, as stated earlier, was not taken literally but represented a broad list of complaints against polluters. Certainly damages and compensation for damages was well recognized within the context of the term "nuisance."

The overall positive effect that litigation has had on improving the environment cannot be disputed. Laws and regulations often followed in the footsteps of litigation and where laws and regulations preceded litigation, often enforcement only came about as a result of litigation.

I believe that the cases given in this chapter, coupled with the references cited, give truth to the comments made in the Introduction. The goal of a pollution free environment can only be approached by the implementation of technology across the board. However, the "guarantee" that available state-of-the-art technology is used and that further technological advances are made often requires a "forcing function" and the forcing function is often litigation, laws, and enforcement, and still more litigation. However, as stated earlier, one cannot "simply regulate advanced technology." Without engineers and scientists developing technological advancements, litigation and laws alone cannot achieve pollution-free goals.

If there is a lesson to be learned from this history it is one of involvement. Clearly, pollution will always manifest itself in some form or another. This is simple reality. However, an ever-alert professional community should, in this writer's opinion, take the lead in recognizing potential environmental problems (hopefully before they manifest themselves) and deal with them within the spectrum of available technology. Too often, certainly true from a historical prospective, lawyers stepped to the plate and took the initiative. The filing of law suits and the passage of legislation to address environmental concerns could not take place without the knowledge extant within the scientific and engineering community. Having said that, it is unfortunate that engineers, with technical solutions in hand, are often placed in the position of having to delay implementation until either litigation or legislation forces implementation. A major excuse preventing the implementation of available advanced technology is "lack of funding." However, when the litigation is over or when laws are enacted, the "money issue" somehow becomes resolved.

A final comment concerns the relationship between law, litigation, and technology. Let it be understood that what technology can accomplish should translate to practice. However, litigation and law cannot accomplish what technology cannot provide, therein lies the rub. Lawyers can sue and laws can be passed but without the available technology the mission cannot be readily accomplished. In any case, hindsight always being 20/20, whereas at one time "the solution to pollution was dilution," perhaps we should have focused on "the solution to pollution is litigation."

Review Questions

1. If, in fact, there is such an extensive list of laws and regulations dealing with pollution, why is it that we still suffer pollution problems today?
2. Since engineering technology exists whereby virtually all discharges of residuals (termed pollution) can be treated (almost to total removal), why is it that these technologies are not universally employed?
3. What prevents full, comprehensive, and rapid enforcement of existing pollution control laws?
4. Just how important is the role of forensic scientists and engineers in litigation, legislation, and enforcement to abate pollution?
5. The "chicken and egg" issue to ponder—what has played the greater role in pollution control, litigation, or legislation?

References

American Canoe Association v. Murphy Farms, Inc., 326 F.3d 505, 509 (4th Cir. 2003).

Anonymous. 1899. "The Smoke Nuisance and its Legal Aspect." *Engineering and Mining Journal* March 10, 67:302–303.

Anonymous. 1904. "Blast Furnaces as a Nuissance." *The Engineering and Mining Journal* LXXVII, January–June.

Anonymous. 1907. "Notes on Smoke Suits." *Mining and Scientific Press* 115:90–91.

Anonymous. 1908. *Engineering and Mining Journal* Butte, 86:1224.

Anonymous. 1910. *Engineering and Mining Journal* 89:650.

Columbus Light and Coke Company v. Jacob Freeland. 1861. 12 Ohio St. 392; (LEXIS 154).

Davenport, S. J., G. G. Morgis. 1954. Air pollution—A bibliography. Bulletin 537, Bureau of Mines.

"Dickey Act." 1949. *Report of the Interim Fact-Finding Committee on Water Pollution.* Randal F. Dickey, Chairman, Assembly of the State of California.

Ebaugh, W. C. 1910. The neutralization and filtration of smelter smoke. *The Journal of Industrial and Engineering Chemistry* September, II (9).

Fulton, C. H. 1915. Metallurgical smoke. Department of the Interior, Bureau of Mines.

Hartman, M. L. 1949. Legal regulation of air pollution. *Industrial and Engineering Chemistry* 41:2391–2395.

Johnson, D. W. 1905. Relation of the law to underground waters. Department of the Interior, United States Geological Survey.

Kennedy, H. W. 1952. The legal aspects of the California air pollution control act. In *Air Pollution.* New York: McGraw-Hill Book Co. Inc., pp. 702–711.

L. A. Darling Company, H.A. Douglas Manufacturing Company and Bronson Plating Company v. Water Resources Commission. No. 86., January 12, 1955 (Westlaw).

Los Angeles Region v. City of Los Angeles; Santa Monica Baykeeper, et al. v. City of Los Angeles, CV 98–09039 RSWL and CV 01–00191 RSWL (1998).

Nemerow, N. L., F. J. Agardy. 1998. *Strategies of Industrial and Hazardous Waste Management.* New York: Van Nostrand Reinhold.

Olds, N. V. 1952. Legal aspects of groundwater contamination. Seventh Purdue Industrial Waste Conference, May 7–9.

P. Ballantine & Sons v. Public Service Corporation of New Jersey. 1908. 76 N. J. L., 358; 70 A. 167; 1908 N. J. Super. (LEXIS 314).

P. Ballantine & Sons v. Public Service Corporation of New Jersey. 1914. 86 N. J. L., 331; 91 A. 95; 1914 N. J. (LEXIS 258).

Peckston, T. 1841. *A Practical Treaty on Gas Lighting.* Hebert: London.

Pensacola Gas Company v. J. C. Pebley. 1889. 25 Fla. 381; So. 593; 1889 Fla. (LEXIS 134).

Pollution Control Legislation. 1949. California Legislature, Regular Session.

Pottstown Gas Company v. Murphy. 1861. 39 Pa. 257; 1861 Pa. (LEXIS 192).

Public Interest Research Group of New Jersey, Inc. v. Powell Duffryn Terminals, Inc., 913 F.2d 64 (3d Cir. 1990), *cert. denied*, 498 U.S. 1109 (1991).

Rivers and Harbors Appropriation Act of 1899, 33 U.S.C. 107.

Sive, D. 1995. The litigation process in the development of environmental law. *Pace Environmental Law Review*, Fall.

Stahl, D., ed. 1958. *Public health laws of Pennsylvania: A study of the laws of the Commonwealth of Pennsylvania relating to public health.* School of Law, University of Pittsburgh.

Sullivan, P. J., F. J. Agardy, R. K. Traub. 2001. *Practical Environmental Forensics: Processes and Case Histories*. New York: John Wiley & Sons.

Swain, R. E. 1949. Smoke and fume investigations—A historical review. Atmospheric Contamination and Purification, 115th Meeting of the American Chemical Society, San Francisco, California. *Industrial Engineering Chemistry* 41.

The Ottawa Gas Light and Coke Company v. James Graham. 1862. 28 Ill. 73; 1862 Ill. (LEXIS 283).

Treadwell, E. E., Meinzer, O. E., Lewis, M. R., and Snow, B. F. 1940. Analysis of legal concepts of subflow and percolating waters—Discussion. *American Society of Civil Engineers* May, 66(5):1020–1030.

CHAPTER 3

Educational Solutions

James R. Mihelcic and David R. Hokanson

Introduction

Currently, the world's population has reached 6 billion and a staggering 80 million people are added each year. By 2025, the world's urban areas will be home to 5 billion, with 90% of growth coming from developing nations. This growth, combined with dramatic increases in per capita resource consumption, contributes to increasingly serious social and environmental problems. These problems will only worsen over the next 50 years as the projected world population approaches 9 billion; developing nations become more industrialized; and people of the world develop personal and household consumption patterns similar to inhabitants of industrialized countries. The global community continues to use finite nonrenewable resources at an ever-faster rate, with little regard for future generations. Facing these facts, we are compelled to ask: are Earth and humankind sustainable?

Global Facts

- Poor environmental quality contributes to 25% of all preventable ill health in the world
- More than 1 billion people lack safe drinking water
- More than 2 billion people lack adequate sanitation
- **Approximately 1/3 of the world's population resides in countries that suffer from moderate to high water stress**[1]
- More than 1 billion people live on less than $1 per day
- As many as 113 million children do not attend school
- Two-thirds of illiterates are women
- Every year 11 million young children die before their 5th birthday

1. Water stress is defined as water consumption being greater than 10% of renewable freshwater resources.

About 47% of the global population currently resides in urban areas, compared to approximately 33% in 1972 and the percentage is expected to grow to 65% by 2050 (UNEP 2002). Urban dwellers are often subject to inadequate sanitation and water supply and are victims of environmental injustices (e.g., urban plant sittings resulting in increased exposure to toxics releases).

The solutions to the world's current and future environmental problems will require that society create ecologically and socially just systems within the carrying capacity of nature without compromising future generations.[2] And importantly, these environmental solutions will have to be sustainable.

In 1987 *Our Common Future* was released by the United Nations. This influential document not only adopted the concept of "*sustainable development*" but also provided the stimulus for the 1992 U.N. Conference on Environment and Development (i.e., Earth Summit). It defined sustainable development as "development which meets the needs of the present without compromising the ability of the future to meets its needs."

Here, sustainable development is defined as:

> the design of human and industrial systems to ensure that humankind's use of natural resources and cycles do not lead to diminished quality of life due either to losses in future economic opportunities or to adverse impacts on social conditions, human health, and the environment (Mihelcic et al. 2003).

In order to achieve a more equitable and thus sustainable world, individuals, communities, governments, and industries worldwide must adopt policies and practices that promote sustainable development. The global community is not on a sustainable path. In fact, despite warnings from numerous authorities related to threats posed by problems such as climate change, loss of biological diversity, and depletion of nonrenewable resources, little progress has been made in implementing changes essential for a sustainable future.

Principle 1 of the Stockholm Declaration[3] states:

2. Carrying capacity refers to the upper limit to population or community size (e.g., biomass) imposed through environmental resistance. In nature this resistance is related to the availability of renewable (e.g., food) and nonrenewable (e.g., space) resources as they impact biomass through reproduction, growth, and survival (from Mihelcic, J. R. 1999. *Fundamentals of Environmental Engineering*. New York: John Wiley & Sons, Inc.).
3. The United Nations (UN) Conference on the Human Environment was held in Stockholm (1972). This conference is significant because for the first time, it added the environment to the list of global problems.

> Man has the fundamental right to freedom, equality, and adequate conditions of life, environment of quality that permits a life of dignity and well-being, and he bears a solemn responsibility to protect and improve the environment for present and future generations.

Principle 2 states:

> The natural resources of the earth including air, water, land, flora, and fauna and especially representative samples of natural ecosystems must be safeguarded for the benefit of present and future generations through careful planning and management, as appropriate.

Society, the environment, and economic/industrial development—the "triple bottom line"—are inherently interconnected, both domestically and worldwide. Without fundamental changes in current educational practices, the future is in jeopardy. Healthy communities require a sustainable future, in which human and industrial systems support an enhanced well-being of all living systems on the planet by recognizing and seeking to understand their interconnectivity. And, change must begin in the place where most change is born—within our households and educational institutions.

One driver of this change in education is related to preparing students for the future. As governments move towards policies that promote an international marketplace, educators need to prepare students to succeed in the global economy and to effectively serve their community. It is clear that in the upcoming decades young people entering the workforce need to play a critical role in the eradication of poverty and hunger and facilitation of sustainable development, appropriate technology, beneficial infrastructure, and the promotion of social change that is socially and environmentally just. It is also clear that to develop and implement environmental solutions to the world's problems, the solutions must be sustainable.

Educators must be proactive in this regard. Graduates of educational programs focused on sustainability need to become influential leaders of their communities and in society. Graduates educated in this new way of thinking will be uniquely suited to meet the challenges of their households and communities, business and government, as well as to pursue research and public service through positions in higher education.

Is the World on a Sustainable Path?

To address whether the world is on a sustainable path, we will simplify the complexities of the world by investigating one region, the North American Great Lakes. When laid out from east to west, the Great Lakes span more than 1,200 km. They also serve as part of the international border between the United States and Canada. They consist of five lakes: Erie, Huron,

Michigan, Ontario, and Superior; they are the world's largest source of fresh surface water, containing approximately 23,000 km^2 of freshwater and cover a total area of 244,000 km.^{2}

When the first Europeans arrived in the 1600s they found a relatively stable ecosystem. Human disturbance was relatively mild, primarily due to hunting and agricultural activities of the native people who had migrated into the area approximately 10,000 years ago. Over 34 million people now inhabit the Great Lakes region and more than 3,500 species of plants and animals have been identified (more than 130 species are considered rare). Figure 3.1 shows the distribution of the population and the boundary of the Great Lakes watershed. One-tenth of the U.S. population and one quarter of the Canadian population resides in the Great Lakes watershed.

It was not until the aftermath of the War of 1812 that "development of the area, from a beautiful, almost inhabited wilderness into a home and workplace for millions began in earnest" (Fuller and Shear 1995). Large-scale modern development in the Great Lakes watershed included agriculture,

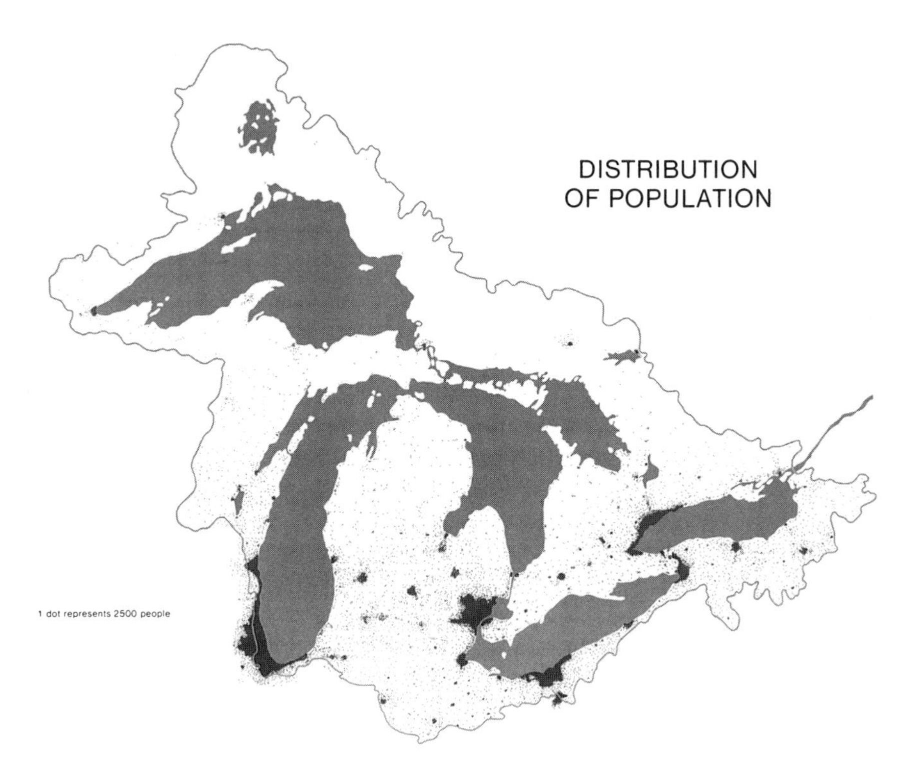

FIGURE 3.1. Distribution of population in the North American Great Lakes watershed. Each dot represents 2,500 people. (Obtained from Fuller and Shear, 1995.)

logging and forest products industry, construction of canals that brought shipping and transportation, exploitation of commercial fisheries, and urbanization and industrial manufacturing.

The economic importance of the Great Lakes watershed is quite significant to the U.S. and Canadian economy. For example, within the watershed is

- US$520 billion per year of personal income
- 25% of world's fresh surface water
- 55% forested land representing 20% of U.S. timberland
- **33% of U.S. manufacturing output**[4]
- 50% of Canadian manufacturing output
- 58% of automobile production in the U.S. and Canada
- US$1 billion sport fishing industry

However, along with this economic activity has come an environmental cost that can be demonstrated with an ecological footprint analysis. An ecological footprint is a determination of the biologically productive area required to provide an individual's resource supplies and absorb the wastes their activities produce. Another way to think of ecological footprint is the ecological impact corresponding to the amount of nature an individual needs to occupy to keep intact their daily lifestyle (Wackernagel et al. 1997). Figure 3.2 clearly shows through an ecological footprint analysis that the North American Great Lakes region is not traveling on a sustainable pathway. This is because while the watershed of the Great Lakes encompasses approximately 580,000 km^2, the current human activity within the watershed requires 8 times more land (approximately 4.6 million km^2) to support its current activities.

On a global level, Table 3.1 shows the ecological footprint of the world's nations. Provided in the table is a country's ecological footprint, available ecological capacity, and ecological deficit on a per capita basis. The ecological deficit is determined by subtracting the footprint from the available ecological capacity. Negative numbers indicate a deficit and positive numbers indicate there is still some remaining ecological capacity.[5] Table 3.1 shows that clearly most of the world's nations are also not following a sustainable path.

4. The U.S. manufacturing output includes 40% of industrial water usage, 23% of farm sales, 70% of steel production, and 301 million tons of shipping per year.
5. 1.7 to 1.8 hectares per person is a typical benchmark for comparing the ecological footprint of the world's population. This number is derived from the fact that on a per capita basis there are 0.25 hectares of biologically productive arable land, 0.6 hectares of pasture, 0.6 hectares of forest, 0.03 hectares of built-up land, and 0.5 hectares of ocean. It is also assumed that at least 12% of land must be set aside to preserve protection of biodiversity.

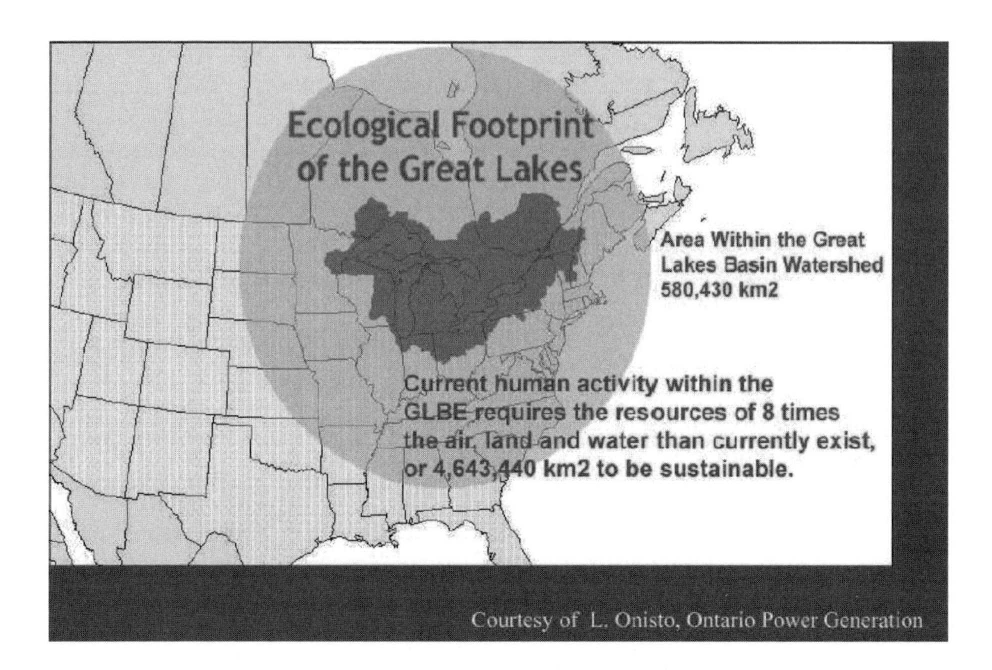

FIGURE 3.2. Ecological footprint of the Great Lakes (from Lickers, F. H., Mohawk Council of the Akwesasne Societal Indicators, presented at the 2000 State of the Lakes Ecosystem Conference, hosted by the U.S. Environmental Protection Agency and Environment Canada, Hamilton, Ontario, www.epa.gov/solec [accessed October 25, 2004]).

The New Revolution, the Sustainability Revolution

Fortunately for the world community, it is at the onset of a new revolution, the sustainability revolution. This revolution is one of several, occurring over the past 10,000 years that have changed how humans interact with nature. Previous revolutions that influenced how humans interacted with nature were the agricultural and industrial revolutions. When the agricultural revolution began 10,000 years ago, the human population was approximately 10 million people. By the year 1750 the world's population had increased to approximately 800 million. An agriculture-based society was not necessarily more productive or efficient. In fact, some believe it was simply the basis to accommodate increasing population. For example, the agriculture revolution resulted in food of a lower nutritional value on a per acre basis; however, it dealt with problems of wildlife scarcity. In addition, the agricultural revolution brought us concepts of land ownership, feudalism, wealth, status, trade, money, power, guilds, temples, armies, and cities (Meadows et al. 2004).

TABLE 3.1
Ecological Footprint of Nations

	Ecological Footprint (units of ha/capita)	*Available Ecological Capacity (units of ha/capita)*	*Difference (ecological deficit if negative) (units of ha/capita)*
WORLD	2.3	1.8	–0.5
Argentina	4.6	3.8	–0.8
Australia	8.1	9.7	1.6
Austria	5.4	4.3	–1.1
Bangladesh	0.7	0.6	–0.1
Belgium	5.0	1.6	–3.4
Brazil	2.6	2.4	–0.1
Canada	7.0	8.5	1.5
Chile	3.5	4.9	1.4
China	1.2	1.3	0.1
Colombia	1.7	1.3	–0.4
Costa Rica	2.5	2.0	–0.5
Czech Rep	4.2	2.5	–1.7
Denmark	5.8	2.1	–3.7
Egypt	1.2	0.6	–0.5
Ethiopia	1.0	0.9	–0.1
Finland	6.3	9.6	3.3
France	5.7	3.8	–1.9
Germany	4.6	2.1	–2.5
Greece	3.9	1.3	–2.6
Hong Kong	2.7	0.5	–2.2
Hungary	2.5	2.0	–0.5
Iceland	9.9	2.5	–7.4
India	0.8	0.8	0.0
Indonesia	1.6	0.9	–0.7
Ireland	6.6	8.3	1.7
Israel	3.1	1.1	–2.0
Italy	4.5	1.4	–3.1
Japan	6.3	1.7	–4.6
Jordan	1.5	0.6	–1.0
Malaysia	2.7	1.7	–1.0
Mexico	2.3	1.4	–0.9
Netherlands	4.7	2.8	–1.9
New Zealand	9.8	14.3	4.5
Nigeria	1.7	0.8	–0.9
Norway	5.7	4.6	–1.1
Pakistan	0.8	0.9	0.1
Peru	1.7	1.5	–0.2
Philippines	2.2	0.7	–1.5
Poland, Rep	3.4	2.3	–1.1

TABLE 3.1 *(continued)*

	Ecological Footprint (units of ha/capita)	*Available Ecological Capacity (units of ha/capita)*	*Difference (ecological deficit if negative) (units of ha/capita)*
Portugal	5.1	2.2	–2.9
Russian Federation	6.0	3.9	–2.0
Singapore	5.3	0.5	–4.8
South Africa	2.6	1.6	–1.0
South Korea	2.0	0.7	–1.3
Spain	4.2	2.6	–1.6
Sweden	5.8	7.8	2.0
Switzerland	5.0	2.6	–2.4
Thailand	2.8	1.3	–1.5
Turkey	1.9	1.6	–0.3
United Kingdom	4.6	1.8	–2.8
United States	8.4	6.2	–2.1
Venezuela	2.6	1.4	–1.2

From Wackernagel et al., 1997.

Over the relatively recent time frame of the industrial revolution, the global population increased to 6 billion individuals. The industrial revolution not only brought us machines, capitalism, roads, railroads, combustion, smokestacks, factories, and large urban areas, it also began a process where in some minds, technology and commerce were elevated above religion and ethics (Meadows et al. 2004). The industrial revolution was also the reason that language added phrases to describe our myriad of modern environmental problems. These phrases now include:

- Non-renewable energy
- Climate change
- Excessive consumption
- Smog and acid rain
- Ozone hole depletion
- Persistent organic pollutants
- Fishing advisory
- Environmental risk
- Environmental justice
- Endocrine disrupters
- Hazardous waste
- Cancer

It is hard to believe that at the onset of the industrial revolution the individuals and governments that promoted this change did not foresee all the social and environmental injustices that were to take place on a global scale over the following centuries. In fact, one might ponder if they had known how their actions would result in common usage of catastrophic phrases such as those listed above, then this revolution would have been designed differently so the final outcome was less harmful, and more sustainable.

Meadows et al. (2004) point out that in order to succeed in implementing the sustainability revolution, members of society will need to imagine what they want, not what someone has taught them to want, and not what they learned to be willing to settle for. The success of this revolution will also require a new set of phrases be not only incorporated into the scientific, educational, and popular language but also valued by members of society. This will obviously require a transformation in the way we educate young and old. Some of these terms include:

- Renewable energy
- Caring, sharing, and compassion
- Resource equity
- Carrying capacity
- Moderation
- Appropriate technology
- Social and environmental justice
- Holistic approaches
- Biodiversity
- Consensus building
- Community-based initiatives
- Values and ethics

Fortunately, educational solutions can be developed to drive the sustainability revolution and in doing so, many of the world's environmental problems may be solved. This effort will require metadisciplinary approaches, an emphasis on appropriate technology, the need for a global exchange of knowledge, realizing the importance of research to education, and also understanding the importance of personal belief systems.

Metadisciplinary Approaches

We have written in the past about the emergence of the new metadiscipline of sustainability science and engineering. The prefix meta refers to "after" and "beyond," thus, a metadiscipline is a higher level discipline. Unlike interdisciplinary approaches that focus on activities at the interface between disciplines, we called this new field a metadiscipline approach because it provides an over-arching framework for adopting and incorporating knowledge across many fields of study to inform human and environmental development and quality (Mihelcic et al. 2003).

Figure 3.3 shows how this metadisciplinary approach will require students and educators to consider sustainability in the context of environmental, societal, and economic/industrial if the world is to achieve a sustainable future. This will require we bring together knowledge from the traditional, more mature fields of the physical sciences, engineering, economics, and human behavioral studies to address educational solutions for sustainability.

Students (and educators) interested in math, science, engineering, and technology, need to understand (and be involved in) social science and public policy issues so they respect the interactions between technology and the environment and society, including the impact of modern technology on local communities. Students majoring in business, humanities, and the social sciences need to have an understanding of science and technology.

Need for Appropriate Technology

All students need to learn about appropriate technology, defined here as the use of materials and technology that are culturally, economically, and socially suitable to the area in which they are implemented. The term

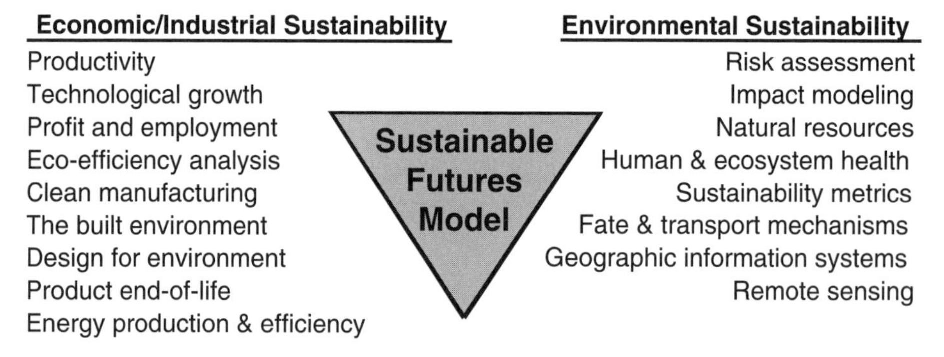

FIGURE 3.3. Sustainability triangle showing the three facets of the metadisciplinary approach for solving environmental problems.

appropriate technology was first used in the mid 1960s as a solution to problems encountered in the developing world. Interestingly, when combined with societal needs, it may have been the precursor of sustainable development.

Educators need to address the use of appropriate technology in the home and work environment. Unfortunately, appropriate technology has mostly been thought of as a term that applies to implementing technology in the developing world. However, numerous examples of unsustainable and harmful technologies have been applied to communities and ecosystems in the industrialized world.

The modern example of personal use of fuel-consuming automobiles and trucks is a non-appropriate technology quickly demonstrating itself to be an unsustainable form of transportation. In fact, this form of transportation is not only adversely impacting the environment, it also contributes to obesity and separation of communities. And even if technological strides are made in reducing environmental impact on a per automobile basis, the automotive infrastructure will remain unsustainable as it is built around a business model that rewards the automotive sector based on sales volume. As automotive companies expand globally into less developed countries (e.g., China, India), it is clear that the current system will break down in terms of short- and long-term sustainability. Sustainable transportation solutions will thus require that educators train students to envision transportation systems that are not focused on single passenger travel in an automobile.

Need for a Global Exchange of Knowledge

Education initiatives will require an understanding of the impact industrialized nations have on communities of the developing world, and incorporate a global perspective and appreciation for the global community. This also means that the industrialized world needs to understand that the challenges of sustainability in the developing world may be different than in the industrialized world. For example, many problems in the developing world are not from manufacturing, but are related to water, soil, agriculture, forestry, and fisheries.

Educators need to realize that many of the world's current environmental problems are being driven by the industrialized world (e.g., climate change, toxic chemical production and release, consumption). Therefore, the diffusion of knowledge that will lead the global community to more sustainable choices may be better derived from the developing world and passed on to the industrialized world, not the other way around as many have suggested.

One example is housing. Families in the developing world typically do not construct excessively large homes for their daily and seasonal needs. They make use of locally made building materials, take advantage of natural architectural features that enhance cooling and ventilation, and more commonly spend time outside their home interacting as a family and a community.

As an example of a nonsustainable method of designing and constructing homes, consider household energy use in the United States. Residential energy use now accounts for 19% of U.S. energy usage (and close to 5% of the world's usage). The percentage of U.S. primary energy use associated with residential housing has been gaining a larger overall percentage of the total energy demand since the 1970s. In the U.S., government data show that an increase in household income results in an increase in the size of a home, and also an increase in the energy consumed by the home (this does not include the "stored energy" associated with purchasing of products to fill up the space in a home). It is also clear from an examination of U.S. data that an increase in household income results in an increase in energy consumption (Table 3.2). In fact, the majority of individuals with greater wealth do not appear to use their extra income to make more sustainable choices related to household energy use.

Thus, one solution to the world's environmental problems is for architects, builders, planners, and policymakers from wealthier nations to be educated about building size and efficiency when designing, building, purchasing, and setting policy initiatives related to home design and construction. In addition, future home owners require education about how their personal choices related to home size and efficiency impact the local and global environment.

For the built environment, the U.S. Green Building Council's Leadership in Energy and Environmental Design (LEED) certification program is having a significant effect on improving environmental impact, energy use, and energy efficiency associated with new and existing buildings. LEED certification programs are available or under development for the construction and operation of new buildings (LEED-NC), for the operation of existing buildings (LEED-EB), and for homes (LEED-H). In effect, the LEED program

TABLE 3.2
U.S. Household Income Influence on Household Energy Consumption

Household Income	*Energy Consumption per Household (millions of BTU)*
<$10,000	65.3
$10,000–$29,999	78.8
$30,000–$49,999	89.4
>$49,999	112

U.S. Department of Energy data from 2001.

represents a form of educational solution that not only improves the sustainability of commercial buildings, but also allows companies to gain competitive advantage by adopting more sustainable ways of doing business. It is in the nascent stages, but perhaps the LEED certification program for homes (LEED-H) will become an educational solution helping to shift the unsustainable energy usage patterns observed in U.S. households, as discussed above, toward sustainable solutions.

Importance of Research to Education

One of the most critical endeavors that must be undertaken in ensuring solutions to the world's environmental problems is in the area of education and human resource development. Educational activities must occur on two levels. On the first level, the technological and environmental awareness of the entire citizenry must be elevated. On the second level, we must begin attracting a sufficient and diverse pool of human talent to solve the problems in sustainability. Early intervention will be a key at both levels (Mihelcic et al. 2003).

In order for the general public to understand environmental problems and solutions to those problems, closer links between scientific research and enhancement of the public's understanding of science must take place. This can include interaction between the research community and their university students, but also precollege students and teachers, museums, and aquariums. The public needs to understand how science affects society, and how their individual choices impact sustainability.

Fortunately, issues related to the environment and sustainability can make excellent examples of how research and educational outreach activities can be connected in mutually supportive relationship between precollege and university educators. For example, the U.S. National Science Foundation now requires that all grant proposals include a discussion about "What are the broader impacts of the proposed activity." This requires that researchers address issues such as:

How well does the activity advance discovery and understanding while promoting reaching, training, and learning?
How well does the activity broaden the participation of underrepresented groups?
To what extent will it enhance the infrastructure for research and education, such as facilities, instrumentation, networks, and partnerships?
What are the expected benefits of the activity to society?

Researchers need to identify a target audience by asking themselves "why they think the public should care about their research" (Avila 2003, p. 11). The target audience could include consumers, students in other disciplines, precollege students and their teachers, members of government, voters, and industrial representatives. Interaction between researchers and precollege teachers has many benefits. Benefits for the researcher include learning new teaching methods and thinking more broadly about the learning process. Precollege teachers benefit because their students may develop a passion for the subject based on the science experience, students may experience the excitement of scientific discovery, and they make become more comfortable with science, and thus take more risks (Avila 2003).

Of course, education needs to continue beyond one's school years. For example, consumers need to be educated about the impact of their activities with regard to the life stage of products they purchase.[6] For example, the global warming impact from manufacturing motor vehicles is smaller relative to other life stages of the vehicle because the "product-use" life stage accounts for a greater amount of greenhouse gas emissions. The U.S. Environmental Protection Agency (2000) estimated that all transportation activities in the U.S. accounted for almost 26% of total U.S. greenhouse gas emissions from 1990 to 1998. For the period of 1990 to 1998, the average greenhouse gas emission associated with the use of passenger cars was estimated to be 184 million metric tons of carbon equivalents per year,[7] and the greenhouse gas emissions associated with use of light duty trucks was estimated to be 92 million metric tons of carbon equivalents per year. The total global warming impact for personal transportation is thus approximately 276 million metric tons of carbon equivalents. However, the global warming impact from industrial use is much less than for transportation. Therefore, in the U.S. the product-use life stage for passenger cars and light trucks has a much higher global warming impact than the premanufacturing and manufacturing life stages.

Education versus Beliefs

The United Nations Environment Programme has an initiative termed "sustainable consumption." This particular area focuses on understanding the forces that drive consumption patterns around the world and how to translate those findings into tangible activities for business and other stake-

6. The five life stages of a product include: (1) the pre-manufacturing step (e.g., mining of raw materials), (2) manufacturing, (3) packaging and transport to consumer, (4) usage by consumer, and (5) end of life (e.g., reuse, recycling, disposal). A life cycle assessment (LCA) is a method to determine the natural resource usage, energy requirements, and pollution generation (i.e., the environmental burden) associated with the total life of a product.
7. The United Nations defines carbon equivalents as the anthropogenic emissions, less removal by sinks, of the greenhouse gases carbon dioxide (CO2), methane (CH4), nitrous oxide (N2O), hydrofluorocarbons (HFC), perfluorocarbons (PFC), sulphur hexafluoride (SF6) weighted according to their greenhouse potential and expressed as CO2 equivalents.

holders. In a previous article (Mihelcic et al. 2003), we discussed the difference between bringing about change through education versus belief systems when an engineer is thinking about green engineering decisions. In that article we wrote

> accordingly, it is very important for engineers to understand individual and household belief systems to help motivate the selection and design of greener products. In this context, it is important to realize that the mechanistic understanding of how pollutants are generated includes not only the production process but also social, economic and political choices and the impact of consumption and population. For example, the "what would Jesus drive?" advertisement campaign of the past year in the U.S. was partially effective because it tapped into consumer beliefs rather than transient wants that are traditionally created by advertising. As an outcome of this initiative, as well as other factors, there has been measurable movement in the past year by consumers to purchase smaller SUVs. Also, some Detroit automakers are now considering broader consumer concerns about SUVs and the cultural trends that influence the buying habits of the next generation of automobile buyers who many believe are shifting away from purchase of extremely low fuel economy vehicles.

This example illustrates that it will not only be education that drives the public's decision-making activities towards more sustainable choices, there must also be examination of how individual's belief systems are influenced. Of course this will also mean that ethical issues related to changing ones beliefs need to be considered.

Putting It All Together: Case Studies of Environmental Solutions

The sustainability revolution is slowly moving forward and change is occurring at all levels of education. At the beginning of this chapter we stated that change must begin in the place where most change is born—within our households and educational institutions. For the remainder of this chapter, we will use examples of sustainability initiatives taking place at our university to demonstrate some of these changes, changes that are beginning to take place all over the world. We note though that similar activities are taking place at other educational institutions on a global scale.

Several years ago, faculty members, staff, and students from a diverse set of disciplines at our university, began collaborating on sustainability issues. Their activity produced the Sustainable Futures Model, which serves as the thematic basis and intellectual focus of its activities (see earlier discussion on *Metadisciplinary Approaches* and Figure 3.3). The sustainable futures model focuses on research and education in four areas: (a) environmental systems, (b) economic/industrial systems, (c) societal systems, and (d) integrative initiatives that join all three of these areas. In 2003, the Sustainable Futures Institute (SFI) was created as an effort to advance the metadiscipline of sustainability science and engineering, which integrates the engineering and technology elements of industrial ecology; the scientific elements of environmental assessment and modeling; and the economic, modeling, and human behavioral elements of the social sciences that support environmental decision making. Table 3.3 describes some of the activities taking place.

We are currently developing and pilot testing middle-school curriculum units on energy and resources, ecosystems, and water quality. The project includes funds for developing and piloting the units in classrooms statewide, and developing Web-based activities for the dissemination of interactive educational materials. The team assembled for this project includes precollege, middle school, and university educators. Specific lesson plans include:

What is trans-boundary air pollution?
What is the potential for climate change in the North American Great Lakes?
How are aquatic ecosystems impacted by changes in population, introduction of exotic species, and input of excessive nutrients?
Where in a watershed do chemical pollutants originate?
How is energy used?
What are the alternatives to traditional energy sources?
What do we do with our by-products and wastes?
How do personal choices impact Michigan, the U.S., and the world?

TABLE 3.3
Examples of University Educational Activities that Can Promote Sustainability

Initiative	*Primary Audience*
Creation of curriculum related to ecosystems, water quality, energy, and resources (that includes pollution prevention/sustainability)	Precollege middle school teachers and their students
Teacher training/workshops	Precollege K—12 teachers
Certificate in Sustainability	Graduate students
Minor in Sustainable Engineering	Undergraduate engineering students
International Senior Design	Undergraduate students
Master's International Program	Graduate students in environmental engineering, civil engineering, geological engineering, geology, forestry
Campus Sustainability Initiatives	Students, faculty, and staff

Examination of a product's life cycle:

Our overall educational goal at the precollege level is to offer a wide variety of programs focused on enhancing the teaching and learning of science, mathematics, and social sciences. Teacher professional development programs help teachers to implement state and national standards and, importantly, provide teachers with tools so they can actively engage their students in learning. A variety of student and community programs stimulate learning in the classroom, apply learning to real world challenges, and involve students in their communities.

Graduate students from all campus disciplines can now combine their specific graduate degree with a "Graduate Certificate in Sustainability" that is managed through the graduate school and Sustainable Futures Institute at Michigan Technological University. The graduate certificate provides formal recognition of a student's curricular breadth in the following three areas: (a) policy and societal systems, (b) environmental systems, and (c) industrial systems, along with some specific core coursework directly related to sustainability.

For undergraduate engineering students, a minor in "Sustainable Engineering" is being developed. The goal of this minor is to enable engineering students to become effective agents for sustainable development by implementing a vision of a better world. This will be accomplished by first encouraging a service learning experience to serve as motivation for understanding the design complexities of community-based solutions. Coursework related to the social, economic, cultural, and philosophical interconnections to engineering will serve as required core concepts of this program. Students will then be allowed to take engineering courses that assist in solving problems in a sustainable manner, either through an emphasis on North American or developing world issues. In all cases, students will be provided a global perspective on sustainable development.

International programs in civil and environmental engineering now provide the opportunity for students to understand the social, economic, and environmental limitations of implementing appropriate technology and sustainable engineering solutions to problems in the developing world. For example, in the past 6 years our Global Sustainable Development Initiative in Civil and Environmental Engineering has involved over 100 undergraduates and 40 graduate students in over 100 projects located in 16 countries.[8] Other graduate international programs in geology, geological engineering, and forestry focus on management of natural hazards and sustainable management of forests.

8. The 16 countries are: Belize, Bolivia, Cameroon, Dominican Republic, East Timor, Honduras, Jamaica, Kenya, Macedonia, Madagascar, Mauritania, Mali, Panama, Philippines, Uzbekistan, and Vanuatu.

These international programs seek graduate students who not only have a strong technical skill set and are skilled at solving complex engineering problems, but also are educated and trained in the many critical, nontechnical skills required of today's engineer. It is also our goal to train engineers to value service to the global community. This educational philosophy can perhaps best be summarized by one of our graduates, who in response to a question about what he learned from this experience, answered that "along with gaining valuable engineering skills, they also learned what it was like to put engineering into practice while taking into consideration the social, economic, and environmental limitations of the developing world."

International Senior Design projects allow students to obtain university credit for working on engineering projects in the developing world. Practitioner involvement has been a key to the success of the projects. Students are obligated by course requirements to address the following issues associated with their design project.

- Economic and health factors
- Social impact
- Appropriate technology
- Constructability
- Sustainability
- Safety and reliability
- Aesthetics
- Ethics
- Environmental limitations

The projects also require that students: develop creativity; solve open-ended problems; develop problem statements; identify and assess alternate solutions; perform feasibility analyses; and develop design drawings and specifications.

Our Master's International programs educate students in nontraditional aspects of their profession, either through coursework or the two plus years of training and service provided by the U.S. Peace Corps engineering experience (Orr et al. 2003; Mihelcic et al. 2004a,b). Some of these skills include the ability to: (1) write and orally communicate; (2) work in diverse teams; (3) build coalitions and consensus; (4) devise strategies and long range plans; (5) secure project funding; (6) budget and manage projects; (7) apply sustainable, appropriate technology; and (8) work in other countries. Coursework is related to traditional graduate engineering education plus nontechnical skills discussed previously. See Figures 3.4 and 3.5.

Most important, students learn that the correct solution is no longer taken directly from a textbook and instead learn to apply sustainable and appropriate technology to solving a problem. As an example, Table 3.4

FIGURE 3.4. Master's International graduate student Jon Annis working with villagers on a water supply project in Madagascar. The three photos show (A) mixing of concrete, (B) the finished spring box that collects water in a mountainous area, and (C) digging of a trench that will distribute water to village via a gravity pipe system. The completed system consists of a spring box, reservoir tank to store water during nightly refilling period, pipes to connect the spring to the village, and several tap stands in the village that provide safe water to several hundred people. Logistically this site is a nightmare because it is only accessible by foot. The path, crossing many small streams and often through thick slippery clay mud, takes $2\frac{1}{2}$ hours on a dry day. The 50 bags of 120-lb cement and all other materials are carried by hand.

provides a few examples of U.S. Peace Corps engineering assignments related to water and sanitation. The importance of water and sanitation to sustainable development is because sustainable development is clearly linked to issues of human health. For example, Principle One of the Rio Declaration states that "*Human beings are at the centre of concerns for sustainable development. They are entitled to a healthy and productive life in harmony with nature.*" Furthermore, the World Health Organization

FIGURE 3.5. Three engineers taking a rest from well installation projects by working on hygiene mural outside the office of a health and AIDS education non-governmental organization in Mali, West Africa. The mural uses the concept of a Baobab tree to show how the roots of peace, health, and unity are activities related to sanitation and water quality.

states that "*health is both a resource for, as well as an outcome of, sustainable development. The goals of sustainable development cannot be achieved when there is a high prevalence of debilitating illness and poverty, and the health of a population cannot be maintained without a responsive health system and a healthy environment. Environmental degradation, mismanagement of natural resources, and unhealthy consumption patterns and lifestyles impact health. Ill-health, in turn, hampers poverty alleviation and economic development.*"

Several years ago, the Environmental Sustainability Committee (ESC) was formed as a cooperative effort among staff, faculty, and students to increase campus action and awareness of issues of sustainability. Although in its early years, the ESC has already begun to make significant strides toward raising awareness of sustainability on and off campus. Its efforts have given way to modest impacts from increased paper recycling on campus, a green landscaping initiative, efforts to educate staff about green

TABLE 3.4
Overseas Projects Performed by Master's International Engineering Students

Area	*Example Projects*
Wastewater treatment	Ventilated improved pit and composting latrines Small-scale treatment systems such was lagoons, evapotranspiration ponds, wetlands
Water supply and treatment	Creation of water boards and water constitutions Construction and/or improvement of wells Gravity fed water transmission lines from springs/streams in mountains to village communities Water storage, filtration, disinfection systems Rainwater harvesting systems Assist health volunteers
Solid waste management	Develop solid waste management plans Implement composting and recycling programs Develop financing in absence of existing tax structure
Water resources	Watershed protection Development of sustainable, land-use watershed management plans Linkage of community health to watershed processes

Adapted from Mihelcic, 2004a.

purchasing decisions, and an initiative to provide alternative transportation through development of a bicycle share program.

The activities of the ESC, operated within the broader scope of the Sustainable Futures Institute, affect many of the economic, environmental, and social aspects of the campus. Their impacts can be measured by use of a carefully selected set of sustainability indicators that represent both the quantitative and qualitative nature of the university's health and vitality. The first step in developing a set of sustainability indicators began with the creation of a social science graduate course entitled "Developing Indicators of Sustainability." This course is one of the listed options for the Graduate Certificate in Sustainability. The goals of this course were to have students become familiar with a variety of processes and tools developed to create sustainability indicators, and then apply them to the campus in a project to create a base set of indicators.

To accomplish a base set of indicators, students decided that their final project was to be the creation of a framework for a "living" sustainability report. To achieve this, they proposed the following plan of action:

1. Collect existing data from reliable sources.
2. Conduct a survey of faculty, students, and staff in order to supplement existing data.

3. Develop indicators from the compiled data that are relevant, logically and scientifically defensible, reliable, leading, policy relevant, and that reflect community values.
4. Determine realistic and tangible short- and long-term goal ranges for each indicator, and note any trends toward or away from those goals.
5. Compile the data into a report that is not only informative to a diverse campus audience with varying levels of interest, but also adaptable to the university's future needs and interests. In other words, develop a framework that is in itself sustainable.

Categories for the selected indicators are related to areas such as the economic vitality of students and employees, quality of education, water and energy usage, transportation, and greening of campus buildings. Table 3.5 provides an example for the category of waste, of how a university goal is developed to an indicator, measure, and short- or long-term target.

Conclusion

There are many educational solutions that can be used to assist solving the world's environmental problems. Ultimately these educational solutions will help drive the world's next revolution, the sustainability revolution. As discussed in this chapter, these solutions will require metadisciplinary

TABLE 3.5
Example Method to Measure the Sustainability of Waste Production at an Educational Institution

Category	*Waste*
Goal	University employs pollution prevention strategies to minimize solid, hazardous, and radioactive waste and seeks to maximize waste recovery (e.g., reuse, recycling, composting, etc)
Indicator	Determine levels of solid, hazardous, and radioactive waste
Measure	Annual per capita waste generation and breakout by type (paper, glass, aluminum, plastics, hazardous, radioactive) and final destination at end of life
Target	University embraces the following principles in purchasing and disposal methods: Pollution prevention techniques in purchasing and use of products and materials Full composting of organics and recycling of appropriate materials as economically and environmentally feasible All hazardous and radioactive waste is properly managed

approaches integrated for a young to old audience, an emphasis on appropriate technology, the need for a global exchange of knowledge, realizing the importance of research to education, and understanding the importance of personal belief systems. Only with this and other knowledge presented at our educational institutions will members of society be able to envision a sustainable future that lives within the means of the carrying capacity of the Earth.

Review Questions

1. What activities in your household can you change to make your community more sustainable?
2. What educational initiatives in your community do you think would make your community more sustainable?
3. What policies should your government enact to encourage more sustainable decision making by households, corporations, and local educational institutions?
4. What indicators would you select to measure the sustainability of an individual household, your community, and your country?
5. Find an ecological footprint calculator on the web. What is your ecological footprint and how does it compare to the world's footprint? What actions can you personally take that would decrease your footprint?

References

Avila, B. K. G. 2003. *Integrating Research and Education: Biocomplexity Investigators Explore the Possibilities*. National Research Council. Washington, DC: The National Academies Press.

Fuller, K., H. Shear. eds. 1995. *The Great Lakes: An Environmental Atlas and Resource Book*. U.S. Environmental Protection Agency and Government of Canada. Chicago: Great Lakes National Program Office.

Meadows, D., J. Randers, D. Meadows. 2004. *Limits to Growth. The 30-Year Update*. White River Junction, VT: Chelsea Green Publishing Company.

Mihelcic, J. R., J. C. Crittenden, M. J. Small, et al. 2003. Sustainability science and engineering: Emergence of a new metadiscipline, *Environmental Science & Technology* 37(23):5314–5324.

Mihelcic, J. R. 2004a. Educating the future's water professional. *Water Environment Technology* 16(9):86–92.

Mihelcic, J. R. 2004b. Educating tomorrow's global engineer through a unique partnership with the U.S. Peace Corps. *Woman Engineer*, Fall, 25(1):30–33.

Orr, B. D., J. R. Mihelcic, T. J. Van Dam. 2003. Engineering help while getting a degree. *IEEE Potentials* 22(2):32–34.

UNEP, *Global Environmental Outlook 3*, United Nations Environment Programme, Nairobi, Kenya, 2002.

U.S. EPA. 2000. *Inventory of U.S. Greenhouse Gas Emissions and Sinks: 1990–1998.* EPA 236-R-00-001, April.

Wackernagel, M., L. Onisto, A. Callejas Linares, et al. 1997. *How much nature do they use?—How much nature do they have?* March 10. http://www.ecouncil.ac.cr/rio/focus/report/english/footprint/ (accessed October 25, 2004).

Part II

Scientific and Technical Solutions

CHAPTER 4

Economic Solutions

Ernest Lowe

Introduction: Solutions for What?

> The stone age came to an end not for a lack of stones, and the oil age will end, but not for a lack of oil. (Saudi Arabia's former oil minister, Sheik Ahmed Zaki Yamani)

Worldwide there is increasing competition for oil, steel, copper, and alumina, cement, and other basic commodities, industrial production and construction of urban infrastructure and industrial plants. In many countries the basic resources of land and water are also at risk. At the same time supply constraints due to warfare and resource depletion are developing, which add significant increases in costs.

Demand for resources is driven not only by new development but also by the need to replace or restore decaying urban, rural, and industrial infrastructure as well as commercial and residential buildings. Industrial rust belts, for instance, exist in developed and developing countries alike.

Sustainable development in both developed and developing countries requires the setting of very high objectives for efficiency in the use of all resources, as one part of a broader strategy to avoid crisis. This calls for systemic planning and action to multiply efficiency, not just make incremental improvements. Gains in efficiency must be achieved across the whole product lifecycle, not just in production. Rapidly increasing consumption is already overwhelming many improvements in manufacturing productivity in a "rebound effect."

Some of the initial responses for achieving such high objectives include the Circular Economy (China) and the Recycling Economy (Germany and Japan), eco-industrial parks and networks, closed-loop and energy generating apartment and office complexes, and sustainable agriculture. All major economic sectors have a role to play in achieving the required gains in efficiency, not just heavy industry. These include land use planning and development; transportation, design and construction of the built environment; commerce; the design and operation of municipal infrastructure; agriculture, and households all have a share in the responsibility. Similarly, gains must be made at every stage of the lifecycle from resource extraction to final recovery or disposal.

There is a close relationship between resource issues and global changes, such as shifts in the chemical balance of the atmosphere, climate change, pollution of rivers and oceans, transboundary pollution, and loss of biodiversity. Even a Recycling or Circular Economy is only a partial economic solution. The deterioration of ecological systems—locally, nationally, and globally—demands economic solutions that will enable their restoration.

When looking at these systems in terms of human welfare, ecological economists call them "natural capital." Natural capital is the source of all natural resources and the sinks that absorb the by-products of human activities. The accounts of natural capital must be recharged through major investments in restoration of forests, grasslands, deserts, farms, watersheds, oceans, and atmospheric balance. This "restoration economy" must also recover the sunk investment in human habitat and infrastructure so as to extend its life. Here too the investments will create massive opportunity for venture development and job creation.

For years economics was defined as the discipline that allocates scarce resources in a society. Mainstream neo-classical economists tended to see resources as substitutable through technological innovation and declined to calculate the limits. Their working assumption was that the stocks and sinks for an economy were virtually infinite. Resource depletion, pollution, degradation of ecosystems and global ecological systems were all counted as value-adding activities that add to gross domestic product.

The failures of neo-classical economics has led the emerging generation of economists to call it "autistic" for its inability to link its elaborate theories and models with the world in which we actually live. The discipline of Ecological Economics in the mid 1990s separated its work from mainstream economics by insisting that our local, national, and global economies must be managed in their ecological context to avoid ever deepening crisis. This means ending the illusion that the critical functions of natural capital must no longer be called "eternalities," as though they had no place in economic analysis. This is true for companies, for cities, and for nations.

Fundamental Economic Transformation

Ecological economists, some environmental economists, and various other theoreticians of sustainable development have argued that a fundamental transformation is required to deal with the challenges of an outmoded production and consumption system. These challenges include increasing competition for limited resources, continued heavy pollution, destruction of ecosystems, loss of biodiversity, and other consequences of not valuing natural capital in neo-classical economics.

At the same time, a number of governments, investment institutions, corporations, and communities have already started this transformation in economic policy and development policy through direct action. The players range from huge insurance companies reeling from losses due to climate change to countries like China, Germany, and Japan responding to constraints on basic resource supplies, and even to teams of workers in Argentina re-opening factories closed down by their bankrupt owners. They are *practicing* economic transformation before the theoretical base is yet in place. They are creating theory on the basis of real world initiatives and projects.

We will review four of the broad initiatives of sustainable economic practice: the German and Japanese Recycling Economy laws; The Chinese Circular Economy initiative; the Service Economy framework in which companies move from selling products to selling the services they deliver; and the foundation of sustainability, restoration of natural capital and the built environment (called the Restoration Economy by one proponent). These initiatives are highly complementary and provide a holistic context in which it may be far easier to implement the kinds of specific solutions covered in the other chapters of Environmental Solutions.

The central objectives of this emerging field of sustainable economic practice include:

- Achieve a very high level of productivity and utility and a very low level of pollution per unit of resource consumed
- Conserve and restore natural capital and the built environment, the dual bases for all economic activity
- Use these first two objectives as drivers of regional economic development, venture development, and job creation
- Share the benefits of such economic development equitably

Achieving these objectives calls for a new level of creative response from humanity. Human systems are natural systems with the emergent quality of self-consciousness and, hopefully, another level of capacity for adaptation. This capacity can be mobilized by our formal and informal institutions taking on the form of learning organizations. The organizations as well as their members learn to learn, to create an organizational memory of successes and failures.

One source of learning is the dynamics of the natural systems in which human systems exist. We can probably best learn from ecosystems how to create a living economy by using models of ecosystem dynamics side-by-side with models of human systems. For instance, develop a system model of the resource recovery industry and its technologies in parallel with models of the detritivore and decomposer subsystems of various ecosystems. Learn from them as independent complex systems and let the ecosystem dynamics inspire new thoughts about the design of resource recovery, not necessarily imitation. This seems a much more profound approach than cherry-picking isolated principles like "the waste of one organism becomes the food of another."

Recycling Economy Legislation

Germany, closely followed by Japan, has taken a legislative approach to improving their utilization of resources under a banner of waste elimination. Their recycling laws have become organizing frameworks for a wide range of policies, regulations and more specific laws regarding categories of waste such as packaging and electronic products.

Germany

In 1996 the German Federal Government passed a new act seeking to move Germany toward a recycling economy: the Waste Avoidance, Recycling and Disposal Act (*KrW-/AbfG*) (often described as the closed-loop economy law). This law seeks change across the lifecycle of production, consumption, and recovery or disposal to reduce the generation of waste in manufacturing and use and to encourage design of products that can easily be reused, repaired, and recycled (Dietmar 2003).

The core themes of the new legislation are:

- The Polluter Pays Principle
- Waste prevention is preferred over recycling
- Thermal and material recycling have equal status, with the possibility of determining a priority in the case of specific waste forms by statutory order
- Producers are responsible for their products (to be implemented in each case by statutory order)
- Opportunities are opened for the privatization of waste disposal

Under this legislation the company that produces, markets, and consumes goods is responsible for reducing waste in production, recycling internally, reusing where feasible, and disposing of wastes that remain in an environmentally sound manner.

Upstream strategies for avoiding waste and enhancing recyclability through design are a very strong part of the German recycling economy legislation. The disciplines of industrial ecology and the practice of cleaner production or pollution prevention identify strategies supported by German law and adopted by advanced corporations such as Volkswagen, BASF, and Bayer. These include:

- Producer responsibility for its products, including product take back infrastructure
- Waste reduction measures at every stage of production and use
- Design of product and production process to optimize utilization of material, energy, and water resources and minimize pollution
- Design to assure that products will have multiple uses, a long life, be repairable, be easily recycled, and disposed of safely at the end of life

The institutes and consulting firms supporting implementation of the German legislation are providing guidance in the use of lifecycle analysis and design for environment (DFE) tools in the industrial sector.

Germany has started a number of eco-industrial parks and by-product exchange projects to support implementation of the Recycling Law. For instance, companies in the Hamburg region have engaged in a process of forming industrial recycling networks to use each others' by-products (Hassler 2004).

Japan

This section is largely based upon an Indigo Development report researched and written by Mari Morikawa (2000).

The Resource Challenge

As a mountainous country of islands, Japan's industrialized economy has generated acute environmental and resource challenges. Japan currently consumes 1,950 million tons of natural resources every year and imports 700 million tons or 35% from overseas (Japan Environmental Agency 1998). Japan managed to achieve its earlier remarkable rates of economic growth over a period of 50 years by maximizing the efficiency with which it uses its limited space and resources. Energy usage per household in Japan (10,724 Mcal) is about half that in Germany and one-third of U.S. consumption. Especially since the escalation of oil prices beginning in 1973, Japanese industry made significant efforts to develop energy efficient production technologies. There is a growing concern, however, that as resources and space for landfill become more scarce and waste disposal costs increase, further economic growth may be hampered by growing resource inefficiencies.

Japan produces a total of 450 million tons of waste per year. This is mostly industrial waste (400 million tons) from the processing of natural resources into industrial products, and the rest is municipal waste. Over 60% of this waste is either incinerated or dumped in landfills. Being a mostly mountainous country with a high population density, Japan faces a serious shortage of landfills. Current estimates predict that remaining landfill capacity will be exhausted by 2007. Capacity for hazardous waste landfills is rapidly disappearing at present (Japan Environmental Agency 1998, 2000; Japan Ministry of Health and Welfare 1999).

While ambient air quality has improved since the enactment of the Air Pollution Prevention Law and its required application of end-of-pipe technologies, Japan still faces severe problems, especially with dioxin and CO_2 emissions. Strong public protest of dioxin emissions from waste incinerators forced the government to develop stricter standards for dioxin emissions in 1998 and a Dioxin law was enacted in 1999. Because nearly 80% of municipal waste in Japan is incinerated and most dioxin emissions are from waste incineration, incinerators are the main target of the regulation. Old and small incineration facilities are being shut down gradually, which puts more pressure on landfills. *The Japanese government has set a target to reduce its waste by half by 2010 to deal with this problem.* This goal calls for a systematic approach across the whole production-consumption-recovery cycle.

According to the Kyoto Mandate, by 2012 Japan must reduce its CO_2 emissions by 6% from 1990 levels. The central government introduced a loan program to encourage the use of green energy, cascading of thermal energy, and materials recycling. The government is promoting eco-industrial projects as a means of integrating such innovations.

The Response

Central and local governments throughout Japan have begun to embrace a comprehensive approach to reducing waste and optimizing resource utilization as the key to sustainable development of the country. There is a growing consensus among policymakers that environmentally friendlier and more sustainable production practices need to be found if the economy is to keep growing in a way that improves the quality of life for society. This motivated creation of a series of innovative resource related laws, leading to passing of the Basic Law for Promoting the Creation of a Recycling-Oriented Society in 2000.

In spite of a difficult economic crisis in the last decade, the government has supported research and community initiatives for improving utilization of resources at a multifactor level. These include development of Eco-town or Eco-city projects throughout Japan (Ministry of International Trade and Industry); research funding programs for environmental technology and engineering (Ministry of Health and Welfare); and creation of the Eco-Factory concept, which seeks systemic changes in design of production facilities, processes, and products. Local governments encourage eco-industrial development by financing environmental ventures or providing infrastructure for recycling businesses.

The Recycling Law and Related Legislation

Japan's government has created a comprehensive program for achieving a Recycling Economy through a series of laws passed in the past decade. The foundation is the Basic Law for Promoting the Creation of a Recycling-Oriented Society and the Law for the Promotion of Effective Utilization of Recycled Resources. This recycling law builds upon and has generated other supporting legislation, including the Law for Promotion of Utilization of Recyclable Resources (1995); the Law Concerning Promotion of Separate Collection and Recycling of Containers and Packaging (2000), which accounts for more than 25% of general waste; and the Specified Home Appliance Recycling Bill (2001).

The foundation for the Basic Recycling Law was a report by the Industrial Structure Council (July 1999) outlining a "Recycling Economy Initiative." This sets objectives for a recycling-oriented economic system: "(1) Minimization of resource and energy input in economic activities (Minimum input) [and] (2) Minimization of emissions of waste and effluents caused by economic activities (Minimum emissions)." Actors at every stage of the production-consumption-recovery life cycle are responsible for achieving these objectives through:

- Reducing resources used in products
- Using products as long as possible before they become waste
- Reusing products again as products and parts after their previous use
- Recycle them as resources for use as raw materials of new products

Eco-factory research is developing a systemic design framework for facility, product, and process in support of realizing these strategies at an optimal level. Eco-town projects seek implementation at the municipal level. Walter Stahel's Service Economy vision provides another important context for understanding the requirements for realization of a Recycling Economy. All are described below. Overall the Basic Recycling Law extends from upstream phases of control on industrial waste and product design to downstream phases of recovery and recycling (Tanabe 2001).

The passage of tighter government regulations concerning waste disposal and recycling practices has been an important driving factor encouraging business and industry to find innovative "waste" management practices. With the enactment of the Disposal and Public Cleansing Law in 1970, Japan's legal framework for waste management began to take shape. For about 15 years under this law, waste treatment was based on "end of pipe" models, which resulted in expansion of incineration and other intermediate processing facilities and disposal sites.

Since 1985 local governments have begun to follow more "up-stream" policies, focusing on the reduction of the volume of waste generated. In 1991, the Waste Treatment Law Amendment and the Law for Promotion of Utilization of Recyclable Resources were enacted. In 1995, the Law Concerning Promotion of Separate Collection and Recycling of Containers and Packaging was enacted to encourage a recycling system for container and packaging waste, which accounts for more than 25% of general waste. In 1997, recycling of polyethylene terephthalate (PET) bottles became obligatory. Recycling of paper and plastic containers and packaging became obligatory in 2000. In addition to these laws, the Specified Appliance Recycling Bill was presented in 1997 and enacted in 2001. This law promotes the product take-back system through promotion of the extended producer responsibility concept, particularly with respect to electronic appliances such as televisions, refrigerators, washing machines, and air conditioners.

Finally, the Realization of Recycling Society Bill was passed in the 2000 Diet (*Asahi Shimbun* 2000). The legal framework has been one of the factors promoting the eco-industrial effort in Japan and it is clear that more recycling and waste management—related laws are going to be enacted (Tanabe 2001). The Ministry of Economy, Trade and Industry (METI) is implementing policies on recycling, with responsibility for setting rules and promoting voluntary efforts in industry.

Japan's Recycling Economy Law encompasses both flows of materials and flows of products and their components. This breadth is important to fulfill the full range of opportunities for achieving resource efficiency along the lifecycle of a product or material.

Eco-Town Projects

Eco-town projects are the strongest community and regional programs for the promotion of the Recycling Economy in Japan. The central government provides both technical and financial support to local governments to establish an area (Eco-towns) where zero-waste is promoted regionally through various recycling and industrial by-product exchange efforts. Once a development plan is approved by the Ministry of International Trade and Industry, local governments are eligible for financing to promote and encourage ecologically sound industrial activity (Morikawa 2000).

Although the Recycling Economy Law and associated research emphasizes a lifecycle approach, many of the Eco-town projects tend to be focused on the resource recovery phase of the lifecycle. This follows from the national policy objective of reducing waste by 50% rather than optimizing overall resource utilization. One Eco-town plan, for Kawasaki City, announced an ambitious plan in 2000 that includes a "zero-emissions" industrial park. Kitakyushu's Eco-town project focuses on an industrial park devoted to a wide variety of recycling companies serving western Japan.

Kawasaki City

The City of Kawasaki is adjacent to Metropolitan Tokyo and has a population of 1.2 million. The Kawasaki Coastal Industrial Area houses over 50 heavy industrial enterprises in a 250-acre area. Its largest tenants consist of oil refineries, steel manufacturers, power generators, and chemical manufacturers. By the 1970s the city and the industrial park were considered one of Japan's most contaminated areas. Residents' complaints and the closing of several plants led to a crisis. To resolve the situation, the government established recycling and material reuse programs between facilities, restrictions on emissions, and higher pollution abatement standards, as well as provision and promotion of logistical support and coordination of material exchange, research and development, and public education (Kawasaki City Homepage; http://www.city.kawasaki.jp/ecotown/ecoen.htm).

Kawasaki has a well-developed transportation infrastructure, a high concentration of Japan's leading large industrial firms, and a large number of medium- and small-size enterprises in the field of resource recycling, and various environment-related facilities. The Industrial Group, Industrial Promotion Section, Division of Industrial Promotion of Kawasaki City aims to create an operationally competitive resource-recycling system. The entire plan is to be completed by 2010.

Some of the processes and technologies currently in place include: cement production using fly ash and bottom ash from incinerator plants as inputs and waste oil used for energy to heat the kilns for production. Electronic appliance recycling provides input for steel manufacturing. A new type of blast furnace utilizing municipal plastic waste as a reducing agent in place of coal was finished by NKK, one of Japan's leading steel makers. This system received funding from MITI as a part of the Eco-town project and is in operation with the capacity of recycling 40,000 tons of waste plastic every year.

A "zero emission"[1] Industrial Park's infrastructure was constructed at the abandoned site of a steel manufacturing plant, in cooperation with Japan Environmental Corporation (JEC), and the first tenants have started operation, including an SDK facility to use scrap plastic to make ammonia. The individual industrial firms within the industrial park will reduce their own emissions but also will use or recycle into usable resources the discards from other facilities located there. The tenants also seek to collectively integrate their energy use to improve overall energy efficiency.

Benefits and Challenges

The Kawasaki Eco-town is a joint effort between government and local business. While still in the early stages of development, it represents a promising example of the industrial area redevelopment model, focusing on environmental technologies and by-product utilization efforts. The city will benefit from the reduced burden of municipal waste treatment by having

1. The terms "zero emission" or "zero pollution" are usually accurate only when one sets the boundary of the system narrowly. A "zero emission" facility may have very low or zero emissions within its space. However, it usually uses energy from an external polluting source, moves goods and people by road and rail transport, and its waste water is often processed by a polluting plant.

an advanced recycling facility on site, and private business can achieve cost savings by utilizing recycled materials, which in turn will result in revitalization of the local economy.

The greatest challenge centers on the coordination of activities necessary to achieve effective by-product and energy exchanges. Bilateral exchanges among the old tenants have always taken place as an efficiency improvement effort in the area. The city then analyzed material flows throughout the area to recruit new tenants that would function to close the material cycle and help achieve the optimal by-product exchange model. This attempt to facilitate new links, however, turned out to be a difficult task. Under the long lasting recession, most small and medium sized enterprises are reluctant to venture into a new investment, and some of the candidates for the Zero Emission Industrial Park decided not to move into the park (personal communication, 2000, T. Miyauchi, Director, Industrial Development Division, Kawasaki City). The plan to cascade the heat energy from a tenant next to the Zero Emission Industrial Park is also facing problems, because this tenant couldn't find any short-term economic benefit from this arrangement.

These examples point out the potential difficulties in a third-party facilitating new arrangements. The current economic circumstances in Japan encourage industries to increase energy efficiency and recycling efforts using an industrial ecology approach, but discourage them from taking risks by making new investments or new arrangements which don't give them immediate tangible economic return. Even though firms are recognizing environmentally-related business opportunities, their financial bottom line continues to be their first priority. Further assistance, such as business incentives or subsidies and education and information sessions conducted by local and central government will be needed to overcome this dilemma.

Kitakyushu

The Kitakyushu Eco-town Project has created a major industrial park for recycling companies that draw inputs from throughout western Japan. The Environment Bureau of Kitakyushu led the development with research and education support from the city's universities and research institutes, such as Fukuoka University, Institute for Resource Recycling & Environmental Pollution Control System, University of Kitakyushu, and Kyushu Institute of Technology.

A "zero-emissions" industrial park includes an aggregation of recycling plants consisting of small, midsize, and start-up businesses and a wind-power plant. The investment of 50 billion yen (around US$450 million) was 64% from private sector and 36% from public. One thousand workers are employed in the resource recovery industries at this site, including:

- Plastic PET bottle recycling to fiber: 18,000 tons/year
- Office equipment recycling: 450 tons/month of copying and fax machines, printers, and PCs are dismantled and recycled. The processes are zero emission recycling through dismantling and separation, not crushing
- Used automobile recycling: 18,000 cars/year are recycled with environmentally sensitive treatment of oils and CFCs and producing quality scrap
- Home appliance recycling: TV sets, refrigerators, washing machines, air-conditioners, and freezers are dismantled and separated to producing quality recycled raw materials. 700,000 units per year
- Fluorescent tube recycling: Glass, fluorescent substances, and caps (aluminum) are collected and raw materials used for recycled fluorescent tubes. 5,270 tons per year (111,600 tubes/day)
- Demolition waste recycling: Mixed waste discarded from construction sites is sorted to concrete, woods, metals, etc. 130,000 tons per year
- Waste wood and plastic recycling: Scrap wood and waste plastic are blended to produce waterproof and weatherproof materials. 5,000 tons/year
- Toner cartridge reusing: Used toner cartridges for printers or copiers are processed through disassembly, cleaning, reassembly, and filling with toner
- Energy from wind-power plant: The energy is sold to Kyushu Electric Co., Inc.
- PCB chemical treatment plant to be operated by JESCO (opening end of 2004)
- Electric supply from composite core facility
- Fuel input to the furnace from the recycling companies, the PCB treatment plant, and other industries
- Power supply about 4,500 kw to recycling companies and 4,500 to the PCB plant
- Resource recovery of slag and metal produced by the furnace

A by-product exchange among some of these companies adds to the level of resources recovered. Each company at the location sends its ordinary discards, such as fluorescent lights, paper, and other materials to the appropriate recycler. There is no water reuse among facilities, a feature often found in by-product exchanges. The materials supplied to the energy plant and the energy delivered to park companies are a major energy component.

Some important discard streams are not processed at the Kitikyushu complex. Wood is the only biomass handled, although sewage and industrial sludge, food processing discards, and household garbage are all important resources to recycle. An associated research complex is working on food discards.

Toner cartridges are the only products that are remanufactured here. Nevertheless, this is a notable instance of implementing the Japanese Recycling Economy and Eco-town concept.

The management of the Zero Emissions Industrial Park acts as facilitator and promoter of the project, seeking a balance between industrial promotion and environmental conservation. It lobbies for both deregulation and strengthening of regulations as appropriate. It acts as a one-stop service to expedite procedures and permitting for companies. Finally, it is responsible for public disclosure and accountability to citizens (based on Sato 2004; see also www.kitaq-ecotown.com).

Eco-Factory

Another allied initiative in Japan has evolved the concept of the "eco-factory" and methods for lifecycle analysis and DFE. Japanese researchers in the Agency of Industrial Science and Technology have offered a logical analysis of the technologies needed to achieve the closed-loop system ideal. The model integrates design of production systems technology—including DFE at product and process levels—with disassembling, reuse, and materials recycling technology. These two large components are then linked to control and assessment technology.

The Eco-factory model describes a total system design for closed-loop manufacturing, with the most detailed technical R&D agenda for resource optimization to appear. The inventory includes innovations in energy, design, production, robotics, materials, systems, and information technologies. Specifics such as robotics for disassembly and sorting, materials recognition systems, information systems for concurrent engineering, and many others clearly add up to major business opportunities for those who create and apply them. Companies that pioneer in this field will realize profound cost savings in their own production and can open global markets for these new technologies (JETRO 1992).

China's Circular Economy Initiative

The laws of thermodynamics and economics make a completely Circular Economy (CE) or closed-loop system impossible. However, it is possible to move toward this ideal and in China it is vitally important that this rapidly growing country succeed in achieving a more Circular Economy.

China's leadership has taken inspiration from the Japanese and German Recycling Economy Laws to form a Circular Economy initiative that has major strategic importance for the whole world, not only China. China's rapid economic growth demands major supplies of all basic industrial commodities, in competition with other nations. Its emissions cross boundaries and oceans, impacting Korea, Japan, and North America. Its contribution to greenhouse gases is rising rapidly, even as its energy crisis becomes more acute.

If China achieves its goal of increasing efficiency of resource utilization by a factor of 8 to 10, this will have global impacts. One critical factor will be the success or failure of Chinese leadership in developing a Chinese model of quality of life before a U.S. style consumerist lifestyle fully emerges and overwhelms gains in productivity through the rebound effect. For foreign producers, China's success in the CE effort would set a new level for competitiveness in the world economy.

The issue of competitiveness gained through resource optimization grows in importance as China rapidly builds regional trade alliances and networks of joint ventures. The CE initiative is unfolding in parallel with purchases of mineral and oil processing facilities in Southeast Asian countries, contracting for major energy and mineral purchases, and creating supply chain joint ventures. China's regional supply and production chains assemble more sophisticated components from South Korea, Taiwan, Japan and Singapore with low-cost labor in China and Southeast Asia. To reinforce the trade alliances, China is now supplying bi-lateral development aid to neighbors such as Burma and Sri Lanka.

The National Context for the Circular Economy

China's rapid industrialization in the last decades has engendered serious problems of depletion of natural resources, degradation of major ecosystems, and pollution extending far beyond its borders (Economy 2004; Yardly 2004). Projections by the country's top leadership have persuaded key officials that continuing this unsustainable model of development is simply not possible. The resources are not available to provide a growing population with higher standards in a Western lifestyle of consumption. The challenge for the Chinese government and people is to create an alternative to Western economic development models. This alternative must enable social and political stability in a time of economic dislocation and growing expectations.

The State Environmental Protection Administration of China and the China Council for International Co-operation for Environment and Development have directed the attention of the top leaders of China, at both national and local levels, to a hard reality: the development target set by the government will not be achieved unless alternative models of economic development are identified and applied.

This ambitious development target is to raise the majority of China's population into "the all-round well-being society." This means that by 2050 a larger population of 1.8 billion would reach a per capita GDP of US$4,000 per year, *5 times the current level*. Some estimate that this increase could occur within the next 30 years. This demands a tremendous increase in production and multiplies pressure on natural resources and the environment. Research by the State Environmental Protection Administration indicates that China's economy will need to achieve at least a 7-fold increase in efficiency of resource use to achieve the goals set for 2050, while maintaining environmental quality. The China Council for International Co-operation for Environment and Development states that an increase as much 10-fold will be required (Shi and Qian 2003).

The need for the parallel objective of reducing pollution is directly experienced in the provinces, cities, and countryside. Five decades of aggressive industrialization has seriously degraded natural resources. The Natural Capital accounts for current and future generations show the massive debits of polluted rivers, cleared mountains, depleted soil, and coal and steel mine sites full of toxic materials.

With China's opening up to foreign investment and increasing inequity in distribution of wealth and income, over a billion poor Chinese are demanding a better life: jobs and higher income as well as a better environment to live in. State ministers, provincial governors and city mayors are feeling the pressure for development. They are acutely aware of China's regular periods of massive instability and unrest. They need to meet the demand for improved quality of life to assure political and economic stability.

To meet the needs for development while restoring the health of ecosystems, the only option is to follow a development path different from the industrialization model of the West. China's leaders see that continuing the present massive exploitation of natural resources and inefficient production practices cannot continue. They also are aware that a U.S. lifestyle emphasizing material possessions is simply not achievable. Their conclusion is to adopt the Japanese and German Recycling Economy approaches and set higher goals than either.

The Response

> China can no longer afford to follow the West's resources-hungry model of development and it should encourage its citizens to avoid adopting the developed world's consumer habits. . . . It's important to make Chinese people not blatantly imitate Western consumer habits so as not to repeat the mistakes by the industrial development of the west over the past 300 years (Pan Yue, Deputy Minister, State Environmental Protection Administration) (New York Times 2004).

Leaders at the highest level are creating a new vision of China's future. National leaders such as Xie Zhenhua, Minister of SEPA, are charting a 50-year plan to achieve sustainability. Several developments in recent years contribute to this planning.

- Growing recognition of the need to create a development path to meet the needs of a growing population at a higher standard of living without following the model of Western consumerism, inefficient resource consumption, and pollution
- Developing a Circular Economy model with high resource efficiency and low pollution
- Passage and implementation of the Cleaner Production Law
- Commitment of US$1.2 billion in science and technology investment for sustainable development by the Ministry of Science and Technology
- Entry to WTO and the need for China's industry to become more competitive
- Acceptance of the nearly universal consensus on climate change, reflected by China's signing of the Kyoto Accords

A Chinese alternative to the outmoded and unfeasible American development path will aim to achieve improved *quality of life* for the Chinese people. Quality of life includes having basic needs for food and shelter met. Access to education, health, and cultural resources is key to such quality. Stable and lively community life is an important context for the lives of families. A healthy environment is the basis for healthy families and a thriving economy. Guaranteeing these aspects of quality of life is the deepest purpose of any economy. Many people in the West are now realizing that economies driven by unsustainable material consumption do not achieve this purpose.

The CE concept has developed in China as a strategy for reducing the demand of its economy upon natural resources as well as the damage it causes to natural environments. The CE concept calls for very high efficiency in resource flows as a way of sustaining improvement in quality of life within natural and economic constraints.

A key leader in creation of the CE concept has been State EPA Minister, Xie Zhenhua, with support coming also from the State Development Planning Commission and the State Economic and Trade Commission. The Ministry of Science and Technology's plan, Science and Technology Framework for Sustainable Development 2001–2010, outlines an agenda for R&D and environmental business development that could support RE enterprises. The China Council for International Cooperation on Environment and Development (CCICED) Taskforce for Promotion of Circular Economy and Cleaner Production has met in Liaoning and Shanghai to review local efforts and produce papers to advance the concept.

In 2003 the National Development and Reform Commission assumed leadership of the implementation of the CE initiative. This mega-agency is also responsible for implementation of the Cleaner Production law, the Renewable Energy Law, and energy efficiency measures.

The Circular Economy Concept

The Circular Economy approach to resource-use efficiency integrates cleaner production and industrial ecology in a broader system encompassing industrial firms, networks or chains of firms, eco-industrial parks, and regional infrastructure to support resource optimization. State-owned and private enterprises, government and private infrastructure, and consumers all have a role in achieving the CE. The three basic levels of action are:

At the individual firm level, managers seek much higher efficiency through the three Rs of Cleaner Production (CP), *reduce* consumption of resources and emission of pollutants and waste, *reuse* resources, and *recycle* by-products. (Sustainable product and process design is important in German and Japanese recycling economy plans but is just emerging as a component of the Chinese CE concept.)

The second level is to reuse and recycle resources within industrial parks and clustered or chained industries, so that resources will circulate fully in the local production system. (The Chinese use the term "eco-chains" for by-product exchanges.)

The third level is to integrate different production and consumption systems in a region so the resources circulate among industries and urban systems. This level requires development of municipal or regional by-product collection, storage, processing, and distribution systems.

Efforts at all three levels include development of resource recovery and cleaner production enterprises and public facilities to support realization of the CE concept. This adds a strong economic development dimension through investment in new ventures and job creation. So the CE opens opportunities for both domestic and foreign enterprises.

Consumers have a role at the household and neighborhood level in applying the CE concept. The majority of the Chinese people still fail to meet all of their basic material needs, including potable water for drinking and sanitation, affordable and good quality food, basic housing, and household equipment. The Circular Economy must support families in achieving these requirements of life. At the same time local initiatives must offer citizens education in the practices of reduce, reuse, and recycle at the home level.

Once basic needs are met, CE leaders are aware of the challenge involved in shifting to less material consumption patterns, one that improves quality of life and avoids the Western lifestyle of wasteful consumption and "throw-away habits." However, given the present level of poverty, the main focus is on meeting basic needs through maximum efficiency of resource-use.

The Circular Economy concept is gaining acceptance due to the urgent need for an alternative approach to achieve both economic development and environmental protection. If the world's most rapidly growing economy continues to follow the conventional approach, which makes environmental protection an add-on to conventional economic development, China will not be able to achieve its linked economic development goals and environmental quality targets.

Moving Toward a More *Circular Economy*
Provinces piloting application of the CE concept include Liaoning, Jilin, Heilongjiang, Jiangsu, and Fujian. Cities include Shanghai, Tianjin, Zhenjiang, and Guiyang, as well as other cities and development zones in these provinces such as Dalian and Shenyang in Liaoning Province and Suzhou, Zhangjiagang, and Changsu in Jiangsu Province.

The Liaoning Plan
The Environmental Protection Bureau of Liaoning Province released the first provincial level Circular Economy plan in the summer of 2002. This Liaoning Circular Economy plan shows a very strong focus on the last part of the materials cycle. However, the major objectives for efficient use of resources set in the document require that measures be taken along the entire lifecycle of resources, products, or services, not just at the end. The CE plan projects implementation of the new Cleaner Production Law by enterprises as the means to create change early in the lifecycle to increase efficiency of resource use and reduce waste produced. Design of products and services using DFE and lifecycle assessment tools is another very important measure. Both public and private sector organizations will need to collaborate to support these and other systems level changes.

Eco-industrial parks, eco-chains (exchange of by-products among industrial plants), and regional infrastructure for materials management are specified by the Liaoning plan. The sites for EIPs include the Dalian, Shenyang, and Yingkou economic development zones and Anshan high-tech industrial park.

Initially implementation of the Liaoning CE plan has been led by provincial, municipal, and development zone Environmental Protection Bureaus, (the local arm of the national-level State EPA). As implementation of the overall CE initiative shifts to the National Reform and Development Commission, its local offices will have an increasing role (Liaoning Province, EPB, 2002).

Eco-Chains in China: The Guitang Group
An important strategy for implementing the Circular Economy was inspired by a Southern China case that evolved over several decades. The term used in CE planning is "eco-chain," where resource use is optimized through a chain of uses for any by-product. The Guitang Group (GG) in Guigang City developed a sugar production based eco-chain internally and now extends it to its competitors in the city.

China produces 10.5 million tons of sugar annually from 539 sugar mills, the majority from sugar cane. Over the last few years, the sugar industry in China has experienced a significant economic decline. This industry has to increase its productivity to remain competitive with Brazil, Thailand, and Australia, three major sugar producing countries. Low prices for sugar on world markets in recent decades have eliminated the industry in former leading countries, including Hawaii and Puerto Rico in the U.S. Sugar production is becoming much less competitive in the Philippines.

The Guangxi Zhuang Autonomous Region, in the far south of China, is the largest source of sugar, producing more than 40% of the national output. The cost of producing sugar has been high in Guangxi. Most farmers have small landholdings, productivity is low, and sugar content of the canes is low. Most refineries are smaller scale and fail to utilize their by-products. This gap causes them to lose secondary revenues and generate high levels of emissions to air, water, and land. The farmers burn the cane leaves every harvest season, generating air emissions. Currently 70,000 families grow sugar in the region and there are 100 sugar mills. One-third of Guigang City's economy depends upon sugar related industries (Zhu and Lowe 2004).

The Guitang Group is a city-owned enterprise formed in 1954 that operates China's largest sugar refinery, with over 3,000 workers. The group owns 14,700 hectares of land for growing cane. Though the sugar industry in China is generally responsible for high levels of emissions, this company has created a cluster of companies in Guigang City to reuse its by-products, thereby reduce its pollution. It has also created supplier relationships with other sugar mills to increase input to its paper mill and alcohol refinery by using their by-products.

The complex includes: an alcohol plant, a pulp and paper plant, a toilet paper and tissue plant, a calcium carbonate plant (recycling waste water from the paper mills), a cement plant, a power plant, and other affiliated units. The goal of the initiative is "to reduce pollution and disposal costs and to seek more revenues by utilizing by-products" (Duan 2001). The GG has established two primary eco-chains, each of which has additional members and some internal feedback loops.

The annual total production of the complex includes sugar (150,000 tons), raw sugar (300,000 tons), pulp (150,000 tons), paper (200,000 tons), alcohol (10,000 tons), cement (330,000 tons), alkali (35,000 tons) and fertilizer (30,000 tons) (Guitang Group 2004). In the late 1990s the secondary products accounted for 40% of company revenues and nearly as large a portion of profits and taxes paid.

The synergy among the Guitang Group's production facilities reduces the market risk for each product and enables an integrated profit for the complex. For example, the sugar refinery suffered a large loss in 1999 because the price of sugar in China decreased by about 1,000RMB (US$122) per ton compared with 1998. However, the GG remained profitable because the prices of other products such as paper remained high and it increased the production of other products (Guigang City and Guitang Group 2000).

The GG is supporting its farmer suppliers in the transition to cultivation of organic sugarcane with higher sugar content, increased productivity, and an extended harvest period. This will reduce the cost of the sugarcane supply per ton as it improves quality and earns a premium price. Eliminating chemical inputs will reduce farmers' costs while higher productivity also increases income (Guigang City and Guitang Group 2000).

The GG's plans for the future include expansion of the production complex and changes in processes at various stages. This innovative plan includes:

- Construct a new beef and dairy farm using dried sugarcane leaves as feed
- Construct a milk processing factory to make fresh milk, milk powder, and yogurt for the local market
- Construct a beef packing house to process beef, oxhide, and bone glue
- Build a biochemical plant to make amino acid based nutrition products and other bio-products using the byproducts from the beef packinghouse
- Develop a mushroom growing company using manure from the new dairy and beef farm
- Process residue from the mushroom base to use on sugarcane fields as natural fertilizer

China's expected entry into the World Trade Organization poses a major threat to the economy of Guangxi. With barriers to lower-priced imports lowered, the economy of this region could be injured profoundly. So the Guitang Group's eco-industrial initiative has strategic importance for this and other sugar producing regions in China.

City of Guigang Plans to Become an Eco-Industrial City

The Guitang Group's example has inspired the town of Guigang to adopt a 5-year plan to become an Eco-industrial city. The heavy dependence of its economy on the sugar industry makes it important to improve the efficiency of its many processing plants. In fulfillment of the plan smaller sugar producers are sending their by-products to Guitang's eco-industrial complex to achieve higher resource utilization. (Targets for the city: "utilization of sugarcane slag reaches more than 80%, use of spent sugar-juice reaches 100%, use of spent alcohol reaches 100%.") The plan also calls for consolidation of cane-growing land into larger holdings. (It will require a transition for small farmers into other crops or into industrial employment.) It includes training of industry and government managers in eco-industrial principles and methods and broader dissemination of Cleaner Production strategies. Some of the long-term goals of this plan are:

- Develop an eco-sugar cane park to enable planting of organic cane, increases in sugar content of canes, increase in production per mu of land, and extend the harvest period
- Enlarge the paper mill with a goal of increasing production to 300,000 tons per year in three phases
- Switch some production from sugar to fructose, which has a strong market
- Build a facility to produce fuel alcohol from spent sugar juice and sugar (capacity 200,000 tons per year). This product will help reduce air pollution from vehicle exhaust
- Adopt low chlorine technology to bleach pulp. Paper made by this technology will be much whiter than the paper made by traditional technologies

Guitang and the leadership of the town are supported by China's State Environmental Protection Bureau (SEPA) and the China National Cleaner Production Center (CNCPC). Financing is from the financial bureau of Guigang City. The local tax administration will return 50% of the agriculture tax to construction of irrigation systems for sugarcane farms.

In India there are at least 20 sugar industry-based complexes similar to what the Guitang Group has created in Guigang. In one of these cases a paper company established sugar cane cultivation and sugar production in order to have bagasse as a by-product feedstock for its mill. Pulp wood prices had risen too high to produce paper competitively with this traditional input. Essentially sugar production in this case can be seen is a "by-product" of paper production (Erkman and Ramaswamy 2003).

See Nelson Nemerow's Chapter 11 for additional information on industrial complexes; as well as his book, *Zero Pollution for Industry*, which includes an alternative pattern for a sugar refinery complex, with part of the bagasse and all of the sludge going to an anaerobic digester to generate methane, which is used as fuel for boilers. The remainder of the bagasse is burnt in another boiler. The material output from the digester is filter cake that goes to farms for fertilizer (Nemerow 1995).

The author visited the Guitang Group complex in April 2001 and interviewed GG managers and city officials. The co-author, Zhu Qinghua, visited the complex in 2004 to update information.

Shanghai

Shanghai is the domestic source and first pilot project of the Circular Economy. Its Environmental Protection Bureau began studying the German and Japanese Recycling Economy, thinking as early as 1995 and incorporating it in the city's Agenda 21 program. By 2000 Shanghai's 10th 5-year plan of economic and social development incorporated Circular Economy goals and programs. Circular Economy principles led to formulation of regulations and laws, as well as policies on water charges and for disposing of sulfur dioxide.

The pilot project in Shanghai has sought to bring together municipal planning, economic, environmental, construction and social agencies to seek an integrated approach to demonstrating a CE. Their efforts have included an energy efficiency law, regulations on construction and reconstruction, planning of industrial areas, and preservation of wetlands and natural habitats (Shi et al. 2003).

Issues with the Circular Economy

Bi Jun, a researcher at Nanjing University, has identified key problems with the Circular Economy, based on his review of national policy, pilot projects, and his own participation in development of provincial and local CE plans. Some of the key points he and his colleagues make include:

- The high expectations of the central government are not matched by the knowledge and experience of local officials and citizens.
- The CE is taking a top-down approach, essentially in the style of China's planned economy.
- However, guidelines for planning are weak or lacking. People attempting to implement CE projects are unclear as to the how this approach differs from standard environmental protection planning.
- The concept is reduced to rigid solutions in some cases, such as company to company exchange, rather than maintaining a broader view.
- Market-based solutions have not been adequately integrated (Bi Jun 2004, Bi Jun et al. 2004).
- Predict the demand of resources in the next 15 years.
- Improve theories for material flow analysis, life cycle analysis, environmentally conscious design, and eco-industrial parks.
- Green construction and building technology in urbanization.
- Eco-farming and agricultural wastes recycling.
- Dematerialization technology.

The System Required for Achieving a More Circular Economy

China's Circular Economy initiative has set a highly ambitious goal of improving the country's productivity of resource use, while reducing pollution. However, initial provincial and municipal pilot projects have tended to focus on industry, primarily heavy industry. These efforts have also tended to emphasize infrastructure for recovery of resources and company to company exchanges of industrial by-products. This heavy industry focus and emphasis on end of cycle solutions is a necessary part of the strategy, but only a part. The four major projects of the 11th 5-year research plan begin the expansion of scope.

Achieving a more Circular Economy in China demands a whole systems guidance framework for policy setting, methods of planning, economic sectors, and stages of the production-consumption cycle (Lowe 2004a). The sectors that must be encompassed include:

- Heavy and light industry, ranging from small to medium enterprises to large scale extraction and production facilities
- Urban master planning, including transportation infrastructure
- Municipal infrastructure for energy, water, discarded materials, transportation, and communications
- Planning, construction and management of the built environment
- Farming, food processing, and agribusiness suppliers
- Households, where critical choices are made impacting resource utilization and resource recovery

This list reminds us that land is a critical resource, along with energy, water, and materials. To achieve as much as a 10-fold increase in efficiency of resource use, profound change in all six sectors is required, since all are major consumers of resources.

A systems framework for this change integrates the following disciplines and professions:

- Eco-industrial development optimizes performance of industry, commerce, and public infrastructure
- Green urban planning and architecture optimizes performance of the built environment and utilization of land
- Integrated transportation planning optimizes movement of people and goods
- Sustainable farming and food processing optimizes use of rural land and increases productivity
- Consumer education and action optimizes performance of households
- Ecosystem restoration renews the natural capital all human systems depend upon for their existence.
- Capacity development optimizes performance of public and private management systems

Achieving a more Circular Economy demands advances within each of these fields and interaction among them. For instance, the long-term damage of a road-based transportation system in China would be its enormous use of scarce land and other material and energy resources, not only the air emissions which are now seen as the major environmental cost. Thus, urban planning, transportation planning, industrial development, and consumer education have to seek transportation solutions that succeed within these major resource constraints. In the U.S. one-third to one-half of urban land is used by streets, roads, freeways, parking structures, and service stations.[2] China does not have the land base needed for a US style transportation infrastructure. A draft of the 11th 5-year research plan in China names 21 programs, including "environment, resources, and circular economy." This draft indicates consideration of expanding the scope and effectiveness of the Circular Economy initiative. A critical change would be making explicit the role of energy, construction, and agriculture in the CE.

It is strategically very important to recognize the economic benefits of the Circular Economy, not just its costs. Market-based mechanisms are crucial to the success of this initiative. They enable the transformation to tap the creative entrepreneurial spirit of the new economy in China, while still utilizing public planning mechanisms to assure balanced development. If the movement toward a more Circular Economy succeeds, China's companies will be more competitive. China's cities and development zones will develop new housing and commercial space in a more affordable way. Entrepreneurs will create new ventures, offering many new jobs. Households will enjoy improved quality of life.

The Economics of the Circular Economy

The broadest economic challenge to achieving a Circular Economy is creation of a strategy for meeting basic needs of the Chinese people, improving quality of life, and moving beyond ever growing material consumption as the measure of quality of life.

However, CE research has focused heavily on technical means and some institutional changes to move the Chinese economy toward a closed-loop system of resource use. Such a major change in resource utilization has profound economic implications. Provincial CE plans, such as the one in Liaoning, call for developing a new economic model but offer little detail on what should be included or how to go about this. The CCICED Circular Economy and Cleaner Production task force research agenda has not explicitly included economic issues.

There are two basic sets of economic issues: (1) how the economic requirements of a CE production and consumption system can be addressed and (2) how the CE fits within a full sustainable economy model.

(1) How can the economic requirements of a CE production and consumption system be addressed?

2. Roads, freeways, and parking garages typically use between one-third and one-half of urban space and a major share of transportation dollars. Interstate highways cover valuable agricultural land while air pollution damages crops and forests. (Renner, Michael. 1988. "Rethinking the role of the automobile." *Worldwatch* Paper 84. Washington, DC: Worldwatch Institute.)

- In what way do the goals of a CE open opportunities for economic development, creation of new enterprises, and new jobs?
- What policies, R&D, financing, and incentives are required for CE business development? What is the role of market-based mechanisms?
- What should be the balance between state ["State" is the word for the national level of government so it should be kept capitalized.] Involvement and market dynamics in financing the investments required by industries to gain high resource efficiency in both public and private sectors?
- Increased resource efficiency may lower demand on public infrastructure such as land fills and water treatment plants. Can deferred public investments in such facilities be used to support resource recovery business development?
- After a certain degree of improvement, investment in increased resource efficiency may not continue to provide an adequate return. Is the answer government subsidy, increased R&D, or accepting the limits to resource recovery?
- How effective can economic instruments, such as greenhouse gas trading credits, be in providing incentives for CE implementation?
- What methods of environmental and ecological economics and industrial ecology should be applied by public and private organizations to calculate true costs and benefits of the CE? Total cost accounting? Lifecycle analysis?
- What system of economic, social, and environmental indicators will provide appropriate feedback to the actors creating the CE in any city or region?

The goals of the Circular Economy must be achieved within the unyielding constraints of resource availability, environmental carrying capacity, and the limits of eco-efficiency. Many nonrenewable natural resources have limited reserves regionally, nationally, and globally. The capacity for renewable resources to regenerate is compromised by pollution and over use, as with ocean fisheries. There are natural limits to eco-efficiency in human systems, including the energy demand of resource recovery, the number of cycles through which materials can be recycled and still maintain value, and dissipative uses of materials where they become unrecoverable by the nature of their use (e.g., particles from the surface of tires dissipate along roadways).

(2) How does the CE fit within a full sustainable economy model? There is an inadequate sense of the boundaries of the CE concept and how it fits within a full sustainable economy planning process. To become a more complete sustainable economy approach it needs to encompass more aspects of the economic system, addressing the following gaps:

- Circular Economy research and projects focus on resource use in the production system, with marketing, distribution, and consumption systems being secondary.
- Transportation, urban planning, building design, and construction are important elements in a full CE model, but are not usually mentioned as such.
- Discussion of energy is weak, except for efficiency targets and mention of energy cascading between processes and firms.
- There is little discussion of renewable energy as a contributor to CE systems.
- Energy and water requirements of resource extraction or recovery are important constraints to consider.
- Renewal of natural capital through ecosystem restoration must be a part of the CE model.

A sustainable economy demonstrates relatively equitable distribution of wealth and income and access to the goods and services providing quality of life. The social issues of worker health and safety and healthy communities are also important to sustainable development. The CE concept does not usually address these important issues.

(Zhao Wei, Environmental Affairs Officer at United Nations Environmental Program and UNEP representative to the CCICED Task force on Cleaner Production and Circular Economy, has greatly contributed to the author's understanding of the CE through conversations and collaboration.)

The Service Economy

Product Life-Extension Leads to the Service Economy

Introduction

Implementation of the Recycling and Circular Economy concepts has emphasized profound advances in product design to achieve much higher efficiency of resource use. Walter Stahel, a Swiss researcher has detailed strategies for lowering demand for energy and materials by designing durable and upgradable products with a long-life span. But he goes on to answer the economically crucial question, *How can manufacturing companies remain profitable if their products are so durable?* Stahel suggests some are already refocusing their mission to delivering customer service (selling results, performance, and satisfaction rather than products) and owning the equipment themselves as the means of providing this service.

The concepts of product-life extension and the service economy are valuable contributions to closing the loop in industrial/consumer systems.

Is Stahel's vision useful only in the long-term? Several major companies are now moving in this direction. Stahel's concepts could possibly be used by networks of entrepreneurs to enter markets now dominated by existing companies.

Summary of Product-Life Extension

Stahel has linked the concerns of industrial metabolism and DFE with a broader level of design: the basic mission of a business. Stahel, a director of the Swiss Product-Life Institute, argues that closing loops through recycling and waste reduction is only a partial solution. It does not slow the rapid and unsustainable flow of materials and goods through economies.

He proposes product-life extension as the necessary complement to recycling. He suggests business strategies for achieving it and the dimensions of a service oriented economy. While his vision implies deep changes, Stahel identifies major corporations, such as Xerox, Schindler (the world's second largest elevator company), Agfa Gevaert, and Siemens, that now demonstrate the concept in practice. Companies like Chemfil and Safety-Kleen have demonstrated that this service model can be applied to supply and recycling of industrial solvents and other chemicals.

Product-life extension implies a fundamental shift from selling products themselves to selling the utilization of products, the customer value they yield. This change in the source of economic value to firms depends on enhancing product life through several key design strategies. Designers would seek to optimize the following product qualities:

- Durable and difficult to damage
- Modular
- Multi-functional[3]
- Sub-components are standardized, self-repairing and easy to repair
- Easy to upgrade
- Components can be reused in new systems
- Units or systems can be easily reconditioned and remanufactured

These design strategies are already part of the DFE toolkit. They would significantly help achieve Recycling/Circular Economy resource optimization objectives of cutting demand on material, water, and energy resources and reducing pollution from manufacturing.

The Xerox Asset Recycle Management Program

The ARM initiative at Xerox Corporation reflects the equipment design strategies recommended by Walter Stahel. The mission: "*Asset Recycle*

3. Several manufacturers have been marketing products integrating a laser printer, scanner, fax, and telephone in one device.

Management provides the leadership, strategy, design principles, operational and technical support to maximize global recycling of parts and equipment, resulting in a major competitive, as well as environmental advantage for Xerox" (Xerox 1995).

This mission has been built into the company's global organization with an ARM Vice President responsible for achieving 100% recyclability of all manufactured parts and assemblies. Remanufacturing to high quality standards and resale to new users will extend the life of equipment several fold and reduce demand on virgin resources. "Xerox chooses to evaluate products on the basis of quality and performance, not on the degree of virgin material used." (Xerox 1995). The initiative is designed to streamline the process by which returned machines are reconditioned, thus increasing return on investment.

The company estimates it has added hundreds of millions of dollars to its bottom line since ARM was formally started in 1991 (Xerox 1995).

Strategies for the Service Economy

As a company moves from maximizing sales of material products to the delivery of customer satisfaction, its long-term source of competitive advantage will become the ability to provide the needed service. Revenues could come from leasing of equipment with long-life; continuing maintenance and service; major upgrading of systems; parts and supplies; service provider training and licensing.[4] Or the company might simplify the transaction by offering one, use-based fee.

Stahel argues that *if the company is compensated on the basis of service provided, its employees will have strong incentives to minimize materials and energy used in the systems that deliver the service to the customer.*

Stahel also considers the larger transition to a decentralized and skill-based service economy that product durability implies. Economic value would be based in utilization (customer satisfaction in the service gained) rather than exchange. Decentralized labor-intensive service centers would create many skilled jobs for workers no longer needed in centralized, automated production units. Resource use would be lowered as products no longer moved rapidly from factory to customer to landfill.

Walter Stahel's work represents design at the level where a company asks, *"What business are we really in?"* Wise decisions at this level will have the greatest impact on a firm's environmental performance.

The Service Economy vision is guiding existing corporations in redefining their mission. In some industries, networks of entrepreneurs

4. Service might be provided through a decentralized service network, licensed and trained by the central company. Economically, this would be an important source of skilled jobs.

could adopt this systems approach in order to compete with established firms. For instance, a home services firm could provide a highly efficient integrated energy and information system to homeowners, owning the equipment and collecting a monthly fee.

Benefits

- Product-life extension is a strategy that promises to make very large reductions in materials and energy use needed to satisfy growing consumer needs. Stahel estimates that it could increase the productivity per unit of resource used ten fold, the goal of the Chinese Circular Economy.
- The strategy includes strong economic incentives for achieving these objectives. Improved resource productivity translates to increased profitability and competitiveness.
- The service economy concept offers a decentralized means of developing skilled jobs.
- Stahel's systems approach could give independent entrepreneurial ventures competitive advantage in entering markets when major corporations who could use it remain focused on selling products.

Challenges

- This approach to sustainability requires long-range vision and major organizational and technological redesign on the part of corporations. (Investment markets' present focus on short-term financial performance does not support such fundamental change.)
- Product-life extension runs the risk of companies making major investments in technologies for service delivery that may become outmoded. To what extent can modular design for easy upgrading offset this risk?
- Are there ways that Stahel's concepts could be applied to more transitional products?

This section on the Service Economy is based upon Lowe et al. 1997; Stahel 2003; Stahel and Giarini 1989/1993, 1994; and the Web site, www.product-life.org.

Restoration: a Fundamental Sustainable Economy Strategy

The restoration and renewal of natural systems, rural areas, and our cities and towns is the most fundamental economic solution. The devastation of natural systems has seriously depleted the natural capital upon which all life depends (Society for Ecological Restoration 2004). *All life* includes every type of ecosystem and its organisms, human economies, industries, and households. Inappropriately designed urban and suburban systems have misused massive volumes of materials, embody huge investments of energy and water, and generally operate with very low resource efficiency.

So investment in restoration of natural systems and the constructed environment is as necessary as the optimization of resources the Circular, Recycling, and Service Economies seek. It actually improves the production of renewable resources, thus adding major economic value. Restoration will play a critical adaptive role as climate change unfolds in the next decades due to greenhouse gas emissions. Every type of ecosystem and all human systems will be impacted by this transition to a warmer, less predictable climate.

Restoration of natural and human systems is already a well-established field of investment, with a value estimated as at least US$1 trillion annually (Cunningham 2002). As environmental crises become more acute, restoration will evolve from scattered projects to regional initiatives integrating renewal of watersheds, forests, grasslands, coastal areas, farm lands, and urban areas (Society for Ecological Restoration International and World Conservation Union 2003). Policy and investment to support restoration will create new opportunities for entrepreneurs, many new jobs (that can't be outsourced), and a strong return in both profits and tax revenues (Cunningham 2002).

A Riparian Restoration Project in Illinois: The Holistic Value of Ecological Restoration

Flood-plain land along rivers is quite fertile and through the centuries has attracted farmers, who in turn receive support from public agencies to minimize flooding and limit or eliminate marshes. Often dikes and pumping keep the land fully available for farming except in extreme flood situations. While contributing to crop production, this widespread practice severely limits the natural services and biodiversity that undeveloped riparian systems provide.

- Flooding is more likely as the absorption of high waters by wetlands is eliminated.
- Habitat for migratory waterfowl and breeding grounds for fish are both reduced.
- Habitat for wildlife including migratory waterfowl, game birds, and land and water species is reduced.
- Farmers who follow the currently conventional approach of applying expensive chemical fertilizer, pesticides, and herbicides, cause severe pollution to rivers.
- Ultimately the soil is degraded by the dependence on chemical inputs and the infrequency of flooding.
- Large animal feedlots and pens add to river pollution through excessive runoff of nutrients.

Reversing these damages through riparian and wetlands restoration brings natural systems back into balance and has both public and private sector economic benefits. For instance, in 1999 The Nature Conservancy (TNC), a large nongovernmental organization, bought a 1,700-acre farm in Illinois, at Spunky Bottom, along the Illinois River. TNC managers discontinued drainage of the site and allowed it to regenerate as wetlands and shallow lakes. Within a few years otters, muskrats, turtles, frogs, and fish populated the former farm. Waterfowl such as osprey and herons search for their meals among the grasses, reeds, and floating plants. It is now a recreational site for people, a buffer for absorbing floodwater, and a purifying system for the river.

Spunky Bottom was a test site for a new project on 7,000 acres, Emiquon Farm 165 miles southwest of Chicago. Nature Conservancy ecologists will stop draining the land over 3 years, allowing a riparian wetlands ecosystem to spontaneously regenerate, as happened in the earlier project. They expect this to also renew the population of fish in the Illinois River, which once was one of the most abundant fishing rivers in the U.S. Migratory birds will find such restored wetlands a welcome stop over on their journeys north and south each year. Ducks Unlimited is an interest group whose members understand the value of ecological restoration as a means of maintaining viable populations of game fowl (Kinzer 2004).

Economic Values Created by This Riparian Restoration

An economist could calculate the economic impact of taking these farms out of agricultural production and allowing them to return to flood plain riparian systems. Some of the values generated include:

- Reduced risk of flooding downstream through the ability of the marshes to absorb high water
- Bio-processing of pollutants from upstream by the marshes and wetlands
- Avoidance of pollution from chemical agriculture
- Sequestration of carbon dioxide to reduce input to climate change
- Productivity of the riparian fishery and increased gamebird population, yielding food and recreational opportunities
- Development of eco-tourism businesses to serve visitors, such as accommodations, restaurants, campgrounds, canoe rentals, tours, and local crafts

A rich variety of enterprises and jobs will be created compared with the relatively few permanent jobs in a modern large-scale farm. The economic investment produces financial profits and rebuilds valuable natural capital, including high biodiversity, sinks for human pollution, and ecological resilience (the ability to adapt to change). "Natural capital" is, of course, a concept centered in the welfare of humans. It is worthwhile to back off from our own self-interest and simply perceive the intrinsic value of viable ecosystems functioning in an evolving balance.

The Dimensions of Restoration

This specific ecosystem restoration project creates a context for exploring the full range of economic opportunities, which are at the same time imperatives for maintaining the vulnerable ecological niche that enables survival of human life. Its special fields of innovation and effort are divided into natural systems and the built environment or constructed systems, however there are many areas of overlap.

Restoration of natural systems:

- Watersheds, including wetlands, streams, rivers, lakes, aquifers, glaciers (and manmade reservoirs and canals)
- Coastal ecosystems, estuaries, dunes, beaches, reefs
- Forests, grasslands, deserts, and avoidance of desertification
- Airsheds and regional atmosphere

Regaining balance in the global "grand cycles" that flow between the atmosphere and other living systems (oxygen, carbon, nitrogen, and sulfur) is a primary objective which requires both forms of restoration.

Restoration of constructed systems:

- Sites of extraction, generation, or collection of natural resources to supply infrastructure and industry
- Urban and rural infrastructure for transportation, energy, water, sewage, waste, telecommunications
- Residential, commercial, and industrial buildings and facilities and their interrelationship in resident usage and resource utilization
- Cities, towns, and suburbs as integrated land use systems in the context of their surrounding ecosystems (including historical restoration, neighborhood renewal, and rehabilitation of infrastructure)
- Natural systems within the built environment, including parks, urban rivers and streams, and wildlife habitat and corridors

We see farms and farming regions as constructed systems that are also natural systems, bridging the two categories. Renewal of farms is intimately related with the restoration of farming towns and villages, and neighboring urban centers. See a following section outlining the ways in which sustainable farming is interdependent with regional restoration initiatives.

Watershed restoration includes reforestation to better hold rain water and prevent landslides, renewal of disrupted hydrogeology of streambeds, renewal of streamside habitats, cessation of grazing near streams, adoption of "best management practices" (BMPs) for animal waste management and manure-to-crops applications, and training of farmers in practices to reduce impacts on waterways and their demand on water.

In urban areas, watershed restoration requires strict measures for pollution prevention and treatment to eliminate flows of chemical and biological pollution into waterways from factories and municipal infrastructure. Naturalizing the channels of urban rivers and streams is often possible when the rest of a regional watershed management is effective in minimizing flood risks. This creates attractive recreational space.

Another vital type of ecosystem restoration, closely linked to watershed management, is **reforestation of mountain ecosystems** denuded by clear-cutting of timber. Clear-cut land usually results in runoff of soil and landslides in hilly and mountainous areas. The goal is to create diverse ecosystems of indigenous plants and animals, not the monocultures of introduced species that often characterize such efforts. Such restoration can have a significant long-term economic impact through sustainable selective harvesting of trees and other forest products. The life of downstream dams is extended by reducing the flow of silt from landslides. The forest contributes to atmospheric purity and, for a significant period, absorbs carbon dioxide.

Restoration of the built environment will become increasingly significant as developed and developing economies deplete supplies of nonrenewable resources. Infrastructure, residential, commercial, and government buildings, and industrial facilities are banks of natural capital in the form of embedded energy and materials. As the price of oil escalates, the economic value of rehabilitating versus new construction will become quite apparent. A relatively basic version of this is conducting resource audits and designing retrofits to improve the efficiency of energy and water use *and* cut the utility bills.

In a documentary titled *The End of Suburbia: Oil Depletion and the Collapse of the American Dream*, James Howard Kunstler, a critic of urban planning called suburban development "the most wasteful use of natural resources in history" (Greene 2004). Nearly all suburban planning assumed the individual automobile would be fueled by cheap oil far into the future. Kunstler and a number of authorities on the petroleum industry lay out a scenario in which continued increases in oil prices will make suburbia's spatial sprawl too costly to maintain present lifestyle. The long commutes of workers and the long "local" trips across a complex road infrastructure to satisfy the basic needs of family life will have to change. The only real uncertainty, they say, is whether oil production has already peaked or will do so in the next 5 to 15 years. Oil prices were already at a record high in the last quarter of 2004.

(See www.endofsuburbia.com/link.htm; and the Web sites of organizations such as the Association for the Study of Peak Oil, www.peakoil.net; Oil Depletion Analysis Centre, www.odac-info.org; and Peak Oil Action, http://peakoilaction.org.)

Once that peak is reached, it will present a major challenge and economic opportunity for urban planners, citizen groups, and engineering and construction firms to retrofit suburbs to adapt to the new economic realities. Infill development, more distributed commerce and services, telecommuting, and light rail transportation systems are a few of the restoration options already emerging. Co-housing is also an option attracting much support. This is where groups of families and extended families set up common services for a cluster of homes. A project in Davis, California has opened their backyards into a shared space for gardens and sports and built a commons building.

Remediation of polluted sites is another area of restoration with the dual value, in best cases, of eliminating a source of continuing pollution and making cleaned up land available for appropriate new uses. Factory sites, exhausted mines and oil fields, and toxic waste dumps are at one increasing scale of difficulty. Much more challenging are the massive military industrial dumps of radioactive and biological weapons waste. Russia and China have similar sites at even greater risk.

Restoration of Farms and Rural Towns

A good example of integration across several fields of restoration is linking the transition to organic farming in a region with watershed, riparian, and soil restoration, as well as village and urban renewal. The following is generalized from a concept for a project on the Yellow River in China's Shandong Province (Lowe et al. 2004).

The business foundation for this initiative would be an agriculturally based eco-industrial park (AEIP) as home to food processing and distribution companies, equipment manufacturers, energy generators, and manufacturers using rural and urban biomass discards. Organic farming requires a set of suppliers and services generally quite distinct from those supporting high-input chemically based agriculture. The AEIP will generate new entrepreneurial and employment opportunities.

An organic agriculture research and training center and a demonstration farm with vegetable, fruit, and medicinal herb cultivation would be located at the site of the AEIP. This center would be responsible for:

- Coordination with watershed management programs to ensure an adequate supply of unpolluted water for crops, to guide selection of crops appropriate to water supplies, and to plan nutrient management.
- Coordination of sustainable farming and eco-park development with regional ecological restoration planning, including restoration of soil quality, constructed wetlands for water treatment, and creation of micro-habitats along riparian corridors and surrounding vegetation.

- Implementation of farming best management training programs to directly assist farmers in adopting the most advanced practices for ecosystem stabilization and soil building to increase the long-term agricultural productivity of the soils and improve water holding capacity.
- Development of business models for organic farming and food processing that help small to mid-size farmers remain financially viable producers.
- Integration of sustainable farming and food processing with town and city programs to use energy, water, and material resources, including urban discards, efficiently.
- Coordination of research between local and foreign organic research centers.

This would be an ambitious initiative, integrating regional rural and urban economic development with restoration of major natural systems. Fortunately development of the agro eco-industrial park offers a business-based center for generating the public sector elements outlined above. It creates value through the real estate development process, which becomes a basis for attracting both private capital and public investment.

It would be critical to attract partners aligned with both the business success of the project and the social goal of preserving small to mid-scale farming as a viable part of the rural economy. Equally, appropriate partners will be those supporting the long-term ecological restoration of the agrarian environment. There is a Chinese program that provides an indigenous approach to marry action toward these three goals.

Chinese Ecological Agriculture (CEA) (Shengtai nongye) is a village-based initiative for reducing the energy intensity and environmental impacts of farming, improving productivity, opening village economic development opportunities, and improving quality of life. The State Environmental Protection Agency has led implementation of the initiative for "a comprehensive agricultural production system which is managed intensively according to the principles of ecology and eco-economics" (Sanders 2000, quoting a Chinese EPA document).

The most successful CEA projects integrate traditional ecologically sensitive farming practices with rural community and economic development. Traditional practices include ones familiar to organic farmers in many countries, such as:

- Crop rotation
- Inter-planting of crops and diversification
- Use of organic fertilizer and minimizing or eliminating use of pesticides
- Combining of crop farming, orchards, and animal husbandry
- Planting of trees

- Utilizing all farm and community by-products to capture resource value
- Working with fields and waterways to avoid soil erosion and conserve water
- Building of greenhouses for intensive growing throughout the year
- CEA villages have also developed
- Bio-gas generation from manure and human night soil at household and village levels
- Use of solar water heating systems and photovoltaic energy
- "Side-line" village industries including value-added food processing and nonfarm-related ventures
- Improvement of housing and community infrastructure

Thus, the ideal of Chinese Ecological Agriculture is a holistic approach that integrates organic farming with ecosystem restoration and village community and economic development. The best cases have achieved notable improvements in the health of the ecological base of farm land and water systems together with significant village economic development and better quality of life (Sanders 2000).

(The author's colleagues, Ivan Weber and Scott Murray, have made major contributions to this section.)

Implications for Economic Development

The fundamental economic initiatives just described seek to convert national, local, and global economies to a sustainable form. This means human needs would be satisfied increasingly with renewable resources, nonrenewable resources would be conserved, natural resources would be restored as sources of natural capital, and both natural and human systems would be restored rather than further degraded.

Achieving these worthy goals will have significant financial costs but far greater economic benefits. Increases in efficiency and productivity at factor 10 levels would help to repay the investments. The early adopters will gain a competitive edge. Each of these new economies will join the earlier "new economy" of information technology and telecommunications. Industrial clusters for resource recovery, renewable resources, sustainable design and construction, ecosystem restoration, environmental cleanup, and renewal of the built environment will expand. Investments in the transformation will create and expand ventures and open new lines of employment at a wide variety of skill levels.

In the U.S. alone estimates of the annual revenues of a narrowly defined "environmental industry" are over $200 billion annually (Gallon 2004). This total is for the following sub-industries:

- Compliance with environmental regulations
- Environmental assessment, analysis and protection
- Pollution control and waste management
- Cleanup of contaminated property
- Provision and delivery of water, recovered materials, and clean energy
- Technologies and activities that contribute to increased energy and resource efficiency, higher productivity, and sustainable economic growth

Developing and developed countries alike are placing a strong emphasis on encouraging environmental technology and service industries and supporting the research to move emerging technologies into market competitiveness. China's Ministry of Science and Technology, for instance, has committed US$1.2 billion to research and venture development in a wide range of technologies. Japan hopes that the waste reduction of the Recycling Economy and development of enterprises will help it break out of its decade long period of economic stagnation (Japan Environmental Agency 2000).

The Infrastructure for Economic Development

A unique set of economic development institutions has emerged in the last decade under the broad term eco-industrial development (EID). The Circular Economy and Recycling Economy efforts have adopted eco-industrial parks (EIP) and eco-industrial networks (EIN) as important means of achieving objectives of resource optimization and waste reduction. The already existing fields of cleaner production and pollution prevention offer many strategies and tools to further support these new economic concepts. The foundation of the Service Economy is a set of design values and methods applicable to a wide range, from specific products to facilities and buildings, including EIPs themselves. This section summarizes the EID institutions and strategies.

Eco-Industrial Parks

The now commonly accepted international definition of "eco-industrial park" is based on the one initially created by an Indigo Development team in 1992 and then expanded for the U.S. Environmental Protection Agency in 1995 (Lowe, Moran and Warren 1997). This definition was refined in an Asian Development Bank publication in 2001 as follows:

> An eco-industrial park or estate is a community of manufacturing and service businesses *located together on a common property*. Member businesses seek enhanced environmental, economic, and social performance through collaboration in managing environmental and

> resource issues. By working together, the community of businesses seeks a collective benefit that is greater than the sum of individual benefits each company would realize by only optimizing its individual performance.
>
> The goal of an EIP is to improve the economic performance of the participating companies while minimizing their environmental impacts. Components of this approach include green design of park infrastructure and plants (new or retrofitted); cleaner production, pollution prevention; energy efficiency; and inter-company partnering. *An EIP also seeks benefits for neighboring communities to assure that the net impact of its development is positive* (Lowe 2001).

This definition is now accepted broadly by innovators in the field of eco-industrial development (Côté 2004; Chertow 2004; Asian Eco-Industrial Estate Network, www.eieasia.org).

This systems understanding implies a mission for EIPs:

> An Eco-Industrial Park will achieve profitable return on investment while demonstrating an environmentally and socially sound form of industrial real estate development. This model of industrial development will be a major hub for sustainable regional development.
>
> Profit for public authorities includes local ventures developed, foreign direct investment attracted, new jobs created, and environmental and social benefits. Nevertheless, publicly owned EIPs should generate sufficient revenue to pay their own operating costs. (Lowe 2004a)

Some academics and consultants have proposed alternative definitions of the term, "eco-industrial park," which are based on a more limited objective of by-product utilization within networks of companies.

An eco-industrial park (EIP) is a "community of companies, located in a single region, that exchange and make use of each other's by-products or energy" (Desrochers 2002).

This constrained understanding of the nature and potential of eco-industrial parks fails to promote many of the benefits of this innovation to industrial and regional resource management. It appears to be popular because it is a simpler concept that can be modeled and communicated relatively easily. See Nelson Nemerow's Chapter 11 for additional ideas on industrial complexes utilizing by-products among companies. His very useful book, *Zero Pollution for Industry*, outlines 15 possible industrial complexes with potential flows of by-products between companies (Nemerow 1995). These complexes usually depend on one major anchor company with large by-product outputs. (The Guitang Group in China is similar to one of Nemerow's hypothetical complexes centered on a sugar refinery.) So the by-product exchange should be seen as one of many strategies for optimizing resource consumption and reducing pollution in an industrial park or region.

EIP Development or Re-development Strategies

There is a rich menu of options for the design and development of EIPs, including ideas for site design, park infrastructure, individual facilities, and shared support services. Some major strategies in planning Eco-parks include:

Natural systems—An industrial park can fit into its natural setting in a way that minimizes environmental impacts while cutting certain operational costs. Industrial facilities, such as the Herman Miller Phoenix Design plant in the U.S., illustrate the use of native plant reforestation and the creation of wetlands to minimize landscape maintenance, purify stormwater run-off, and provide climate protection for the building. These and other natural design concepts can be used throughout an industrial park.

At another level, design choices in materials, infrastructure, and building equipment, plant design, and landscaping can reduce an EIP's net emissions of greenhouse gases and consumption of nonrenewable resources.

Park management and support services—As a community of companies, an EIP needs a more sophisticated management and support system than a traditional industrial park. Management or a third-party supports the exchange of by-products among companies and helps them adapt to changes in the mix of companies (such as a supplier or customer moving out) through its recruitment responsibilities. Management may maintain links into regional by-product exchanges and a site-wide telecommunications system. The park may include shared support services such as a training center, cafeteria, daycare center, office for purchasing common supplies, or transportation logistics office. Companies can add to their savings by sharing the costs of these services.

Sustainable design and construction—EIP planners design buildings and infrastructure to optimize the efficient use of resources and to minimize pollution generation. They seek to minimize ecosystem impacts by careful site preparation and environmentally sensitive construction practices. The whole park is designed to be durable, maintainable, and readily reconfigured to adapt to change. At the end of its life, materials and systems can be easily re-used or recycled.

Energy—More efficient use of energy is a major strategy for cutting costs and reducing burdens on the environment. In EIPs, companies seek greater efficiency in individual building, lighting, and equipment design. For example, flows of steam or heated water from one plant to another can be used (energy cascading) and these can also be conducted into district heating or cooling systems. (In power plants and many industrial processes, the majority of heat generated goes up the stack rather than producing value.)

In many regions, the park infrastructure can use renewable energy sources such as wind and solar energy.

Materials flows—In an eco-park, companies perceive wastes as products they have not figured out how to re-use internally or market to someone else. Individually, and as a community, they work to optimize use of all materials and to minimize the use of toxic materials. The park infrastructure may include the means for moving by-products from one plant to another, warehousing by-products for shipment to external customers, and common toxic waste processing facilities. Companies in the EIP also enter into regional exchanges.

Water flows—In individual plants, designers specify high efficiency building and process equipment. Process water from one plant may be re-used by another (water cascading), passing through a pre-treatment plant as needed. Park infrastructure may include mains for several grades of water (depending on the needs of the companies) and provisions for collecting and using stormwater run-off.

Integration into the host community—Relations of EIP developers with neighboring communities should compensate the many benefits to the park from government services, educational systems, housing, etc. The project can return value to the community through such institutions as a business incubator to support new businesses or expansion of existing ones in the community. Some will become tenants and others may provide essential services or supplies to tenants. Training programs will build a stronger workforce in the community and strengthen the local economy, beyond the needs of the park. A major return from this collaborative approach is the potential formation of a public private partnership to assume financing of some aspects of an EIP's design.

An EIP operates in a regional context of *eco-industrial development*, which is defined as: "Networks of businesses that work with each other and in conjunction with their communities to efficiently share resources (information, materials, water, energy, infrastructure and local habitat), leading to economic gains, environmental quality gains, and the equitable enhancement of human resources for the business and local community" (Côtá 2004).

A basic EIP principle is to design the project in relationship to the characteristics of the local and regional ecosystem. Equally important is the EIP development concept must match the resources and needs of the local and regional economy. The development team needs to work in an inquiry mode, discovering the right solutions for the specific site and community within the broad principles of industrial ecology, sustainable planning, architecture and construction. The turbulence of the time will open many new opportunities for those who keep asking, *"How do we create an eco-industrial park suitable to this community and this ecosystem in this amazing time?"*

Challenges in EIP Development

Review of eco-industrial projects in Asian countries, as well as Western countries, suggests that two primary challenges must be addressed: (1) dominance of public sector in the planning and implementation and (2) ineffective management of the projects. In most Asian countries, eco-industrial park or network projects have been managed by public authorities with insufficient private sector input or participation. Often political influences have determined decisions, rather than a clearly defined economic development requirement and strategy. In many areas this has resulted in excessive investment in development of industrial sites with little chance of fully using the land and gaining adequate return on this public investment. In China, for instance, the Ministry of Land and Resources cancelled 4,800 development zones (24,900 square kilometers of land) which were found to be unfeasible (*China Daily*, August 24, 2004).

The second issue is that management of eco-industrial projects has often failed to overcome the inertia and lack of capacity of bureaucratic structures. In Thailand, the Industrial Estate Authority of Thailand launched a major initiative in 2000 with support from the German bi-lateral aid organization, GTZ. The goal was to eventually make all 28 estates eco-industrial estates, beginning with five pilot estates. GTZ withdrew its support a year early because only one of the pilot estates had demonstrated any significant progress. The project suffered from changes in management of the other estates.

Around the world, innovations that seek to bridge the perceived gap between economic development and environmental protection require highly effective management. Tables 4.1 and 4.2 summarize the advantages and disadvantages of exclusively public or private sector development.

Public Private Partnerships in Eco-Industrial Land Development

Eco-industrial parks seek to achieve social, economic, and environmental benefits in an appropriate balance. The economic innovations described in this chapter set very ambitious goals for improvement of efficiency of resource use. Thus public sector involvement in EIP projects is important to assure gaining these benefits to society. Private sector developers bring very important strengths to eco-industrial development projects. Some form of public private partnership in industrial land development could capture the advantages of the two sectors and neutralize their disadvantages.

The balance and roles of each sector may shift over time from the development phase to an operational phase. In North America, authorities developing industrial parks on public land may hire a development company on a consulting basis. In other cases the public sector invests the land and joint ventures with the private firm, which invests the planning and land development costs. The public and private sector then collaborate

TABLE 4.1
Advantages and Disadvantages of Public Sector Led Eco-Industrial Development

Advantages	*Disadvantages*
With protection of the public interest: economic, social, and environmental outcomes	Bureaucracies often slow and fragmented
Can directly utilize government policy, R&D, and incentives	Changes in administration and agency leadership may interrupt/end projects
Not required to make a profit, so long as projects return their investment	Weak sense of ownership
The "profit" of real estate development includes this whole package of economic, social, and environmental benefits	May lack understanding of business needs and values
Long-term view	Focus on regulations, forgetting economic rationale
Leverage for inter-firm cooperation	Managed as cost (not-profit) center
Has enforcement ability	Financial motivation unclear due to lack of accountability
Tax increment financing flexibility	May be subject to corruption
Can set own regulations	
Can leverage tax revenues	

TABLE 4.2
Advantages and Disadvantages of Private Sector Led Eco-Industrial Development

Advantages	*Disadvantages*
Strong in financial analysis of real estate investment	Disadvantages
Greater innovation	Little experience of private industrial real estate development in China
Profit-motivated	Difficult to raise pre-development funds (high-risk money encourages developer to follow old models)
Forced to develop value concept	Seeking maximum return on investment may eliminate many features of EIP design
Share risks with businesses	Limited funds beyond those required for basic infrastructure and recruitment
Understands corporate management	Estate manager makes money at cost of tenants—-using environmental issues
Tenants as shareholders	Integrity questionable
Estate manager can be better motivated by financial incentives	
Better integration into business routine	

in recruitment of companies. Once developed, the development company, a management company, or the government may manage an industrial park. Development authorities and companies should explore what form of public private partnership best fits the practices in their country and test this approach in eco-industrial parks.

Test projects would need to be sure they include leadership at a high level of the organizations involved; measures to maintain continuity of key managers in the projects; a clear statement of responsibility for the success of the EIP projects in staff job descriptions; incentive structures to reward managers for performance; and improved interagency coordination (Lowe 2004).

Eco-Industrial Networks (EIN)

An eco-industrial network extends collaboration among companies in a region to achieve a broad agenda for improvement of business, environmental and social performance. An EIN may include eco-industrial parks or it may be a network of stand-alone firms. In the Philippines the PRIME project (led by the Ministry of Industry in 1999–2002), for instance, recruited five industrial estates to collaborate as an eco-industrial network to create a regional by-product exchange and to assess the feasibility of a common resource recovery system and a business incubator (PRIME 2000). Eco-industrial networks may include quite a broad variety of programs for member companies, community service programs, and many other joint programs (Cohen-Rosenthal and McGalliard 1999).

The development of a regional eco-industrial network may grow out of interfirm collaboration, without EIP involvement. Or it may be led by the management of an eco-industrial park, building upon the network of suppliers and/or customers of park tenants in surrounding industrial areas. An EIN extends such collaboration across a number of activities and institutions.

Eco-Industrial Network Supporting Institutions

Formation of an eco-industrial network may entail creation of a number of supporting institutions:

- An integrated resource recovery system
- A system for encouraging and managing the exchange of by-products between companies
- Training and services in all aspects of eco-industrial development
- A network management/coordinating unit and working groups
- A community enhancement office to manage projects with neighboring communities
- One or more business incubators (for small-to-medium size enterprises, or SMEs)

- Public sector support in R&D, policy development, access to investment, and information management

The development of **by-product exchanges and integrated resource recovery systems (IRRS)** are closely linked in their management of by-product energy, materials, and water. Some company-to-company exchanges develop with little effort, others become complicated and costly transactions. So the exchanges between pairs of companies is only one part of a system for optimizing by-product utilization. A network of companies specializing in collection, re-use, recycling, and remanufacturing offer comprehensive by-product management to industry. This system may be created by an industrial park, a service business or utility, an independent entrepreneur, or possibly by a public industrial development agency. The resource recovery companies could be distributed throughout the region or concentrated in a resource recovery park at a central location.

Training enables EIP or EIN members to gain the technical and business skills required for eco-industrial development. Course opportunities range from reviews of state of the art technology in energy management, by-product utilization, green chemistry, or hazardous materials recycling to broad themes like sustainable industry and industrial ecology. Training in basic cleaner production techniques like source reduction, materials substitution, and efficient process design allow factory managers to cut the overall volume of by-products generated at their plants, thus avoiding overload of the IRRS. A number of Asian countries, including Thailand, the Philippines, and China, have created national and regional "cleaner production centers" that will provide training and other guidance in this realm.

Services can include consulting to support projects in most of the subjects just listed for training. For instance, a comprehensive audit of materials, energy, and water flows through a facility will prioritize opportunities for increasing efficiency and lowering pollution. It will also evaluate the feasibility of internal reuse, exchange with neighboring plants, or recovery through the IRRS.

Other services in an EIN enable members to cut costs through common procurement of goods and services or by integrating employee transportation management. For example, a service firm could do bulk purchases of office supplies or commonly used process chemicals. Environmental service firms could handle permitting, monitoring, and reporting requirements for SMEs who cannot afford their own environmental staff.

A network management/coordinating unit and working groups—An eco-industrial network needs a management and coordination system to do needs assessment, set priorities, conduct organizing events, and oversee common projects. Management units can draw from industrial park or estate managers, managers of stand alone factories, trade associations, and local government representatives. The focus of control should be in the region so that projects are well suited to the needs and capabilities of network members.

A community enhancement unit to manage relations with neighboring communities—Initially this could be one of the working teams of the management committee, with the mission of two-way communication, community project development, and enlistment of stand alone firms into the EIN. Thai industry is discovering that it needs to go beyond public relations to active partnering with neighboring towns. Companies and industrial park management itself can participate in workforce training, contribution of surplus IT equipment to schools, provision of employee housing located in the community, and other projects enhancing the quality of life in their region.

Business incubators for small to medium enterprises—An EIP or EIN team can enhance the success rate of new businesses by setting up an incubator through public/private collaboration. This resource can play a vital role in developing suppliers to anchor tenants, filling out a theme cluster such as resource recovery or renewable energy, or helping to fill niches in the by-product exchange network.

Transportation services—Van lines, shuttle services from terminals, and car-pooling offices can all reduce the environmental burdens of single-driver auto transport while easing employee travel needs.

Environmental management—Small-to-mid-size firms particularly may benefit from outsourcing aspects of their environmental management tasks to consulting and training companies. Permitting, training, and reporting are some of the duties that open business opportunities for local firms. A cleaner production center could provide a site for such businesses as well as one-stop shops for permitting and other regulatory functions.

Integrated Resource Management System

Achieving the goals of the Recycling and Circular Economy initiatives is an ambitious and complex undertaking. Success calls for changes in all aspects of life, all sectors of society. Success must be defined by a strengthening of the regional economy, by improvement in quality of life, and by a much higher level of efficiency and lower level of pollution in utilization of resources. Achieving success in these four realms requires a system-wide approach to planning and development for moving toward a more closed-loop economy. This calls for a higher level of integration along several dimensions of resource planning and management:

- Short- to long-term time scales
- Interactions among water, materials, energy, air, and land resources
- The lifecycle of resources and products, from extraction through production, consumption, and recovery or disposal

- The collaboration of public and private sector organizations that control resource flows and their utilization (agencies, companies, associations, and public interest groups)

Increasing integration in these four realms is a continuing process, with a more integrated resource management framework as a goal. Full integration is unlikely and probably not desirable. The process toward greater integration allows managers in different organizations and from different disciplines to minimize the errors of fragmented planning. Integration is a way of seeing and thinking that promotes decisions with understanding of the broader system opportunities, constraints and requirements. (*See Chapter 12 by Erkman and Ramaswamy for applications of materials flow analysis that support this planning process.*)

Some general principles to guide the process sketch some of the dimensions of integration required and outline some innovative institutional forms that could promote greater integration.

The first of the principles suggests "a decentralized, self-organizing process of innovation." A more integrated approach to resource management does not mean fusing organizations with diverse missions into one mega-organization. It is much more a matter of having stronger channels for communication and interaction and capacity building for their managers and staffs in systems thinking.

Design Principles for a More Integrated Resource Management System

Application of a number of principles will support design of a more integrated system for resource management and planning:

- Emphasize a decentralized, self-organizing process of innovation and management, with strong information links across systems and levels of system
- Engage the points of social and economic control in the process (financial institutions, policy, regulatory, and development agencies, and standards setting organizations, among others)
- Link the staffs of public agencies responsible for environmental protection, economic development, infrastructure, and urban planning in a shared systems approach
- Support capacity building so that managers in both public and private organizations can learn to work with integrative tools of systems management and industrial ecology (lifecycle assessment, industrial metabolism, green supply chain management, etc.)
- Improve coordination of agencies in their policy making, their planning, and their operation of public facilities

- Encourage collaboration among industry sectors and enterprises to achieve increased internal efficiency and to build resource chains for using by-product materials
- Encourage both government and industry's leadership role in changing consumer behavior and expectations
- Design incremental changes in the context of visions for broader transformative change
- Plan short-term actions with awareness of their interactions and their potential long-term impacts
- Maintain strong communication among people and institutions implementing the CE and scientists researching it

Components of a Regional Resource Recovery System

Resource recovery is a vital activity in a resource management system and a major economic development opportunity. A municipality or broader authority may sponsor an integrated regional resource recovery system or an eco-industrial park may serve as the coordinating hub to such a system, including these elements:

- Coordinating firm or agency that integrates management of industrial, residential, commercial, and government "waste" streams for maximum resource recovery. (This entity should also coordinate with waste reduction, reuse, and recycling programs within plants, as well as DFE programs)
- By-product exchange networks among plants
- A resource recovery industrial cluster
- R&D to improve value of recovery and include more materials
- An investment capital network committed to this industry (both public and private sources)

The resource recovery industrial cluster may often be more important than plant-to-plant exchange networks in achieving a high level of efficiency in regional resource use. This cluster includes:

- Niche collectors to separate and preserve the value of materials
- Recycling firms to manage processing and distribution
- Manufacturing firms to utilize recycled materials
- Re-use firms to sell used equipment, materials, and products
- *Re*-manufacturing firms to rehabilitate used equipment
- Composting firms to process biomaterials for nutrients
- Energy companies to generate fuel or power from biomass
- A firm that supports companies in creating by-product exchanges and eco-industrial networks and integrates their efforts into the larger system

An eco-industrial park can benefit its whole community by hosting such a cluster of resource recovery companies, as well as other firms promoting cleaner production and DFE to increase efficiency and productivity of major companies and government facilities. Eco-industrial parks should be designed to support implementation of this whole system.

Limitations of Company-to-Company By-Product Exchange

There are outstanding examples of the effectiveness of the company-to-company by-product exchange strategy, but it may only be feasible in a broader eco-industrial network or region. The original inspiration was the case of Kalundborg in Denmark, where materials, energy, and water exchanges evolved among a group of companies in a region and their host community. Here a large power plant's outputs were the basis for a series of interfirm deals to use by-products economically among a refinery, a wallboard company, a pharmaceutical company, a fish farm, and community district heating. However, this "symbiosis" depended on the size and diversity of the companies, which were not located in a formal industrial park. *The network of companies at Kalundborg is not an eco-industrial park.*

The Guangxi Guitang Group mentioned above is a notable example of using sugar production by-products, first within a single city-owned company, and then in a broader network including other sugar producers in the city of Guitang and the farmers growing cane. This success was possible because the first investments in infrastructure and plants were all within a single corporate group, not between separate firms. From this single company, an *eco-industrial network* has evolved, including other sugar producers and the farmers growing the sugar cane.

There are serious limits to company-to-company exchanges achieving the resource efficiency goals of the Circular Economy:

- Often plants located together in an industrial park or development zone are clustered by industry and may generate very similar by-products. A cluster of electronics firms will find relatively few ways to use each other's secondary product outputs.
- Even when there are diverse by-products available, firms have limited time to negotiate the transactions required for one plant to use another's un-marketed products. The cost of reaching a deal (and minimizing the risks) may be greater than the value of the material, water, or energy utilized.
- The investment in negotiation needed results in a focus on the higher volume and higher value by-products and avoids dealing with many resource flows of lower value or volume.

- By-product exchange is an end-of-pipe solution that may diminish efforts to redesign process or product design to eliminate the by-product in the first place.

Conclusion

Sustainable development may be seen as humankind learning to live within the current income the Earth provides, rather than overdrawing the accounts of natural capital. This chapter, Economic Solutions, has surveyed four pathways to a sustainable economy: Germany and Japan's Recycling Economy, China's Circular Economy, the Service Economy, and restoration of natural and constructed systems. These are interrelated initiatives that seek to meet the challenge of depletion of natural resources and degradation of natural capital. Finally, we have discussed the forms of eco-industrial development that supply infrastructure for these pathways.

The German and Japanese Recycling Economy laws grew out of issues of waste management and reduction, not directly from perception of limits for the global supply of natural resources. Their goals are set in terms of waste reduction, although many of the strategies and tools adopted will also lead to more optimal utilization of resources. Both countries lay strong emphasis upon major improvements in product and process design, not just recycling at the end of the line.

What is unique about the Chinese Circular Economy concept is the setting of a very high objective for optimization of resource use in all sectors. Chinese leaders understand that achieving higher resource efficiency at the level of 8 to 10 factors is fundamental to achieving the goal of developing an "all round well-off society." This is also a vitally important means of becoming even more competitive in global markets. Other countries cannot afford to ignore that advantage. Another notable goal in the CE initiative is defining a Chinese quality of life distinct from the western consumerist model. While the broad vision of the Circular Economy is a strong one, China needs to fill significant gaps in its program of implementation.

The Service Economy concept seeks to answer the question, how will companies benefit from producing products with long lives, rather than the present throw-away models. This question is important to making economic sense out of a set of strategies for product life extension that cut resource consumption but threaten the need of companies for continuing sources of profits. Walter Stahel answered the question by identifying how companies such as Xerox, Schindler Elevators, and Agfa-Gaevert have moved to a service mode. They sell the outputs of their equipment and own the equipment, thus gaining a self-interest in building highly durable, repairable, and recyclable products.

Restoring natural and human physical capital rounds out the transformation to a more sustainable economy. Natural capital is embodied in the various realms of our physical environment and in the diversity of our ecosystems. It is the ultimate source of inputs of physical resources to the economies of nations. It includes the sinks that absorb the discarded material, fluid, and atmospheric outputs of our economies. Restoration of natural capital is a major industrial sector that will require continually increasing investment to provide the renewable resources required for development and to adapt ecosystems to climate change.

Restoration of the built environment may be seen as product life-extension for infrastructure, housing, and commercial or industrial buildings and sites. It builds upon the sunk investment in place across our countries, extending its useful life. This renewal process will draw upon resources at a lower rate than totally new construction. Restoration of farms and farm communities is transitional, involving both natural capital and the built environment.

This chapter concludes with discussion of the different facets of eco-industrial development or applied industrial ecology. This provides essential infrastructure for implementing the economic transformations described earlier. Developers of eco-industrial parks or redevelopers of conventional industrial parks take a whole systems approach to creating a sustainable home for industries. They practice environmentally sensitive site design, management to support the environmental, social, and financial performance of tenant companies; encouragement of high standards of resource utilization in each plant; exchange of by-products, when feasible; and creative interaction with neighboring communities.

Eco-industrial networks are groups of companies and support agencies that collaborate to achieve similar high performance across a broader region. One of their functions may be to support creation of an integrated resource management system, a public private partnership working toward optimization of resource use and reuse in a city or region.

Eco-industrial development and cleaner production are the two cornerstones for Recycling and Circular Economy initiatives. The Service Economy identifies a business and economic strategy for optimizing design of products without bankrupting the companies that make highly durable, long-lived products. This is vitally important to making cleaner production's environmentally sensitive design strategy a reality.

Integrating restoration into sustainable economy concepts and action programs supplies an important missing piece. Sustainable development seeks to enable humanity to learn to live well within the inevitable constraints set by Earth as a living system. Destruction of natural capital and wasteful design and use of the built environment tighten the constraints. Restoration recovers value at a level of great strategic importance. We must remember the further requirement (not discussed here except in connection with China's Circular Economy) for achieving more equitable distribution and redefining quality of life so it is not dependent upon ever continuing increases in material consumption.

Review Questions

1. How would the skills and knowledge you have (and are learning) contribute to implementation of the Recycling Economy, Circular Economy, Service Economy, and restoration as a foundation for sustainable development?
2. Can the concepts being applied in Germany, Japan (two developed countries), and China (a rapidly developing nation) be adapted to the economies of developing countries?
3. What are the likely consequences if the U.S. economy fails to act to improve efficiency of resource use at a multifactor level, as China's Circular Economy proposes?
4. How does restoration of natural and constructed environmental systems add value to national and regional economies? How would you justify investment in restoration activities in economic terms?
5. What major transformations and strategies other than those described in this chapter are required in order for a truly sustainable economy to evolve?

References

Asahi Shimbun, April 8, 2000.

Bi, Jun. 2004. Development of Circular Economy in China. Presentation at Yale University Industrial Symbiosis Symposium, January 2004, New Haven, CT.

Bi, Jun, Jie Yang , Zengwei Yuan, Juan Huang. 2004. Circular economy: An industrial ecology practice under the new development strategy in China. Center for Environmental Management & Policy, Nanjing University.

Chertow, Marian. 2004. Industrial Symbiosis and Eco-Industrial Parks. Yale University School of Forestry & Environmental Studies. Presentation delivered by Reid Lifset at Partnership for the Future: 2nd Annual Conference and Workshop for Eco-Industrial Development, Eco-Industrial Estates Asia Network, Bangkok, Thailand, March 11–12, 2004.

China Daily. 2004. "China cancels 4,800 development zones," August 8, 2004.

Cohen-Rosenthal, Edward N., Thomas McGalliard. 1999. *Eco-Industrial Development–Prospects for the United States.* Work and Environment Initiative, Cornell University, Ithaca, NY.

Côté, Raymond. 2004. The Industrial Park as an Ecosystem. Industrial Ecology Research Group, Dalhousie University, Nova Scotia, Canada: http://www.mgmt.dal.ca/sres/research/rlnpark.htm.

Cunningham, Storm. 2002. *The Restoration Economy*. San Francisco: Berrett Koehler. (See www.restorationeconomy.com; see also Revitalization Institute, www.revitaliz.org.)

Desrochers, Pierre. 2001. Eco-Industrial Parks, the Case for Private Planning. The Independent Review, v.V, n.3, Winter 2001.

Dietmar, Rolf. 2003. Approaches to a circular economy in Germany. Proceeding of Workshop on Circular Economy in Guiyang, Guiyang, August.

Duan, Ning. 2001. Make sunset sunrise: Efforts for construction of the Guigang eco-industrial city. Draft paper from Chinese Research Academy of Environmental Sciences, Beijing. Presented at April 2001 workshop: New Strategies for Industry, Manila.

Economy, Elizabeth C. 2004a. China's environmental challenges. Congressional testimony: Subcommittee on Asia and the Pacific, U.S. Congress, House International Affairs Committee, September 22, 2004. (http://www.cfr.org/pub7391/elizabeth_c_economy/congressional_testimony_chinas_environmental_challenges.php; accessed September 30, 2004.)

Economy, Elizabeth. 2004b. *The River Runs Black*. Ithaca, NY: Cornell University Press, Council for Foreign Relations. [A critical view of China's environmental crisis.]

Erkman, Suren, Ramesh Ramaswamy. 2003. *Applied industrial ecology—A new strategy for planning sustainable societies*. Bangalore: AICRA Publishers.

Gallon Environment Letter. 2004. *"Enabling the Environment Industry"* Canadian Institute for Business and the Environment, Ontario, Vol 8, No. 4.

Greene, Gregory. 2004. *The End of Suburbia: Oil Depletion and the Collapse of the American Dream*, documentary produced by The Electric Wallpaper Co. Toronto. web site http://www.endofsuburbia.com.

Guigang City and Guitang Group. 2000. Initial planning for industrial eco-park demonstration project in the Guitang group. Compiled by Technology Center of the Guitang Group, August 1, 2000.

Guitang Group. 2004. *The brochure of the Guitang Group*, May.

Hassler, Arnulf. 2004. "Research Projects About Recycling Networks in Europe," Proceedings of International Seminar on Taiwan Environmental Science and Technology Park, Industrial Technology Research Institute, Taipei.

Japan Environmental Agency. 1998. *Ippan Haikibutsu no Life Cycle to Mondaiten (Life Cycle and Problem of Waste)*. Tokyo.

Japan Environmental Agency. 2000. *Kankyo Hakusho (Environmental White Paper)*. Tokyo.

Japan Ministry of Health and Welfare. 1999. *Nihon-no Haikibutsu Shori (Waste Management in Japan)*. Tokyo.

JETRO. 1992. Ecofactory—Concept and R&D themes. *New Technology*, Special Issue, FY, Japan External Trade Organization, Tokyo. [Report based on work of the Ecofactory Research Group of the Mechanical Engineering Laboratory, Agency of Industrial Science and Technology.]

Kinzer, Stephen. 2004. Future of Illinois farm may lie in swampy past. *New York Times*, September 27, 2004.

Liaoning Province, Environmental Protection Bureau. 2002. Pilot project plan for development of a recycling economy in Liaoning Province, Shenyang.

Lowe, Ernest. 2001. *Eco-industrial park handbook for Asian developing countries*. Report prepared for Environmental Department, Asian Development Bank. (http://www.Indigodev.com/ADBHBdownloads.html.)

Lowe, Ernest. 2004a. Defining eco-industrial parks: The global context and China. Report prepared for the Policy Research Center for Environment and Economy, State Environmental Protection Administration, China. Oakland, CA: Indigo Development.

Lowe, Ernest. 2004b. The system for achieving a more circular economy. Paper prepared for the Policy Research Center for Environment and Economy, State Environmental Protection Bureau. Oakland, CA: Indigo Development.

Lowe, Ernest A., Stephen R. Moran, Douglas B. Holmes. 1997. *Eco-Industrial Parks: A Handbook for Local Development Teams*, Indigo Development. Emeryville, CA: RPP International. [This publication was developed as *The Eco-Industrial Park Fieldbook* in 1994–1995 under a U.S. EPA Cooperative Research Agreement. The present handbook replaces it.]

Lowe, Ernest, Scott Murray, Ivan Weber. 2004. An agro eco-industrial park: an opportunity to promote sustainable farming and food production in Shandong Province, China. The International Center for Sustainable Development [ICSD, a U.S.-based NGO active in China] and the Environmental Education Media Project for China (EEMPC) in Beijing.

Morikawa, Mari. 2000. *Eco-Industrial developments in Japan.* Indigo Development Working Paper No. 11. Oakland, CA: Indigo Development Center.

Nakajo, H. 1998. Recycling society. *LookJapan.* (www.lookjapan.com)

Nemerow, Nelson, L. 1995. *Zero Pollution for Industry: Waste Minimization Through Industrial Complexes.* New York: John Wiley & Sons.

PRIME. 2000. Website for the Philippine Board of Investments eco-industrial network project (www.iephil.com. See also www.eieasia.com.)

Sanders, Richard. 2000. *Prospects for Sustainable Development in the Chinese Countryside: The Political Economy of Chinese Ecological Agriculture.* Aldershot, UK: Ashgate Press.

Sato, Meiji. 2004. "Kitakyushu Eco-Town Project," Proceedings of International Seminar on Taiwan Environmental Science and Technology Park, Industrial Technology Research Institute, Taipei.

Shi, Lei, Yi Qian. 2003. *Strategy and mechanism study for promotion of circular economy in China.* Department of Environmental Science and Engineering, Tsinghua University, Beijing, 100084. [Manuscript received December 6, 2003. This summarizes recommendations of the CCICED Task Force on Cleaner Production and Circular Economy.]

Shi. Lei, Jining Chen, Tianzhu Zhang, Yi Qian. 2003. *Circular economy: International experiences and China's experimentation.* Department of Environmental Science and Engineering, Tsinghua University, Beijing, 100084. [Manuscript received December 6, 2003.]

Society for Ecological Restoration. 2004. *Natural capital and ecological restoration.* An occasional paper of the SER Science and Policy Working Group (1), April 2004. (http://www.ser.org/content/Naturalcapital.asp; accessed August 2004.)

Society for Ecological Restoration International (SERI) and World Conservation Union (IUCN). 2003. *SERI and IUCN's Draft Global Rationale for Ecological*

Restoration. (http://www.ser.org/content/Globalrationale.asp; accessed August 2004.)

Stahel, Walter. 1994. The utilization-focused service economy: Resource efficiency and product-life extension. In Allenby, B. and Richards, D., *Greening of Industrial Ecosystems.* Washington, DC: National Academy of Engineering.

Stahel, Walter. 2003. The functional society: the service economy. In Dominique Bourg, Suren Erkman, eds. *Perspectives on Industrial Ecology.* London: Greenleaf Publishers.

Stahel, Walter, Orio Giarini. 1989/1993. *The Limits to Certainty: Facing Risks in the New Service Economy.* Dordrecht and Boston, MA: Kluwer Academic Publishers.

Tanabe, Yasuo. 2001. *Environmental initiatives undertaken by Ministry of Economy, Trade and Industry towards building society oriented to recycling of resources.* Ministry of Economy, Trade and Industry, Japan. ECP Newsletter No. 18, September. [Environmentally conscious products newsletter (www.jemai.or.jp/english/e-ecp/ecp_no18.htm).]

Xerox. 1995. CSS/ISC Asset Recycle Management. June, Revision 2.0.

Yardley, Jim. 2004. "Bad Air and Water, and a Bully Pulpit in China." *NY Times,* September 25, 2004 article on State EPA Deputy Director, Pan Yue.

Zhao, Wei, Ernest Lowe. 2003. *The circular economy in China: An introduction.* Prepared for the Professional Association for China's Environment Circular Economy Forum.

Zhu, Qinghua, Ernest Lowe. 2004. Integrated chain management for green marketing and regional sustainable development: a case study of Guitang Group, China. (Unpublished.)

Recommended Resources

Balmford, A., A. Bruner, P. Cooper, et al. 2002. Economic reasons for conserving wild nature. *Science* 297:950–953.

Cairns, Jr., J. 1993. Ecological restoration: replenishing our national and global ecological capital. In: D. Saunders, R. Hobbs, P. Ehrlich, eds. *Nature Conservation 3: Reconstruction of Fragmented Ecosystems.* Chipping Norton, NSW, Australia: Surrey Beatty & Sons, 193–208.

Chertow, Marian R. 1999. The Eco-Industrial Park Model Reconsidered. *Journal of Industrial Ecology* 2(3):8–10.

Chertow, Marian. 2000. Industrial symbiosis: A review. *Annual Review of Energy and the Environment* Vol. 25:313–337.

Costanza, R. R., R. R. d'Arge, S. de Groot, et al. 1997. The value of the world's ecosystem services and natural capital. *Nature* 387:253–260.

Fleig, Anja-Kathrin. 2000. *Eco-industrial parks as a strategy towards industrial ecology in developing and newly industrialised countries.* Report prepared for The Deutsche Gesellschaft fär Technische Zusammenarbeit (GTZ) GmbH

(German Technical Cooperation), Berlin. [The GTZ is a federally owned limited liability company responsible for planning and implementation of projects and programs in the field of technical co-operation with developing countries; www.gtz.de.]

Guitang Group. 2002. The Group's Web site (www.guitang.com; accessed March 23, 2002).

Guitang EIP Office. 2004. *Developing recycling economy in the Guigang City*, May 20.

Ide, Tsugio, Kotaro Kimura, Tohru Sasaki. 2003. Efforts towards the construction of a recycling-oriented society in Japan, Proceeding of Workshop on Circular Economy in Guiyang. Guiyang, August.

Koenig, Andreas. 2000. *Development of eco-industrial estates in Thailand, project development and appraisal, June to December 2000.* Overheads for Roundtable Meeting, GTZ. Bangkok, November 27. 2000.

Lowe, Ernest A. 2000. *Prospects for eco-industrial development in Thailand.* Report of short-term expert on eco-industrial development to GTZ. Bangkok, November 28.

Lowe, Ernest. 2003. Report to the Dalian development zone on the eco-planning process. (http://www.indigodev.com/DDZReport.html.) [In support of the Zone's response to the Liaoning Province Circular Economy Plan.]

Society for Ecological Restoration International Science & Policy Working Group. 2002. *The SER Primer on Ecological Restoration.* (www.ser.org; accessed August 2004.)

Stahel, Walter. 1986. Product-life as a variable: The notion of utilization. *Science and Public Policy* August 13(4).

Stahel, Walter R. 1995. The functional economy and cultural and organizational change. In Deanna J. Richards, 1996. *The Industrial Green Game: Implications for Environmental Design and Management.* Washington, DC: National Academy Press.

Stahel, Walter, G. Reday. 1981. *Jobs for Tomorrow: The Potential for Substituting Manpower for Energy.* Commission of the European Community. New York: Vantage Press.

Product Life-Extension, www.product-life.org.

Revitalization Institute, www.revitaliz.org [devoted to research, consulting, and implementation of Restoration Economy projects].

Society of Ecological Restoration, *Restoration Ecology Journal,* http://www.ser.org/content/restoration_ecology.asp.

University of Wisconsin Arboretum, *Ecological Restoration Journal,* http://ecologicalrestoration.info.

CHAPTER 5

Environmental Engineering Solutions

Ram Tewari

Introduction

As pointed out in the introduction of the book, times have changed and so have the past practices and approaches for environmental management. This revolutionary paradigm shift has taken place because of the concerns and realizations among environmental engineering community about enhanced value of all forms of life, environmental consciousness, endangered planet Mother Earth (Earth's future is in our hands), long-term survival of humanity (time to anticipate about our common future), dilemma of globalization (pollution does not have political boundaries), ensuring environmental justice and social equity (meet multicultural needs—a profound issue and dilemma), willingness to adjust our wasteful lifestyles (natural needs versus acquired needs), and the urgent need for sustainable development (from destructive, exploitative philosophy to one that fosters long-term protection of the environment and its inhabitants). As a result of all this, the role, commitment, and responsibilities of an environmental engineer (EE) have also changed.

The mission of an EE is to protect public health, safety, welfare (well-being), and to safeguard the natural environment—air, water, and land—on which life depends. In doing so, an EE has to use a holistic approach for waste management, in place of a fragmented single-medium approach of the past, for harmonious environmental protection. An EE has to apply waste management technology that mimics natural processes, conserves resources and energy, promotes sustainability, and achieves a delicate balance among sustaining and assimilative capacities of our biosphere (interconnectedness of an ecosystem working in tandem to sustain life on Earth). It means a dramatic transition from past development practices when technological developments were guided by efficiency, productivity, profitability, and other economic criteria. Now added to those parameters are health and environmental impacts, resource and energy conservation, waste management and its social impacts. An EE has to recognize that the importance of creative problem-solving is essential to achieving innovative, precedent setting solutions.

Let us consider the limitations and the framework within those, an EE has to work for well-rounded creative solutions to complex environmental problems, which is environmentally balanced, resource optimized, socially and ethically acceptable, economically viable and universally beneficial:

1. There are only three sinks, with finite capacity, on this planet, namely the air, the land, and the oceans.
2. The first law of thermodynamics is that nothing disappears. Matter can neither be created nor destroyed but that it can be changed in form (law of conservation of mass-materials balance). Matter may change from one form, substance, or concentration, but it does not simply disappear. The second law of thermodynamics states energy spontaneously disperses from being localized to becoming spread out if it is not hindered. Pollution disperses.
3. Waste is something that everyone produces, but no one wants.
4. Need a global perspective—*think globally but act locally.*
5. Pollution problems must be solved by using a holistic, organized, systematic, and multi-media approach.
6. Sustainability and accountability to the public should remain an integral part and absolute necessity for any environmental management project.
7. Essential prerogative of a solution to environmental pollution is a provision of a safe, environmentally compatible, resource optimized (conserving and recovering material and energy resources), and at a least possible cost (value engineering).

The next logical question is what should be the basis of these engineering decisions? Decisions should be based on technical analyses, cost-effective analysis, cost/benefit analysis, risk analysis (levels of risk and liability), an environmental impact analysis, and ethical analysis (increasingly important these days). While doing environmental protection, we are forced to confront public, political and ethical perceptions and concepts concerning environmental justice and equity. The three important issues of human health, human welfare, and ecological concerns should always governing our planning.

Our mission is to solve pollution problems and enhance the environment. In any scenario, some level of uncertainty, risk, and liability always exists. But the final choice has to be made which has an acceptable long-term risk, because there cannot be any choice which has zero-risk. Also, flexibility should be an integral part of solutions because no single approach can satisfy all the stakeholders.

Systems Analysis Approach

Using a systems approach can solve complex environmental problems. This approach consists of a logical series of steps that can lead to the optimum solution of an environmental problem. A systems approach provides a framework for solution, but the quality of the solution depends on assumptions, definition of the system, goals and objectives, controlling factors, limitations, individual judgment, development of alternative solutions, and selection criteria of an alternative.

What is a system? A system is a collection of things that function together. There are obvious advantages in treating environmental problems as systems. These problems can be considered in their totality, and most the effective solution can be obtained by looking at all interrelated parts, functions, and their effects on one another. Real systems may be simplified with certain assumptions in such a way the model provides a fair approximation of what is happening in an actual environment. It is a duplication of principles observed in natural purification processes in engineered system. Thereafter, generated systems for environmental problem can be transformed into mathematical relationships (numerical accounting/mathematical models) and those can be manipulated to evaluate the effects of differing alternatives.

Mathematical models, which are approximations or abstractions of the actual system, are essentially experimental tools with mathematical description of objectives, component interactions, and management methods. These models help to generate and evaluate alternative solutions to the problem. Selection of an alternative can be done by application of optimization techniques. Though quantification does not altogether eliminate subjectivity, it minimizes vagueness.

Three environmental systems: the water resources (surface and groundwater) management system, the air resource management system, and solid waste (land resource) management system. Since many important environmental problems cross the air—water—land boundary, these are known as multimedia problems, meaning that the problems encompass more than one medium. We have to learn from Mother Nature's design—that no one component in an ecosystem can operate independently from the other.

Our experience tells us that it is risky to develop models that are too simplistic, instead use the multimedia approach (interaction and collective effects). To obtain the best and optimum solution to environmental pollution is waste minimization—if waste is not produced, it does not need to be treated or disposed of. Pollution prevention should be the first and foremost target for handling any pollution problem. Remember **4 Rs**—**r**educe, **r**euse, **r**ecycle, and then **r**esidue disposal. What is pollution? It is an unreasonable interference with another beneficial use. Waste not, want not. Think before you throw. Live in harmony with Nature. You get back what you paid for. Nothing in life is free. Nearly everything we do leaves behind some kind of waste, so we cannot escape from waste. We must understand and acknowledge that we are all part of the problem; therefore, we must change our attitude and life style. We can transition to a society that views wastes as inefficient uses of resources and believes that most wastes can be considered misplaced resources and can be reused and eliminated. Eliminating wastes will contribute to environmental, economic, and social vitality.

There are three main processes observed in the natural environment: physical, chemical, and biological (see Chapter 6 for a complete presentation of biological solutions to environmental problems). This chapter provides an overview of physical and chemical processes (natural systems), which can be considered for environmental management.

Physical Processes: flow equalization, mixing, dilution, deep well injection, ultraviolet radiation disinfection, fluoridation, gravity, solids separation-screening, grit removal, settling-sedimentation-clarification, filtration (land application, percolocation, slow, rapid, direct, pressure sand, high rate), microstraining, membrane filtration, dewatering-thickening, compaction, solidification, thermal and solar evaporation, evapotranspiration, adsorption-ion exchange, absorption, solidification and stabilization, trees, wetlands, aeration (air stripping, air sparging), incineration, symbiotic processes (stabilization, polishing ponds), surface impoundments, dilution, phytoremediation, phytodegradation, phytoextraction, phytostabilization, trees and plants (as a part of landscaping, barriers, aesthetics, noise prevention, energy conservation, wetlands), and natural attenuation.

Chemical Processes: conditioning and stabilization, coagulation and flocculation, chemical disinfection-chlorine, ozone, dechlorination, phosphorus removal, chemical oxidation (see Chapter 8 for a more detailed presentation of chemical treatment solutions).

A problem well-defined is a problem half solved is very true in the case of waste management. We need to emphasize that one should embark not on a project but rather on a process. Choose the most effective solution based on your specific situation—write a strategic plan. An innovative and integrated (reduction-conservation and preservation, pollution prevention, waste minimization, substitution of less polluting material, reuse, recycle, residue disposal) approach is essential for overall regulatory compliance and environmental protection.

Pollution prevention (P2) is simply a way of eliminating wasteful practices. P2 measures such as source reduction, recycling, and treatment can reduce waste disposal needs; minimize impacts across all environmental media; reduce the volume and toxicity of waste; and ease some of the burdens, risks, and liabilities of waste management. A methodical and rigorous assessment helps to identify the types of wastes currently being generated and how those wastes are being managed. Instead of focusing on just treatment and disposal, concentration is on why these wastes were created in the first place and how they can be either eliminated or minimized at the source. Pollution prevention techniques can be in various forms: process and/or design improvement, materials substitution, energy or water conservation, inventory control, good housekeeping and preventive maintenance, employee training, and waste exchange or pollution trading.

Ideas have been introduced at concept levels and the aim is to raise awareness of complexities and how to manage potential problems. It is our hope that this encourages appropriate positive effort in establishing useful projects and avoiding potential pitfalls and failures. This should raise the level of appreciation that successful projects arise from a sound understanding of site, technologies, local, social, environmental and economic issues, all developed in a systematic manner.

Selection Guide for Treatment Strategies: Approaches to Environmental Problem-Solving

Decisions should be made on the basis of what is to be achieved and only then can a treatment flow scheme be proposed to accomplish the desired objective. For engineers, that is a major responsibility because they have to meet the obligation of their customers, public and/or private. Action is to be carefully and methodically decided. Remember that in the engineering profession, there is no single best solution, and one definition of pollution is "unreasonable interference with another beneficial use." Therefore, pollution prevention, reuse, wastes exchange, and so forth, should be studied first. The general approach goes like this:

1. Getting started: Understanding and defining the problem, characterization, simplifying assumptions, analyzing the problem—origin, magnitude, and short- and long-term impacts (air, surface water, groundwater, ecological risks and liabilities, etc.) of the problem, monitoring, compliance.
2. Identify environmentally friendly options: Goal is to study performance-based alternatives that are scientifically and technologically sound, cost-effective, fair, holistic, ethical, flexible, and effective in achieving environmental goals. This will need a thorough study of most current and credible available options (e.g., literature review, practices, professional publications, design manuals) for addressing the problem; this may require evaluating environmental technologies, changes in management practices—integrating pollution prevention.
3. Comprehensive review: After completion of preliminary analysis, unit processes and unit operations, materials balance (conservation of mass), reactions, reactors, energy consumption (use gravity whenever possible because gravity always wins).

4. Recognizing critical role: The widest possible discussion with administrators and regulators and actively involve stakeholders, interested parties, and public in developing a consensus.
5. After incorporating diverse comments, suggest and check potential treatment scheme systematically and realistically, then do a cost estimates (capital, operating, maintenance, monitoring, repairs, etc.) and value engineering of the chosen option before finalizing recommendations.

From an environmental engineering point of view, substances/solids can exist in one of the three classifications: suspended, colloidal, or dissolved. Particle size increases from dissolved to colloidal to suspended stage. A dissolved substance is one that is truly in solution. The substance is homogeneously dispersed in the liquid; therefore, only one phase is present. Thus for removal of dissolved substances, a phase change process is used. Distillation, precipitation, adsorption, extraction, are some of the processes which have been found to be useful. Physical methods such as filtration, sedimentation, or centrifugation cannot remove dissolved substances. Varying degrees of treatments are generally used for liquid—solid separation process. Useful parameters are: particle size (volume), particle shape, surface area, mass, particle density, fluid density, and fluid viscosity. Whenever the ratio of surface area to the mass of a particle is low, mass is dominant and, therefore, gravity force predominates, as is the case with the plain sedimentation. On the other hand if the ratio of surface area of particle to the mass is high, surface phenomenon (electrostatic repulsion and hydration) is important. The treatment objective is to get particles to settle at the highest velocity. To fulfill this requirement, it is necessary to have large particle volumes, compact shapes, high particle and low fluid densities, and low viscosities.

Microfiltration: It may remove turbidity, algae, cryptosporidium, giardia, bacteria, etc.

Ultrafiltration: Asbestos, virus, organic micromolecules, are likely to be trapped.

Nanofiltration: This method, which complies with stricter federal standards, uses a system of membranes to screen out most impurities. The membranes remove contaminants causing color, smell, and taste. For example, organic chemicals, hardness, color, radionuclides are separated.

Reverse osmosis (RO): Metal ions and salts are eliminated, making water suitable for drinking. Advances in reverse osmosis technology are creating new sources of potable water for a growing number of communities.

Desalination is a viable water supply solution that will only become more attractive and cost-effective (lower cost and higher efficiency) in the near future. Fifty-four percent of the population lives near coastal regions and desalination is being implemented to balance shrinking supplies and growing demands for drinking and the environment. In 1993, there were

134 desalination plants in the United States; by 2003 that number had grown to 234. As the contamination (pollution, salt water intrusion) expands, it puts tremendous pressure on the available water supplies. Industry experience, standardization, and competition will continue to reduce desalination costs, making it a competitive and cost-effective solution. For example, presently desalination of brackish water costs $325 to $650 per acre-foot, compared to $500 acre-foot, in some areas, for imported water. The costs for water recycling (reclamation) and conservation are $400–$800 and $350–$500 per acre-foot, respectively.

Optimizing brine disposal: RO desalination has its own waste disposal problem—disposal of the high salt concentrate (brine) left after the RO process. On average the volume of brine can be about 20%. In some situations, disposal can account for as much as 50% of desalination costs. The brine, residue left by desalinated water is a concentrated slurry containing some water and a high level of dissolved solids. The traditional disposal strategies are in sewers or surface waters, such as rivers, canals, lakes, or oceans; deep-well injection into underground aquifers; land disposal methods, such as spray irrigation and lined evaporation ponds. However, all of these strategies can present some problems. There is a definite need for new promising solutions for concentration management and ways to reduce the energy required in thermal processes. Some creative, collaborative, and innovative planning can be done to increase efficiency and reduce costs. For example, selective removal of marketable salts (like high quality form of calcium carbonate) from the concentrate and sell them to off set disposal costs, increasing the size of desalination plants, co-locating these desalination plants with waste water treatment facilities or/and with power plants using a once-through cooling system. This approach can be mutually beneficial because of joint use of feed water and discharge infrastructure, reduced electric transmission and line losses, use of power plant once through cooling water to dilute brine concentrate before disposal.

Indicator Pollutants For Drinking Water Supply Source

Nutrients, including phosphorus and nitrogen compounds that can stimulate algae growth in receiving waters. For example, in Florida's fragile Everglade's ecosystem, phosphorus is a pollutant of concern and the South Florida Water Management District (SFWMD) is planning and implementing source controls, treatment facilities, and regional programs to reduce phosphorus loads.

Taste and odor compounds that are commonly produced in drinking water reservoirs by blue-green algae.

Total organic carbon (TOC), including live algae and decaying plant and algal material that can lead to the production of suspected carcinogens.

Total dissolved solids, or salinity that can lead to unpalatable mineral tastes, higher water treatment costs, and physiological effects in consumers.

Suspended solids and turbidity from runoff that can reduce the efficiency of water treatment plants.

Pathogens from human contact, livestock, wildlife, urban runoff, septic systems, and wastewater discharges.

Coagulation process: destabilization, mixing, and flocculation: Suspended particles in the colloid range are too small to settle in a reasonable time period and too small to be trapped in the pores of the filter, and have to be treated for removal. Particulate suspensions commonly removed with coagulation process are: clay and silt-based turbidity, natural organic matter, and other associated constituents, such as microbial contaminants (bacteria), iron and manganese, color (dissolved—true and colloidal-apparent), synthetic organic chemicals, toxic metals, etc. For colloids to remain stable they must remain small (the larger the surface charge, the more stable suspension), in a state of hydration—chemical combination with water, and possess surface electric charge. Most colloids are stable because they possess a negative charge that repels other colloidal particles before they collide with one another. The colloids are continually involved in Brownian movement, which is merely a random movement. Since colloids are stable because of their surface charge, in order to stabilize the particles, charge has to be neutralized.

The purpose of coagulation is to alter the colloids so that they can adhere to each other. During coagulation a positive ion is added to water to reduce the surface charge to the point where the colloids are not repelled from each other. This process promotes aggregation of small colloidal particles into larger particles that can subsequently be remove by sedimentation and/or filtration. The coagulation process is completed in three sequential steps: coagulant formation, particle destabilization, and particle aggregation. While selecting a coagulant, three key properties are to be considered: it should be insoluble in the neutral pH range, nontoxic, and a trivalent cation. Most common coagulants are aluminum and ferric salts. Organic polymers (synthetic coagulants—polyelectrolytes) have also been found to be cost-effective in a narrow range of turbidity. To assist in the coagulation—flocculation process, certain additives and coagulant aids have been successfully used. The additives are weighting agents, adsorbents, oxidants, and so on. Four basic types of coagulants aids (interparticle bridgers) are: pH adjusters, activated silica, clay, and polymers (anionic, cationic, nonionic, or polyamphotype).

To sustain growth and future developments adequate water supply has to be provided. Planning for management of water resources, to bridge the demand—supply gap, and to ensure an adequate reliable, safe long-term supply for multiple uses, protection (by using natural systems) of the quality of runoff that fills the reservoirs and recharges the groundwater basins, water conservation, reclamation, underground water (lower grade) storage, and change in our lifestyles (change in mindset), and other sensitive steps and practices have to be initiated. Remember thinking and creativity are the most important human resource. Because of the explosion in technology, globalization and rapid growth, environmental challenges and quality of life issues will always remain with us.

Physical and/or engineering solutions should focus on natural processes wherever and whenever possible. Some examples are a rain garden for storm water management. A man-made rain garden is an attractive landscaping feature planted with perennial native plants (bioretention area) to filter (nutrient removal) treat/polish storm water. The rain garden concept is based on the hydrologic function of the forest habitat. Rain gardens that minimize the volume and improve the quality of water, are suitable for any land use.

Ecologically engineered systems should be preferred over conventional treatment systems. An ecologically engineered system design combines environmental engineering concepts with other disciplines such as biology, microbiology, chemistry, biogeochemistry, and ecology. Thus an ecologically engineered system is a more completely integrated system that takes all aspects of the desired treatment process into consideration and melds them into a cohesive unit. It uses sound engineering in coordinating the interaction of appropriate ecosystems to degrade and remove waste materials from the wastewater. For example, wetlands, the areas of transition between the land and water, are one of the world's most important natural resources and amongst the most productive ecosystems in the world. Wetlands provide values no other ecosystem can, and their underlying natural processes have been successfully duplicated in wastewater treatment and polishing.

The practice of using engineered tree caps to cover landfills is beginning to grow roots in the waste industry. The sponge and pump method (provided by tree cap) relies on rootable soil, amendments, and vegetation that are placed over waste. Trees and grasses provide the pump by taking water up through their root systems and releasing it into the atmosphere via evapotranspiration. Using hydrologic models, engineers can determine which sponges are best-suited to store water during the trees' growing and dormant season. Tree caps prevent water from infiltrating waste, intercept and treat leachate plumes, capture odors and wind blown waste, and mitigate irrigated leachate. These also can help in controlling erosion, reduce subsurface landfill gas movement offsite and sequester greenhouse gases. Since 1990, more than 15 tree caps on a demonstration basis have been installed at U.S. landfills.

Bioengineered treatment systems have been used in environmental engineering projects.

Conclusion

We, environmental engineers, have a unique opportunity and role and should strongly believe that we can make a major difference in the global community and on this planet Earth. Our mission should be to protect human health and environment by promoting sustainability and safely managing all wastes by properly selecting the treatment methodologies. Overall goals for us should be efficient use of resources, creating a healthy environment and protecting and/or restoring the natural environment whenever possible. Our environmental challenges for the 21st century should build on past successes, capitalize on technological advances, and work on environmental improvements that respect the importance of social and economical health. A long-term objective is a sustainable global civil society. This necessitates integration of environmental considerations with economic policy.

We should also accept that wastes will continue to be a part of everyday lives, because everything we do leaves behind some type of waste. Globalization and technological advances are rapidly changing our lives and thinking. We should assume leadership roles in setting goals, policies, priorities, and implementation. We should be willing to learn policies and projects, which are environmentally sound and economically practical. Aim to apply best practices and innovations of international community.

Review Questions

1. What is the current mission of environmental engineering solutions?
2. What are the major limitations of implementing these solutions?
3. What are engineering decisions mainly based on?
4. Of what value is the systems approach in deriving engineering solutions? List the three major ones.
5. What are the 4 Rs of pollution prevention and list the major treatment systems under the three main engineering processes.

Recommended Resources

www.webdirectory.com
www.worldwatch.org
www.watermatters.org
www.epa.gov
www.worldwater.org
www.wri.org
www.cnie.org
www.water.usgs.gov

www.unep.or.jp
www.wasteage.com
www.smartgrowth.org
www.planning.org
www.webdirectory.com
www.WasteXchange.org
www.wes.army.mil
www.eaarth911.org
www.webdirectory.com
www.awwa.org
www.wef.org
www.pollutiononline.com
www.floridacenter.org
www.ajph.org
www.swana.org
www.apwa.org
www.asce.org

American Water Works Association. 2000. *Water Quality and Treatment Handbook*. New York: McGraw-Hill.

Davis, M. L., D. A. Cornell. 1998. *Introduction to Environmental Engineering*, 3rd ed. New York: McGraw-Hill.

Design Manual, Municipal Wastewater Stabilization Ponds. October 1983. Center for Environmental Research Information, EPA 625/1-83-015, Cincinnati, OH.

Reynolds, T. D., P. A. Richards. 1996. *Unit Operations and Processes in Environmental Engineering*, 2nd ed. Boston: PWS Publishing.

Vesiland, P. A. 1997. *Introduction to Environmental Engineering*. Boston: PWS Publishing.

Wastewater Treatment Plant Design. 2003. Water Environment Federation and IWA Publishing.

Selected Manuals of Practice (MOPs) from the Water Environment Federation:
Clarifier Design (ISBN 0943244-61-7).
Design of Municipal Wastewater Treatment Plants (ISBN 0-7844-0342-2).
Energy Conservation in Wastewater Treatment Facilities (ISBN 1-57278-034-7).
Operation of Municipal Wastewater Treatment Plants (ISBN 1-57278-040-1).
Operation of Natural Systems for Wastewater Treatment (ISBN 1-57278-165-3).
Water Reuse (ISBN 0-943244-45-5).

CHAPTER 6

Biological Solutions

Daniel B. Oerther

Introduction

Environmental engineers are interested in microorganisms for at least three different reasons: (1) pathogenic microorganisms cause disease and need to be eliminated to protect public health; (2) catalytic microorganisms transform pollutants and need to be encouraged to clean up the environment; and (3) nuisance microorganisms hinder the stable operation of environmental engineering systems and need to be monitored and controlled. In each of these applications, environmental engineers are interested in knowing three things about specific microorganisms (1) identity, (2) abundance, and (3) activity.

For example, to provide clean drinking water, environmental engineers need to know the identity of putative pathogenic microorganisms in the source water. Viruses, bacteria, and protozoa show different responses to disinfection strategies (i.e., while viruses and bacteria are often easily disinfected using free chlorine, protozoa can demonstrate significant resistance to chlorine; similarly, protozoa and bacteria can be disinfected using ultraviolet irradiation while viruses can demonstrate significant resistance to UV disinfection). Thus, knowing the identity of a particular microbial contaminant is important to engineering suitable protection of public health.

The abundance of specific microbial populations is important to know to calibrate mathematical models used to size unit operations in environmental engineering. For example, molecular biology-based forensic tools can provide a robust measure of the abundance of suspended growth nitrifying bacteria in a sewage treatment system designed to oxidize ammonia. By providing an independent experimental measurement of $X_{autotrophs}$, environmental engineers can better calibrate models to specific treatment plants.

Finally, knowing the activity of particular microbial populations is useful when developing a control strategy for a nuisance population, such as filamentous bacteria causing poor biosolids separation in the clarifier of a municipal activated sludge wastewater treatment plant. Respirometry provides critical kinetic information used to develop kinetic selectors (e.g., a baffled section of an aeration basin placed at the head of a full-scale activated sludge system) to encourage the growth of floc-forming bacteria and reduce the growth of filamentous bacteria.

Biological solutions to problems in environmental engineering often involve engineers integrating apparently disjointed biological knowledge and tailoring this knowledge to address specific engineering challenges. Environmental engineers often work with biologists to adapt tools and techniques from the field of biology to the field of environmental engineering. Often this work is challenging because the fields of biology and environmental engineering are rich with specific terminology requiring a certain level of "translation" to move between the fields. Environmental engineers working at the interface between biology and engineering typically have backgrounds in the emerging discipline of environmental biotechnology where molecular biology, mathematical modeling, and bioprocess engineering are integrated to develop biological solutions to environmental engineering challenges. In this chapter, examples of biological solutions to environmental engineering challenges are provided to show how the emerging discipline of environmental biotechnology contributes to the field of environmental engineering.

Public Health

Risk Assessment

Microorganisms are a form of pollution that threatens public health. Nearly 600 years ago, the Black Death killed one-quarter of the global human population. Bacteria, specifically *Yersinia pestis*, were present in the fleas of rats infesting densely populated urban centers. The solution to this problem was a systematic program for rodent control including the critical component of regular solid waste disposal. To this day, an outbreak of disease of this magnitude has been avoided due to concepts such as weekly garbage pickup.

In 1854, cholera was once again sweeping through urban London. In the first example of public health epidemiology, Dr. John Snow mapped the location of homes containing individuals sick with cholera and the wells where the homes collected their drinking water. Snow observed that contaminated well water provided the fecal—oral route for the spread of the bacteria *Vibrio cholerea*, the etiological agent of cholera. The solution to the problem was simple. Snow removed the pump handle from the Broad Street pump, and deaths from cholera dramatically declined.

In 1993, a dozen children in Washington state died from a gastrointestinal illness brought on by infection with *Escherichia coli* H7:O157. Subsequent epidemiology studies demonstrated that hamburger meat provided to the Jack-in-the-Box food chain was contaminated with *E. coli* H7:O157. Coupled with poor food safety practice, undercooked hamburgers were served, resulting in wide-spread infection and a tragic number of deaths. The solution to this problem was an improved quality control process for inspecting raw meat and the enforcement of proper food safety practice.

These examples demonstrate the nature of the risk posed to human health from microorganisms as pollutants. Environmental engineers play a significant role in wide-spread protection of public health from microorganisms as pollutants by producing significant quantities of potable water at reasonable costs. As part of the Safe Drinking Water Act of 1974, regular monitoring for coliforms (total and fecal) as well as *E. coli* is required. The purpose of this testing is to protect public health. The strategy for this testing is based on the biological principle of "indicator species." An indicator species is an "easy to assay" microorganism serving as a surrogate for other microorganisms. Coliforms were originally suggested as an appropriate indicator species to monitor in drinking water because it was believed that coliforms were present in feces; coliforms survived in the environment in a manner similar to fecal pathogenic microorganisms; and coliforms were susceptible to physical removal and chemical disinfection in a manner similar to fecal pathogenic microorganisms. With time, environmental engineers have systematically moved from testing total coliforms, to testing fecal coliforms, to testing *E. coli*, and most recently to testing *Enterococci* as indicator species to assess the possible threat to human health from potable drinking water *and* to monitor the efficacy of drinking water treatment unit operation to reduce the levels of putative pathogenic microorganisms.

Assessing the risk to human health and determining the effectiveness of environmental engineering design decisions to reduce this risk to an acceptable level for the least possible cost represents a significant challenge.

This challenge is being met using "biological solutions." First, environmental engineers have begun to recognize that not all microorganisms are created equal. Size, shape, and surface properties play a role in determining the best strategy for physical removal of microorganisms (e.g., filtration). Microbial physiology and biochemistry play a role in determining the best strategy for chemical removal of microorganisms (e.g., disinfection). Molecular biology provides state-of-the-art forensic tools to identify and enumerate specific microorganisms alleviating the need for "easy to assay" indicator species and providing environmental engineers with a biological solution to track specific pathogenic microorganisms such as those listed on the Contaminant Candidate List (CCL) (Adenoviruses, *Aeromonas hydrophila*, Caliciviruses, Coxsackieviruses, Cyanobacteria (and their toxins), Echoviruses, *Helicobacter pylori*, Microsporidia, and *Mycobacterium avium intracellulare*).

The development of biological solutions to assess the risk to human health and to determine the effectiveness of environmental engineering design decisions to reduce this risk is an active area of research that will continue to play out over the next decade. At the present time, at least two observations can be made with some degree of certainty, namely: (1) environmental engineers increasingly recognize that the current strategy of employing coliform-like microorganisms as indicators is potentially inadequate to successfully protect public health; and (2) molecular biology-based forensic tools appear to be the next-generation technique that will be used to identify and enumerate a broad spectrum of putative pathogenic microorganisms such as those found on the CCL.

Forensic Epidemiology

As discussed under Risk Assessment, microorganisms represent a form of pollution that must be removed during the production of potable water. For environmental engineers, the design and operation of a drinking water production facility begins with the selection and protection of the best possible source water. Source water quality, whether surface water or ground water, depends significantly upon the types of pollutants discharged upstream. Microbiological pollutants are no exception, and the microbiological quality of source water is a major concern for the safe and economical production of potable water.

Based upon the Clean Water Act Section 303d list of impaired waters, nearly one-half of all waterways in the United States are impaired due to microbiological pollution. States have until approximately 2015 to complete total maximum daily load (TMDL) (i.e., selecting a pollutant to consider, determining assimilative capacity, estimating sources of pollution, predicting pollutant levels, and allocating allowable pollution) plans for the waters listed on their respective Section 303d lists, including microbiologically impaired waterways. The significant challenge in meeting this requirement is to identify the sources of microbiological pollutants in waterways. For example, measuring fecal coliforms along a branch of a stream provides an assessment of microbiological water quality indicating an impaired waterway. But, this assessment does not provide definitive evidence of the source of the impairment. Are migratory bird populations responsible? What about overflow from septic systems? Are wild animals or domestic animals responsible? What is the impact of sanitary or combined sewer overflows? Answers to these difficult questions are needed before a plan can be developed to cost effectively reduce the microbiological pollution and eliminate the impairment.

Molecular biology-based forensic tools are providing a biological solution to this environmental engineering challenge. Molecular biology is a relatively new field in biology, existing for arguably the past thirty 30 years (i.e., since the discovery of recombinant DNA technology in 1972). Molecular biology involves the study of molecular-scale phenomena in biological systems. Molecular biology represents the first example of nanotechnology. Processes occurring naturally within biological systems are researched and subsequently industrialized using biotechnology to produce products for further research and analytical diagnostics. Molecular biology operates at the level of individual nucleotides of DNA and molecules of protein.

For environmental engineers, molecular biology offers the promise of dramatically improved analytical capability to identify and enumerate microorganisms in a variety of environmental matrices. Molecular biology-based forensic tools increasingly are used by researchers in environmental engineering to address the problem of identifying the source of microbiological pollution for Section 303d waters. This emerging field of microbial or bacterial source tracking (MST or BST) often relies upon molecular biology-based assays to identify specific microorganisms and to link environmental microbiological pollution to its source (i.e., to answer the question, "Did this *E. coli* originate from human feces, or domestic animal feces, or wild animal feces?").

Although still under development, MST is being used increasingly by state agencies to develop required TMDLs. In some instances, MST has been shown to be an effective tool for correctly identifying the source of microbiological pollution, yet in other demonstration projects MST has performed poorly. The continued development of molecular biology-based forensic tools to identify and enumerate specific microorganisms is expected to remain an area of active research for at least the next decade as environmental engineers develop these tools as an effective biological solution to the problem of assessing the microbiology quality of source water.

Biocatalysis

Application of Biocatalysis to Reduce the Impact of the Discharge of Organic Pollution

Historically, high-quality water has been readily available throughout most of the United States. The dependence of life on a sufficient quantity of potable water cannot be overstated. As densely populated urban centers have developed in the U.S. over the last two centuries, the availability of high-quality water has been reduced. Dense urban populations release large quantities of highly concentrated pollutants that can quickly overcome the assimilative capacity of receiving bodies (i.e., the environment can deal with a certain degree of pollution—termed the assimilative capacity—but when this level is exceeded, deterioration of water quality is assured).

Nearly a hundred years ago, Streeter and Phelps developed a set of empirical equations to predict the impact of the discharge of organic pollution on the levels of dissolved oxygen (DO) in waterways. DO was selected as the primary measure of water quality for a number of reasons including the necessity to maintain sufficient levels of dissolved oxygen for aquatic life. The Streeter-Phelps equations predicted the critical DO sag, or the point downstream at which the levels of DO would reach their minimum. By properly allocating pollution discharge within the watershed and incorporating engineering works to enhance re-aeration, the critical DO sag could be improved. As the levels of pollution continued to increase, it became increasingly difficult to eliminate dangerously low DO sag points. To solve this problem, environmental engineers embarked on the process of wastewater treatment plant design.

Wastewater treatment plants are a biological solution to the problem of highly concentrated organic pollution. In a wastewater treatment plant, the processes which occur naturally in the environment (e.g., microbial degradation of organic waste with biomass production following by sedimentation) are encouraged to occur in a highly controlled environment. The two most critical processes in biological wastewater treatment of organic pollution include cost-effective oxygen transfer and biomass capture.

Trickling filters were one of the first successful biological solutions to the problem of wastewater. Wastewater containing organic pollution is allowed to flow by gravity over a bed of packed media. Counter-current airflow provides oxygen transfer through the thin film of water to the substratum-attached biomass where the organic pollution is consumed. Because of their simplicity of design and low costs for oxygen transfer, trickling filters were a successful early biological solution to the problem of organic pollution in wastewater. However, the large physical footprint needed for trickling filters, operational difficulties associated with freezing temperatures and biomass sloughing, and nuisance species such as flies and snails represent significant concerns when evaluating trickling filters as a biological solution to wastewater treatment.

In parallel to the development of attached growth treatment systems, environmental engineers were interested in the development of suspended growth biological treatment systems. The suspended growth activated sludge system is perhaps the most widely used biological solution to the problem of organic pollution in wastewater. In a conventional activated sludge design, flocculated biomass is mixed with influent wastewater and oxygen transfer is accomplished through bubbling compressed air in what is known as the aeration basin. In the aeration basin, the biomass consumes both oxygen and the organic pollution resulting in the net growth of microorganisms and the release of carbon dioxide. Thus, dilute soluble organic pollution is converted into particulate biomass. A secondary clarifier follows the aeration basin, and it is used to separate the particulate biomass from the purified water. Employing gravity settling, the flocculated biomass is removed from the bottom of the clarifier while purified effluent water is discharged through weirs at the surface of the clarifier. A portion of the biomass is eliminated from the system (e.g., waste-activated sludge, WAS) and a portion of the biomass is returned to the head of the aeration basin (e.g., return-activated sludge, RAS).

In general, activated sludge systems require a significantly smaller physical footprint as compared to trickling filters with equivalent capacity. Furthermore, activated sludge systems provide environmental engineers with a greater degree of control over the composition of the community of microorganisms consuming the pollutants in the wastewater. The trade-off for these advantages is the increased cost of oxygen transfer and the requirement to manage a significant stream of particulate biomass (e.g., WAS).

Since their inception nearly one hundred years ago, the biological solutions of trickling filters and activated sludge systems have been used successfully to remove broad spectrum organic pollution (measured as biochemical oxygen demand, BOD) from concentrated waste streams to prevent dangerously low critical DO sags. In the twenty-first century, the focus of wastewater treatment increasingly is moving from broad spectrum pollution control toward protecting human health and the environment from specific compounds of interest. In the next two sections, the removal of the macronutrients nitrogen and phosphorus are discussed. In addition to macronutrients, environmental engineers have recently become intensely interested in the removal of "emerging micropollutants" including synthetic food additives (e.g., Olestra), increasingly dangerous microorganisms (e.g., multiple-antibiotic resistant bacteria), pharmaceutical products, and compounds that are suspected to impact endocrine systems (e.g., pseudo-estrogen having an impact at the part per billion concentration). In the next decade, environmental engineers can expect regulatory agencies to begin to ask if existing wastewater treatment systems are effective at removing very specific compounds.

Application of Biocatalysis to Reduce the Impact of the Discharge of Nitrogen Pollution

Nitrogen is a critical macronutrient required for the synthesis of biomass. Although dinitrogen gas is abundant in the atmosphere, only a limited spectrum of organisms is capable of "fixing" nitrogen into a useable form (e.g., organic nitrogen). Nitrogen is applied as a fertilizer and taken up by plants where it is assimilated into plant biomass. Plants form the base of the human food chain, so human waste contains a significant level of excess nitrogen (i.e., the major pollutant in urine is reduced forms of nitrogen). As a component of wastewater, nitrogen is primarily found in reduced forms including organic nitrogen (e.g., proteins) and free ammonia. If left untreated, the discharge of reduced forms of nitrogen can have significant impacts on aquatic life where ammonia irritates the gills of fish resulting in mucous production that ultimately suffocates the fish. In addition to specific problems associated with reduced forms of nitrogen, the discharge of significant levels of total nitrogen is highly correlated to enhanced eutrophication of slow moving surface waters including lakes and estuaries.

Since the 1950s, regulatory agencies increasingly have required wastewater treatment plants to solve the problem of discharging reduced forms of nitrogen. In response to this challenge, environmental engineers have employed a biological solution wherein microorganisms produce enzymes to hydrolyze organic nitrogen releasing free ammonia. Subsequently, ammonia is used by nitrifying bacteria as an energy source. Autotrophic nitrifying bacteria transfer electrons from reduced nitrogen using the energy to fix carbon dioxide yielding new biomass. Because of the high energy costs of autotrophic growth and the relatively low energy yield of nitrogen oxidation, nitrifying bacteria grow slowly. This characteristic slow growth represents a challenge to environmental engineers designing wastewater treatment plant.

Biochemically, nitrification occurs as a step-wise process where ammonia is initially oxidized to nitrite by ammonia oxidizing bacteria (AOB). Subsequently, nitrite is oxidized to nitrate by nitrite oxidizing bacteria (NOB). To evaluate the extent to which conventional wastewater treatment plants required modification to incorporate successful nitrification, environmental engineers collaborated with microbiologists to identify model populations of AOB and NOB. *Nitrosomonas europaea* was originally determined to be the most abundant AOB found in wastewater treatment plants. *Nitrobacter winogradski* was originally determined to be the most abundant NOB. Studies of pure cultures of each of these model microorganisms provided environmental engineers with kinetic information required to design wastewater treatment plants to successfully oxidize reduced nitrogen. Because the kinetics of nitrite oxidization were determined to be faster than the kinetics of ammonia oxidization, many modeling approaches simplified the step-wise process of nitrification with a single overall reaction with kinetic parameters similar to *N. europaea*. The outcome of these efforts resulted in a standard for design with requirements for nitrifying activated sludge systems to employ a solids retention time (SRT) roughly 5 times larger than a conventional activated sludge system (e.g., SRT of 10 days for nitrification as compared to 2 days for the removal of organic pollution).

Recently, environmental engineers and microbiologists have employed molecular biology-based techniques to re-examine the identity of the predominant AOB and NOB in wastewater treatment systems. Although *N. europaea* and *N. winogradski* continue to be found in many treatment systems, it has become clear that these may not be the predominant AOB and NOB in all systems. This discovery has lead environmental engineers to question some of the underlying assumptions regarding the design of nitrifying wastewater treatment systems. For example, modeling efforts have evolved to include nitrification as a step-wise process with explicit reactions for ammonia and nitrite oxidation. Furthermore, the kinetic parameters used to describe nitrification often are measured for a specific system rather than assumed based on literature values.

Once reduced forms of nitrogen have been oxidized, then the oxidized nitrogen must be reduced to dinitrogen gas for successful removal of total nitrogen to the atmosphere. If left untreated, the discharge of oxidized forms of nitrogen can have significant impacts on human health where nitrate consumed by infants and toddlers can be reduced to nitrite that strongly binds myoglobin reducing the ability of the blood to carry oxygen resulting in suffocation of the child (e.g., "blue baby" syndrome). As mentioned above, in addition to human health concerns the discharge of significant levels of total nitrogen can stimulate eutrophication.

To meet this challenge, environmental engineers have turned to a biological solution by modifying the design of activated sludge wastewater treatment systems. Oxidized nitrogen is a readily consumed electron acceptor that can be used as a substitute for DO in the aeration basin of an activated sludge system. Specifically, environmental engineers have developed a number of systems designed to mix nitrified wastewater (containing high levels of oxidized nitrogen) with fresh wastewater (containing organic pollution) and flocculated biomass. The biomass consumes both oxidized nitrogen and the organic pollution resulting in the net growth of microorganisms. Thus, two pollutants are treated simultaneously as the organic pollution is converted into particulate biomass and the oxidized nitrogen is converted into dinitrogen gas.

Although successful for removing total nitrogen from wastewater, the sequential complete oxidization and then partial reduction to dinitrogen gas suffers from obvious inefficiencies from the perspective of redox chemistry.

In the past 10 years, environmental engineers have collaborated with microbiologists to develop alternative technologies for total nitrogen removal that avoid some of the inefficiency of nitrification followed by denitrification. The alternative biological solution to total nitrogen removal is known as anaerobic ammonia oxidization (ANAMMOX). In the ANAMMOX process, specific populations of microorganisms couple the reduction of nitrite to the simultaneous oxidization of ammonia to produce dinitrogen gas. This process avoids the need to fully oxidize and subsequently partially reduce nitrogen by combining partially oxidized nitrogen (e.g., nitrite) with fully reduced nitrogen (e.g., ammonia). As with the process of nitrification, the biomass responsible for the ANAMMOX process is slow growing and therefore requires modification of conventional wastewater treatment plant design. Furthermore, the process appears to be sensitive to the presence of oxygen requiring careful control of the bioreactor environment.

In the future, biological solutions to the problem of total nitrogen discharge will continue to be developed by environmental engineers as regulatory agencies continue to set lower discharge permits to protect the environment from accelerated eutrophication.

Application of Biocatalysis to Reduce the Impact of the Discharge of Phosphorus Pollution

Phosphorus is an important macronutrient found in nucleic acid and membrane lipids. In addition, phosphorus has been used extensively as a builder to produce more effective detergents. Phosphorus in wastewater is derived from biological material and detergent. If left untreated, the discharge of phosphorus to the environment can significantly accelerate the process of eutrophication causing difficult-to-reverse damage to sensitive aquatic ecosystems.

For environmental engineers, the challenge of removing phosphorus from wastewater can be addressed with a biological solution, namely, enhanced biological phosphorus removal (EBRP). By accident, the EBPR process was originally described nearly 40 years ago when effluent wastewater from an activated sludge treatment system was used to irrigate a crop of rice. Phosphorus deficiency in the wastewater was noted, and subsequent analysis indicated that a condition of organic overloading at the front of the aeration basin resulted in the formation of an anaerobic zone encouraging the growth of bacteria that could accumulate phosphorus in excess of anabolic needs. Subsequent years of collaborative research by environmental engineers and microbiologists identified a number of microorganisms as candidate bacteria responsible for EBPR. These putative phosphorus accumulating organisms (PAOs) included *Acinetobacter* spp. among others.

With the advent of molecular biology-based techniques to monitor microorganisms in the environment, it became apparent that *Acinetobacter* spp. were not the predominant PAO. Rather, bacteria phylogenetically related to *Rhodocylcus* spp. appeared to predominate in some EBPR systems. To date, no pure culture has been shown to demonstrate the PAO metabolism. Despite this lack of fundamental biochemical information regarding PAOs, full-scale installations of EBPR have been used successfully for many years to solve the problem of phosphorus discharge.

In an EBPR activated sludge system, influent wastewater containing organic pollution and phosphorus is mixed with flocculated biomass containing PAOs. Because compressed air is not included in this mixing zone, fermentation occurs transforming the organic pollution into short chain volatile fatty acids such as acetate and propionate. The PAOs degrade internal stores of polyphosphate to release energy that is used to import and store the volatile fatty acids. In a subsequent zone of the aeration basin, compressed air is introduced to provide oxygen transfer and increase the level of DO. In the presence of oxygen, the PAOs metabolize the internal stores of volatile fatty acids to release energy that is use for bacteria growth and the important and storage of polyphosphate. Thus, PAOs alternatively store polyphosphate or volatile fatty acids depending upon the presence of oxygen.

As a biological solution to the problem of phosphorus pollution, EBPR is an effective technology used by environmental engineers. In the future, it is anticipated that continuing research will eventually isolate a pure culture of a PAO so that the biochemistry of phosphorus metabolism can be more fully understood and the information can be used by environmental engineers to improve the reliable performance of EBPR systems.

Sludge Stabilization

As described in the three prior sections, the biological solution often employed by environmental engineers when dealing with aquatic pollution is to convert diluted soluble pollution into concentrated particulate pollution (e.g., converting organic pollution into biomass). Although this approach significantly reduces the volume of pollution, identifying an appropriate solution to deal with particulate biomass is a challenge. In general, environmental engineers continue to use biology as a way to further reduce the problem of particulate organic pollution.

Biomass decay is one of the most commonly employed biological solutions to the problem of particulate organic pollution. When microorganisms are mixed with pollutants the outcome includes the net growth of biomass. However, growth of biomass is an exergonic process with a significant degree of entropy. In other words, mass and energy is lost as biomass grows. Thus, environmental engineers take advantage of the inherent inefficiency of biological systems to solve the problem of particulate organic pollution.

Predation is one biological solution employed by environmental engineers. In general, the treatment of soluble organic pollution and macronutrients is carried out by single cell bacteria organized into flocculated biomass or adhered biofilm. These bacteria are susceptible to predation from both larger and smaller microorganisms. Protozoa and higher eukaryotes graze on bacteria in wastewater treatment systems. Alternatively, bacteria phage (also known as bacteria viruses) infect and kill bacteria in wastewater treatment systems. By encouraging both forms of predation, environmental engineers create a more complex trophic structure (i.e., predator and prey) that takes advantage of the inefficiency of biological systems resulting in the further elimination of pollution.

Alternatively, environmental engineers can expand the trophic structure of the artificial ecosystem by controlling the overall redox state. Anaerobic digestion and stabilization of biomass is a widely applied biological solution to the challenge of particulate organic pollution. In a typical anaerobic treatment system, biomass from a mainstream aerobic system is transferred to an anaerobic vessel. Deprived of DO, many of the microorganisms die. The products of microorganism death and decay include large organic molecules that are initially hydrolyzed into smaller molecules and subsequently fermented to release volatile fatty acids (e.g., propionate and acetate). The volatile fatty acids are metabolized syntrophically by bacteria through the cooperation of methanogenic archaea that reduce the levels of dissolved dihydrogen and produce methane gas. Thus, through a step-wide biochemical process the biomass harvested from the aeration basin of an activated sludge system (e.g., the waste activated sludge, WAS) is converted into methane, carbon dioxide, and the net growth of syntrophic bacteria and methanogenic archaea.

Syntrophs and methanogens derive only small amounts of energy from their catabolic reactions. Thus, syntrophs and methanogens growth slowly and yield low levels of biomass (i.e., similar to the nitrifiers discussed above). By expanding trophic structure to include anaerobic metabolism in addition to aerobic metabolism, environmental engineers have found a suitable biological solution to the challenge of organic pollution.

Although sludge stabilization has been a technology available to environmental engineers for many years, research to optimize predator—prey interactions or anaerobic treatment systems generally lags research on less complicated aerobic systems designed for the removal of organic and macronutrient pollution. In the future, as the cost of disposal of particulate biomass continues to increase (i.e., tipping fees at land fills rise and agricultural application of biomass is restricted), environmental engineers are expected to invest significant effort to further develop innovative biological solutions to deal with particulate pollution.

Bioprocess Control

Difficulties with Solids Separation

In a conventional suspended growth-activated sludge process, dilute soluble organic pollution is converted into concentrated particulate biomass. Ultimately, this biomass must be separated from the purified wastewater before discharge to the environment can occur. In general, a secondary gravity clarifier is the most commonly employed technology for biomass separation. In a clarifier, the biomass settles to the bottom of the vessel under the influence of gravity while the purified wastewater exits the vessel from the top over the weirs. The ability of flocculated biomass to be separated in a secondary clarifier represents a competition of forces between gravity settling (pulling downward) versus drag and buoyant forces (pushing upward). When the force due to gravity is insufficient to overcome drag and buoyant forces, the flocculated biomass settles poorly resulting in biomass in the effluent of the treatment system.

To solve the problem of poor biomass separation, environmental engineers have collaborated with microbiologists to develop a biological solution. The conceptual model of successfully flocculated biomass includes a balance between floc-forming bacteria and filamentous bacteria. When floc-forming bacteria significantly outnumber filamentous bacteria, pinpoint floc formation occurs and gravity force is significantly reduced because of the limited mass incorporated into individual flocs. When filamentous bacteria significantly outnumber floc-forming bacteria, filamentous sludge bulking occurs and drag force is significantly increased because of the large equivalent surface area of individual flocs. Therefore, maintaining an appropriate balance between floc-forming and filamentous bacteria is a critical component to successful operation of a secondary clarifier.

A fundamental understanding of microorganism metabolic kinetics is needed to control the balance between floc-forming and filamentous bacteria. In general, floc-forming bacteria tend to have higher specific growth rates and lower affinity for substrates as compared to filamentous bacteria. In other words, floc-forming bacteria represent rapid-growth members of the activated sludge ecosystem (e.g., weeds and rodents) while filamentous bacteria represent slow-growth members of the activated sludge ecosystem (e.g., trees and elephants). When bioreactor configurations are selected to encourage significant gradients in substrate concentration (e.g., from high-influent substrate levels through low-effluent substrate levels), floc-forming bacteria predominate. When bioreactor configurations are selected to discourage gradients in substrate concentration (e.g., completely stirred tank reactor systems), filamentous bacteria predominate. By controlling the overall hydraulic mixing conditions in an activated sludge wastewater treatment system, environmental engineers can influence the balance between floc-formers and filamentous bacteria resulting in control over the settleability of the biomass.

In addition to hydraulic considerations, a number of other factors are known to influence poor biomass settling including oxygen limitation, nutrient deficiency, long SRT, and power input due to mixing equipment. In each of these scenarios, environmental engineers benefit from knowing the identity of the predominant filamentous bacteria present in the system. Thus, part of the biological solution for the challenge of poor biomass settling is related to the use of molecular biology-based techniques to identify and enumerate specific populations of microorganisms. In the past 10 years, the use of molecular biology-based techniques to identify filamentous bacteria has exploded with a wide variety of tools now available to diagnose poor biomass separation attributed to filamentous bacteria. In the future, environmental engineers are expected to employ these molecular biology-based tools increasingly to solve problems of poor biomass separation.

Application of Bioaugmentation to Enhance Biocatalytic Capability

Increasingly, knowledge of applied microbial ecology is becoming a useful tool for environmental engineers. Engineered bioreactors and bioremediation scenarios rely upon microorganisms to degrade and transform pollutants, but microorganisms do not exist in isolation. Rather, for environmental engineering applications microorganisms exist in complex communities interacting with pollutants, one another, and the abiotic environment. Approximately 30 years ago, with the advent of recombinant DNA technology there was an effort to design "super-bugs" capable of accelerating the clean-up of environmental pollution. When these custom-made microorganisms were introduced into the environment, the typical outcome was that the natural flora out-competed the exogenous microorganisms. With improvements in research, it has become increasingly clear that communities of microorganisms share greater complexity beyond the straightforward metabolic link between genetically modified microorganisms and pollutants.

For environmental engineers, microbial ecology represents a future biological solution to the problem of optimizing biocatalytic capacity. In place of developing super-bugs capable of enhanced degradation, the path forward lies in the application of fundamental ecosystem understanding to engineer communities of microorganisms to more effectively transform pollutants. As described above, metabolic competition for growth limiting nutrients—as it relates to understanding the competition between floc-formers and filamentous bacteria—represents an initial step down this path.

Environmental engineers will continue to develop a more thorough fundamental appreciation for microbial ecology which should become manifest in successful efforts of bioaugmentation. The process of bioaugmentation involves the addition of an exogenous population into an ecosystem. In the simplest case, bioaugmentation is performed by adding an abundance of a microorganism originally isolated from the environment of interest, prepared in a separate cultivation scheme, and amended to the original system. The advantages of bioaugmentation include increasing the abundance of a key population of microorganisms in the ecosystem of interest. In the future, environmental engineers may apply the concept of bioaugmentation to enhance the performance of existing biological treatment systems for the treatment of increasingly specialized environmental pollutants (e.g., pollutants discussed above in Biocatalysis).

Review Questions

1. Why are environmental engineers interested in microorganisms?
2. How can bacterial source tracking be used as a biological solution to the problem of impaired water quality?
3. How have the results of molecular biology-based tools changed the manner in which environmental engineers approach models of nitrifying wastewater treatment systems?
4. What are the two main changes in trophic structure that can be used as biological solutions by environmental engineers to address the challenge of particulate biomass disposal?
5. Beyond considerations of floc-formers and filamentous bacteria, what properties of wastewater would influence engineering design decisions regarding bioreactor hydraulic configurations?

CHAPTER 7

Hydrogen Energy Solutions

T. Nejat Veziroglu, S. A. Sherif, and Frano Barbir

Introduction

Energy carriers are a convenient form of stored energy. Electricity is one type of carriers that can be produced from various sources, transported over large distances, and distributed to the end user. Hydrogen is another type of energy carrier. If produced from clean sources, it constitutes a clean, efficient and versatile carrier that supplements electricity well. Together, these two carriers may satisfy most of human's energy needs and form an independent energy system (Bockris 1975, 1985, 1991).

Hydrogen has some unique properties that make it an ideal energy carrier (Veziroglu and Barbir 1992):

- It can be produced from and converted into electricity at relatively high efficiencies.
- Raw material for hydrogen production is water, which is available in abundance. Hydrogen is a completely renewable fuel, since the product of hydrogen utilization (either through combustion or through electrochemical conversion) is pure water or water vapor.
- It can be stored in gaseous form (convenient for large-scale storage), in liquid form (convenient for air and space transportation), or in the form of metal hydrides (convenient for surface vehicles and other relatively small scale storage requirements).
- It can be transported over large distances through pipelines or via tankers (in most of the cases more efficiently and economically than electricity).
- It can be converted into other forms of energy in more ways and more efficiently than any other fuel, i.e., in addition to flame combustion (like any other fuel) hydrogen may be converted through catalytic combustion, electro-chemical conversion and hydriding.
- Hydrogen as an energy carrier is environmentally compatible. Since its production (from electricity or directly from solar energy), its

storage and transportation, as well as its end use do not produce any pollutants (except small amounts of NO_x if hydrogen is burned with air at high temperatures), greenhouse gases, or any other harmful effects on the environment.

Figure 7.1 shows a global energy system in which electricity and hydrogen are produced from available energy sources and used in applications where fossil fuels are being used today. In such a system, electricity and hydrogen are produced in large industrial plants as well as in small, decentralized units, wherever the primary energy source (solar, nuclear, and even fossil) is available. Electricity is used directly or transformed into hydrogen. For large-scale storage, hydrogen can be stored underground in ex-mines, caverns and/or aquifers. Energy transport to the end users, depending on distance and overall economics, is either in the form of electricity or in the form of hydrogen. Hydrogen may be transported, by means of pipelines or super tankers. It is then used in transportation, industrial, residential, and commercial sectors as a fuel. Some of it may be used to generate electricity (via fuel cells), depending on demand, geographical location or time of day. Fuel cells may be available in MW power plant size or as individual devices (several kW) suitable for distributed power generation.

This chapter will discuss the use of hydrogen energy in an overall hydrogen-based economy. The topics covered will involve hydrogen pro-

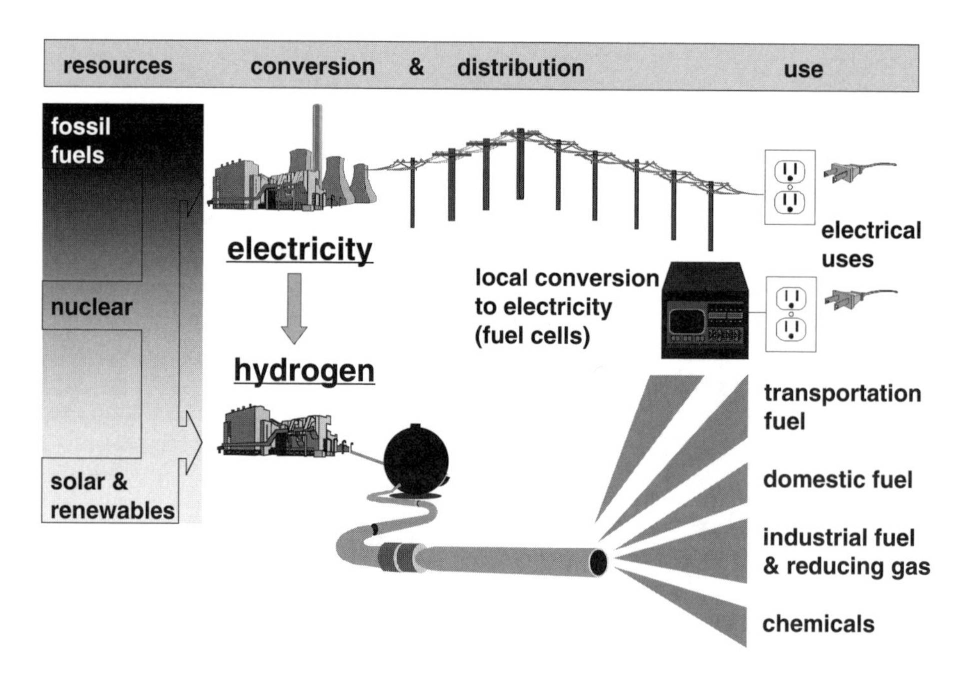

FIGURE 7.1. Hydrogen/electricity energy system.

duction, storage, distribution, conversion, and safety. Other alternative energy solutions will be discussed later in this book.

Properties of Hydrogen

Hydrogen is an odorless, colorless gas. With a molecular weight of 2.016, hydrogen is the lightest element. Its density is about 14 times less than air (0.08376 kg/m^3 at standard temperature and pressure). Hydrogen is liquid at temperatures below 20.3 K (at atmospheric pressure). Hydrogen has the highest energy content per unit mass of all fuels—higher heating value is 141.9 MJ/kg, almost 3 times higher than gasoline. Some important properties of hydrogen are compiled in Table 7.1. It exists in three isotopes: protium, deuterium, and tritium. A standard hydrogen atom (protium) is the simplest of all the elements and consists of one proton and one electron. Molecular hydrogen (H_2) exists in two forms: ortho- and para-hydrogen. Both forms have identical chemical properties, but due to different spin orientation have somewhat different physical properties. At room temperature hydrogen consists of approximately 75% ortho-hydrogen and 25% para-hydrogen. Since para-hydrogen is more stable at lower temperatures its concentration increases at lower temperatures, reaching virtually 100% at liquid hydrogen temperatures.

TABLE 7.1
Properties of Hydrogen

Molecular weight		2.016
Density	kg/m^3	0.0838
Higher heating value	MJ/kg	141.90
	MJ/m^3	11.89
Lower heating value	MJ/kg	119.90
	MJ/m^3	10.05
Boiling temperature	K	20.3
Density as liquid	kg/m^3	70.8
Critical point		
Temperature	K	32.94
Pressure	bar	12.84
density	kg/m^3	31.40
Self—ignition temperature	K	858
Ignition limits in air	(vol. %)	4–75
Stoichiometric mixture in air	(vol. %)	29.53
Flame temperature in air	K	2,318
Diffusion coefficient	cm^2/s	0.61
Specific heat (c_p)	kJ/(kg·K)	14.89

Hydrogen Production Methods

While hydrogen is the most plentiful element in the universe, making up about three-quarters of all matter, free hydrogen is scarce. The atmosphere contains trace amounts of it (0.07%), and it is usually found in small amounts mixed with natural gas in crustal reservoirs. A few wells, however, have been found to contain large amounts of hydrogen, such as some wells in Kansas that contain 40% hydrogen, 60% nitrogen, and trace amounts of hydrocarbons (Goebel et al. 1984). The earth's surface contains about 0.14% hydrogen (10th most abundant element), most of which resides in chemical combination with oxygen as water. Hydrogen, therefore, must be produced. Logical sources of hydrogen are hydrocarbon (fossil) fuels (C_XH_Y) and water (H_2O). Presently, hydrogen is mostly being produced from fossil fuels (natural gas, oil, and coal). However, except for the space program, hydrogen is not being used directly as a fuel or an energy carrier. It is being used in refineries to upgrade crude oil (hydrotreating and hydrocracking), in the chemical industry to synthesize various chemical compounds (such as ammonia. methanol, etc.), and in metallurgical processes (as a reduction or protection gas). The total annual hydrogen production worldwide in 1996 was about 40 million tons (5.6 EJ) (Heydorn 1998). Less than 10% of that amount was supplied by industrial gas companies; the rest was produced at consumer owned and operated plants (so-called captive production), such as refineries, and ammonia and methanol producers. Production of hydrogen as an energy carrier would require an increase in production rates by several orders of magnitude.

A logical source for large-scale hydrogen production is water, which is abundant on earth. Different methods of hydrogen production from water have been, or are being developed. They include electrolysis, direct thermal decomposition or thermolysis, thermochemical processes, and photolysis. These are described below.

Electrolysis

Electrolysis appears to be the only method developed to date, which can be used for large-scale hydrogen production in a post-fossil fuel era. Production of hydrogen by water electrolysis is a mature technology, based on a fundamentally simple process, it is very efficient, and does not involve moving parts.

The following reactions take place at the electrodes of an electrolysis cell filled with a suitable electrolyte (aqueous solution of KOH or NaOH or NaCl) upon the application of a potential:

$$\text{Cathode reaction: } 2H_2O\ (1) + 2e^- \rightarrow H_2\ (g) + 2\ OH^-\ (aq) \tag{7.1}$$

$$\text{Anode reaction: } 2OH^-\ (aq) \rightarrow 1/2\ O_2\ (g) + H_2O\ (1) \tag{7.2}$$

$$\text{Overall cell reaction: } H_2O\ (l) \rightarrow H_2\ (g) + O_2\ (g) \tag{7.3}$$

The reversible decomposition potential ($E_{rev} = \Delta G/nF$) of the above reaction is 1.229 V at standard conditions. The total theoretical water decomposition potential is 1.480 V corresponding to hydrogen's enthalpy (since $\Delta H = \Delta G + T\Delta S$). Due to irreversible processes occurring at the anode and cathode, including the electrical resistance of the cell, the actual potentials are always higher, typically between 1.75 and 2.05 V. This corresponds to the efficiencies of 72% to 82%.

Several advanced electrolyzer technologies are being developed such as:

Advanced alkaline electrolysis, which employs new materials for membranes and electrodes that allow further improvement in efficiency—up to 90% (Bonner et al. 1984; Dutta 1990).

Solid polymer electrolytic (SPE) process, which employs a proton-conducting ion exchange membrane as electrolyte and as membrane that separates the electrolysis cell. This type of electrolyzers can operate at very high current densities (up to 2 A/cm^2 which is about one order of magnitude higher than standard electrolyzers with alkaline liquid electrolyte). The water to be dissociated does not require dissolved electrolytes to increase its conductivity, and is added solely to the anode side (Dutta 1990; Wendt 1988).

High temperature steam electrolysis, which operates between 700° and 1,000°C, uses oxygen ion-conducting ceramics as electrolyte. Electrical energy consumption is reduced since part of the energy required for water dissociation is supplied in the form of heat. The water to be dissociated is entered on the cathode side as steam, which forms a steam-hydrogen mixture during electrolytic dissociation. The O_2-ions are transported through the ceramic material to the anode, where they are discharged as oxygen (Liepa and Borhan 1986).

An electrolysis plant can operate over a wide range of capacity factors and is convenient for wide range of operating capacities, which makes this process interesting for coupling with renewable energy sources, particularly with photovoltaics (PV). Photovoltaics generate low voltage-direct current, exactly what is required for the electrolysis process.

The performance of photovoltaic-electrolyzer systems has been studied extensively both in theory and in practice (Hancock 1984; Carpetis 1984; Siegel and Schott 1988; Steeb et al. 1990). Several experimental PV-electrolysis plants are currently operating all over the world, such as:

- Solar-Wasserstoff-Bayern pilot plant in Neunburg vorm Wald in Germany (Blank and Szyszka 1992)
- HYSOLAR project in Saudi Arabia (Grasse et al. 1992)
- Schatz Energy Center, Humboldt State University, Arcata, California (Lehman and Chamberlain 1991)

- Helsinki University of Technology, Finland (Lund 1991)
- INTA Energy Laboratory, Huelva, Spain (Garcia-Conde and Rosa 1992)

Direct Thermal Decomposition of Water (Thermolysis)

Water can be split thermally at temperatures above 2000K. The overall thermal dissociation of water can be shown as (Baykara and Bilgen 1989):

$$H_2O \rightarrow a\,H_2O + b\,OH + c\,H + d\,O + e\,H_2 + f\,O_2 \qquad (7.4)$$

The degree of dissociation is a function of temperature: only 1% at 2,000K, 8.5% at 2,500K, and 34% at 3,000K. The product is a mixture of gases at extremely high temperatures. The main problems in connection with this method are related to materials required for extremely high temperatures, recombination of the reaction products at high temperatures and separation of hydrogen from the mixture.

Thermochemical cycles

Thermochemical production of hydrogen involves the chemical splitting of water at temperatures lower than those needed for thermolysis, through a series of cyclical chemical reactions which ultimately release hydrogen. Since the mid 1960s, research has been performed to investigate a number of potential thermochemical cycles for the production of hydrogen, and some 2,000 to 3,000 cycles have been invented. After examining their practicability in terms of reaction and process technology, only 20 to 30 remained applicable for large-scale hydrogen production. Some of the more thoroughly investigated thermochemical process cycles are listed herein below (Wendt 1988; Engels et al. 1987; Yalcin 1989):

- Sulfuric acid—iodine cycle,
- Hybrid sulfuric acid cycle,
- Hybrid sulfuric acid—hydrogen bromide cycle,
- Calcium bromide—iron oxide cycle (UT-3),
- Iron chlorine cycle

Depending on the temperatures at which these processes are occurring, relatively high efficiencies are achievable (40–50%). However, the problems related to movement of large mass of materials in chemical reactions, toxicity of some of the chemicals involved, and corrosion at high temperatures remain to be solved in order for these methods to become practical.

Photolysis

Photolysis (or direct extraction of hydrogen from water using only sunlight as an energy source) can be accomplished by employing photobiological systems, photochemical assemblies, or photoelectro-chemical cells (Bull 1988; Willner and Steinberger-Willner 1988). Intensive research activities are opening new perspectives for photo-conversion, where new redox catalysts, colloidal semiconductors, immobilized enzymes and selected microorganisms could provide means of large-scale solar energy harvesting and conversion into hydrogen.

Hydrogen Production from Biomass

Hydrogen can be obtained from biomass by a pyrolisis/gasification process (National Hydrogen Association 1991). The biomass preparation step involves heating of the biomass/water slurry to high temperatures under pressure in a reactor. This process decomposes and partially oxidizes the biomass, producing a gas product consisting of hydrogen, methane, CO_2, CO, and nitrogen. Mineral matter is removed from the bottom of the reactor. The gas stream goes to a high temperature shift reactor where the hydrogen content is increased. Relatively high purity hydrogen is produced in the subsequent pressure swing adsorption unit. The whole system is very much similar to a coal gasification plant, with the exception of the unit for pretreatment of the biomass and the design of the reactor. Because of the lower calorific value per unit mass of biomass as compared to coal, the processing facility is larger than that of a comparably sized coal gasification plant.

Gaseous Hydrogen Storage

Because of hydrogen's low density, its storage always requires relatively large volumes and is associated with either high pressures (thus requiring heavy vessels), or extremely low temperatures, and/or combination with other materials (much heavier than hydrogen itself). Table 7.2 shows achievable storage densities with different types of hydrogen storage methods. Some novel hydrogen storage method may achieve even higher storage densities, but have yet to be proven in terms of practicality, cost, and safety.

Depending on storage size and application, several types of hydrogen storage may be differentiated:

1. Stationary large storage systems: These are typically storage devices at the production site or at the start or end of pipelines and other transportation pathways.

TABLE 7.2
Hydrogen Storage Types and Densities

	kg H2/kg	*Kg H2/m3*
Large volume storage (10^2 to 10^4 m^3 geometric volume)		
Underground storage		5–10
Pressurized gas storage (above ground)	0.01–0.014	2–16
Metal hydride	0.013–0.015	50–55
Liquid hydrogen	~1	65–69
Stationary small storage (1 to 100 m^3 geometric volume)		
Pressurized gas cylinder	0.012	~15
Metal hydride	0.012–0.014	50–53
Liquid hydrogen tank	0.15–0.50	~65
Vehicle tanks (0.1 to 0.5 m^3 geometric volume)		
Pressurized gas cylinder	0.05	15
Metal hydride	0.02	55
Liquid hydrogen tank	0.09–0.13	50–60

2. Stationary small storage systems: At the distribution or final user level, for example, a storage system to meet the demand of an industrial plant.
3. Mobile storage systems for transport and distribution: These include both large-capacity devices, such as a liquid hydrogen tanker-bulk carrier, and small systems, such as a gaseous or liquid hydrogen truck trailer.
4. Vehicle tanks: These are used to store hydrogen used as fuel for road vehicles.

Large Underground Hydrogen Storage

Underground storage of hydrogen in caverns, aquifers, depleted petroleum and natural gas fields, and man-made caverns resulting from mining and other activities is likely to be technologically and economically feasible (Taylor et al. 1986). Hydrogen storage systems of the same type and the same energy content will be more expensive by approximately a factor of 3 than natural gas storage systems, due to hydrogen's lower volumetric heating value. Technical problems, specifically for the underground storage of hydrogen other than expected losses of the working gas in the amount of 1% to 3% per year are not anticipated. The city of Kiel's public utility has been storing town gas with a hydrogen content of 60% to 65% in a gas cavern with a geometric volume of about 32,000 m^3 and a pressure of 80 to 160 bar at a depth of 1,330 m since 1971 (Carpetis 1988). Gaz de France (the French national gas company) has stored hydrogen-rich refinery by-product gases in an aquifer structure near Beynes. Imperial Chemical Industries of

Great Britain stores hydrogen in the salt mine caverns near Teeside, United Kingdom (Pottier and Blondin 1955).

Above Ground Pressurized Gas Storage Systems

Today, pressurized gas storage systems are used in natural gas business in various sizes and pressure ranges from standard pressure cylinders (50 L, 200 bar) to stationary high-pressure containers (over 200 bar) or low-pressure spherical containers (>30,000 m^3, 12 to 16 bar). This application range will be similar for hydrogen storage.

Vehicular Pressurized Hydrogen Tanks

Storage of hydrogen in automobiles has been enabled by the development of ultra-light but strong new composite materials. Pressure vessels that allow hydrogen storage at pressures >200 bars have been developed and used in automobiles (such as Daimler-Benz NECAR II). A storage density higher than 0.05 kg of hydrogen per 1 kg of total weight is easily achievable (Mitlitsky 1996).

Metal Hydride Storage

Hydrogen can form metal hydrides with some metals and alloys. During the formation of the metal hydride, hydrogen molecules are split and hydrogen atoms are inserted in spaces inside the lattice of suitable metals and/or alloys. In such a way, effective storage comparable to the density of liquid hydrogen is created. However, when the mass of the metal or alloy is taken into account, then the metal hydride gravimetric storage density is comparable to storage of pressurized hydrogen. The best achievable gravimetric storage density is about 0.07 kg of H_2/kg of metal, for a high temperature hydride such as MgH_2 as shown in Table 7.3, which gives a comparison of some hydriding substances with liquid hydrogen, gaseous hydrogen, and gasoline (Veziroglu 1987).

During the storage process (charging or absorption) heat is released which must be removed in order to achieve the continuity of the reaction. During the hydrogen release process (discharging or desorption) heat must be supplied to the storage tank. An advantage of storing hydrogen in hydriding substances is the safety aspect. A serious damage to a hydride tank (such as one that could be caused by a collision) would not pose fire hazard since hydrogen would remain in the metal structure.

Novel Hydrogen Storage Methods

Hydrogen can be physically adsorbed on activated carbon and be "packed" on the surface and inside the carbon structure more densely than if it has

TABLE 7.3
Hydriding Substances as Hydrogen Storage Media

Medium	*Hydrogen Content kg/kg*	*Hydrogen Storage Capacity, kg/Liter of vol.*	*Energy Density kJ/kg*	*Energy Density kJ/Liter of vol.*
MgH_2	0.070	0.101	9,933	14,330
Mg_2NiH_4	0.0316	0.081	4,484	11,494
VH_2	0.0207		3,831	
$FeTiH_{1.95}$	0.0175	0.096	2,483	13,620
$TiFe_{0.7}Mn_{0.2}H_{1.9}$	0.0172	0.090	2,440	12,770
$LaNi_5H_{7.0}$	0.0137	0.089	1,944	12,630
$R.E.Ni_5H_{6.5}$	0.0135	0.090	1,915	12,770
Liquid H_2	1.00	0.071	141,900	10,075
Gaseous H_2 (100 bar)	1.00	0.0083	141,900	1,170
Gaseous H_2 (200 bar)	1.00	0.0166	141,900	2,340
Gasoline	—	—	47,300	35,500

been just compressed. Amounts of up to 48 g H_2 per kg of carbon have been reported at 6.0 MPa and 87 K (Schwartz and Amankwah 1993). The adsorption capacity is a function of pressure and temperature, therefore at higher pressures and/or lower temperatures even larger amounts of hydrogen can be adsorbed. For any practical use, relatively low temperatures are needed (<100 K). Since the adsorption is a surface process, the adsorption capacity of hydrogen on activated carbon is largely due to the high surface area of the activated carbon, although there are some other carbon properties, which affect the capability of activated carbon to adsorb hydrogen.

Researchers from Northeastern University in Boston, Massachusetts, have recently announced that they have developed a carbon storage material that can store as high as 75% of hydrogen by weight (Chambers et al. 1998). This material, apparently some kind of carbon nanotubes or carbon whiskers, is currently being researched in several laboratories. The best results achieved with carbon nanotubes to date confirmed by the National Renewable Energy Laboratory is hydrogen storage density corresponding to about 10% of the nanotube weight (Dillon et al. 1996).

Hydrogen can be stored in glass microspheres of approximately 50 μm diameter. The microspheres can be filled with hydrogen by heating them to increase the glass permeability to hydrogen. At room temperature, a pressure of approximately 25 MPa is achieved resulting in storage density of 14% mass fraction and 10 kg H_2/m^3 (Rambach and Hendricks 1996). At 62 MPa, a bed of glass microspheres can store 20 kg H_2/m^3. The release of hydrogen occurs by reheating the spheres to again increase the permeability.

Researchers at the University of Hawaii are investigating hydrogen storage via polyhydride complexes. Complexes have been found which catalyze the reversible hydrogenation of unsaturated hydrocarbons. This catalytic reaction could be the basis for a low temperature hydrogen storage system with an available hydrogen density greater than 7% (Jensen 1996).

Liquid Hydrogen

At present, liquid hydrogen is primarily used as a rocket fuel and is predestined for supersonic and hypersonic space vehicles primarily because it has the lowest boiling point density and the highest specific thrust of any known fuel. Its favorable characteristics include its high heating value per unit mass, its wide ignition range in hydrogen/oxygen or air mixtures, as well as its large flame speed and cooling capacity due to its high specific heat which permits very effective engine cooling and cooling the critical parts of the outer skin (Brewer 1982; Winter and Nitsch 1988). Liquid hydrogen has some other important uses such as in high-energy nuclear physics and bubble chambers. The transport of hydrogen is vastly more economical when it is in liquid form even though cryogenic refrigeration and special dewar vessels are required. Although liquid hydrogen can provide a lot of advantages, its uses are restricted in part because liquefying hydrogen by existing conventional methods consumes a large amount of energy (around 30% of its heating value). Liquefying 1 kg of hydrogen in a medium-size plant requires 10 to 13 kWh of energy (electricity) (Winter and Nitsch 1988). In addition, boil-off losses associated with the storage, transportation, and handling of liquid hydrogen can consume up to 40% of its available combustion energy. It is therefore important to search for ways that can improve the efficiency of the liquefiers and diminish the boil-off losses.

Liquid Hydrogen Production

Production of liquid hydrogen requires the use of liquefiers that utilize different principles of cooling. In general, hydrogen liquefiers may be classified as conventional, magnetic, or hybrid. Many types of conventional liquefiers exist such as the Linde-Hampson liquefiers, the Linde dual-pressure liquefiers, the Claude liquefiers, the Kapitza liquefiers, the Heylandt liquefiers, and those liquefiers utilizing the Collins cycle, just to name a few. Conventional liquefiers generally comprise compressors, expanders, heat exchangers, and Joule-Thomson valves. Magnetic liquefiers, on the other hand, utilize the magnetocaloric effect. This effect is based on the principle that some magnetic materials experience a temperature increase upon the application of a magnetic field and a temperature drop upon lifting the magnetic field. The magnetic analog of several conventional liquefiers includes the Brayton liquefiers, the Stirling liquefiers, and the active magnetic regenerative (AMR) liquefier. Additional information on liquid hydrogen production methods can be found in Huston 1984.

Liquid Hydrogen Storage

Hydrogen is usually transported in large quantities by truck tankers of 30 to 60 m^3 capacity, by rail tank cars of 115 m^3 capacity, and by barge containers of 950 m^3 capacity (Sherif et al. 1989). Liquid hydrogen storage vessels are usually available in sizes ranging from one liter dewar flasks used in laboratory applications to large tanks of 5,000 m^3 capacity. The National Aeronautics and Space Administration (NASA) typically uses large tanks of 3,800 m^3 capacity (25 m in diameter). The total boil-off rate from such dewars is approximately 600,000 LPY (liters per year), which is vented to a burn pond.

The contributing mechanisms to boil-off losses in cryogenic hydrogen storage systems are: (1) ortho-para conversion; (2) heat leak (shape and size effect, thermal stratification, thermal overfill, insulation, conduction, radiation, cool-down), (3) sloshing, and (4) flashing. These mechanisms will be described below.

At high temperatures, hydrogen is an equilibrium mixture of 75% ortho-hydrogen and 25% para-hydrogen. Hydrogen of this composition is termed normal and is in equilibrium at room temperature and above. At zero Kelvin, on the other hand, all the molecules must be in a rotational ground state at equilibrium, therefore all the hydrogen molecules at equilibrium should be in the para state.

By cooling hydrogen gas from room temperature to the nbp (normal boiling point) the ortho-hydrogen converts to para-hydrogen spontaneously. The conversion of ortho-hydrogen to para-hydrogen is an exothermic reaction. The heat of conversion is related to the change of momentum of the hydrogen nucleus when the direction of spin changes. The amount of heat given off in this conversion process is temperature dependent. The heat of conversion is greater than the latent heat of vaporization of normal and para-hydrogen at the nbp. If the unconverted normal hydrogen is placed in a storage vessel, the heat of conversion will be released within the container, which leads to the evaporation of the liquid. Because of these peculiarities in the physical properties of hydrogen, the boil-off of the stored liquid will be considerably larger than what one would be able to determine from calculations based on ordinary heat leak to the storage tank. In order to minimize the storage boil-off losses, the conversion rate of ortho-hydrogen to para-hydrogen should be accelerated with a catalyst that converts the hydrogen during the liquefaction process (Newton 1967a,b; Baker and Shaner 1978; Hands 1986).

The use of a catalyst usually results in a larger refrigeration load and consequently in an efficiency penalty primarily because the heat of conversion must be removed. The time for which hydrogen is to be stored usually determines the optimum amount of conversion. For use within a few hours, no conversion is necessary. For example, large-scale use of liquid hydrogen as a fuel for jet aircraft is one of those cases where conversion is not necessary since utilization of the liquid is almost a continuous process and long-term storage is therefore not needed (Baker and Shaner 1978). For some other uses, a partial conversion might be required to create more favorable conditions. It should be noted that for every initial ortho concentration there exists a unique curve for boil-off of hydrogen with respect to time.

The heat of conversion at higher temperatures is small (at 300 K the heat of conversion is 270 kJ/kg) and increases as the temperature decreases, where it reaches 519 kJ/kg at 77 K, the nbp for liquid nitrogen. For temperatures lower than the nbp of nitrogen, the heat of conversion remains constant (523 kJ/kg). However, the equilibrium percentage of para-hydrogen at the nbp is 50%. Therefore, in order to produce equilibrium liquid hydrogen, 50% of hydrogen must be converted from ortho to para at temperatures lower than liquid nitrogen temperatures. Since the heat of conversion is almost constant for temperatures lower than those of liquid nitrogen, the refrigeration load required to remove the heat of conversion should be almost constant at different temperatures. Ideal refrigeration requires that the energy needed for the same amount of refrigeration load be temperature-dependent. This suggests that in order to operate at an optimum power, it is necessary to have a continuous conversion, especially for temperatures lower than those of liquid nitrogen. By continuous conversion, up to 15% energy saving may be achieved.

The heat leakage losses are generally proportional to the ratio of surface area to the volume of the storage vessel (S/V). The most favorable shape is therefore spherical since it has the least surface to volume ratio. Spherical shape containers have another advantage. They have good mechanical strength since stresses and strains are distributed uniformly. Storage vessels may also be constructed in other shapes such as cylindrical, conical, or any combination of these shapes. Cylindrical vessels are usually required for transportation of liquid hydrogen by trailers or railway cars because of limitations imposed on the maximum allowable diameter of the vessel. For normal highway transportation, the outside diameter of the vessel cannot exceed 2.44 m. From an economics standpoint, cylindrical vessels with either dish, elliptical or hemispherical heads are very good, and their S/V ratios are only about 10% greater than that of the sphere (Scott 1962).

Since boil-off losses due to heat leak are proportional to the S/V ratio, the evaporation rate will diminish drastically as the storage tank size is increased. For double-walled, vacuum-insulated, spherical dewars, boil-off losses are typically 0.3% to 0.5% per day for containers having a storage volume of 50 m^3, 0.2% for 103 m^3 tanks, and about 0.06% for 19000 m^3 tanks (Ewe and Selbach 1987). Obviously, the larger the size of the dewar, the smaller the cost per unit volume of storage. Interestingly enough, the rate of evaporation does not substantially decrease with increasing the size of the container for cylindrical vessels of constant diameter.

Additional boil-off may be caused by thermal stratification in liquid hydrogen in which case the warmer upper layers evaporate much faster than the bulk liquid (Scott 1962). One way of decreasing boil-off losses due to stratification and thermal overfill is by employing high-conductivity plates (conductors) installed vertically in the vessel. The plates produce heat paths of low resistance between the bottom and top of the vessel and can operate most satisfactorily in eliminating temperature gradients and excessive pressures. Another way is to pump the heat out and maintain the liquid at subcooled or saturated conditions. An ideal refrigeration system to perform this task can be an efficient magnetic refrigerator. The magnetic refrigerator is very suitable for this job because of its relatively higher efficiency, compactness, lower price, and reliability (Barclay 1982, 1983, 1984; Barclay and Sarangi 1984; Barclay and Stewart 1982; Barclay and Steyert 1982; Barclay et al. 1983, 1985).

Liquid hydrogen containers are usually of three types: double-jacketed vessels with liquid nitrogen in the outer jacket, superinsulated vessels with either a reflecting powder or multilayer insulation, and containers with vapor-cooled shields employing super insulation. Although multilayer insulation (MLI) provides for a low boil-off rate, the addition of a vapor-cooled shield (VCS) will lower the boil-off losses even further. A VCS is a type of insulation that takes the vapor boil-off and passes it past the tank before being re-liquefied or vented. Published data indicate that a reduction of more than 50% in boil-off losses may be achieved for a 100,000 lb liquid hydrogen cryogenic facility with a VCS than without one. Brown (1986) showed that locating the VCS at half the distance from the tank to the outer surface of a four-inch MLI of a 100,000 lb facility would reduce the boil-off by ten percent. A dual VCS system on the tank would improve the performance by 40% over a single VCS. Brown (1986) also showed that the preferred locations for the inner and outer shields in a dual VCS system are 30% and 66% of the distance from the tank to the outer surface of the MLI.

Mechanical supports that connect the inner and outer vessel of double-walled vacuum tanks are an integral part of the insulation problem, primarily because of the heat influx that can be conducted to the inner vessel. One way to diminish these boil-off losses is to continuously cool the support employing the produced hydrogen vapor in a counter-flow fashion.

Another process that leads to boil-off during liquid hydrogen transportation by tankers is sloshing. Sloshing is the motion of liquid in a vessel due to acceleration or deceleration. Due to different types of acceleration and deceleration, there exist different types of sloshing. Acceleration causes the liquid to move to one end and then reflect from that end, thus, producing a hydraulic jump. The latter then travels to the other end, thus transforming some of its impact energy to thermal energy. The thermal energy dissipated eventually leads to an increase in the evaporation rate of the liquid (Scott 1962; Hands 1986). The insertion of traverse or anti-slosh baffles not only restrains the motion of the liquid, thus reducing the impact forces, but also increases the frequency above the natural frequency of the tanker (Barron 1985; Hands 1986).

Another source of boil-off is flashing. This problem occurs when liquid hydrogen, at a high pressure (2.4 to 2.7 atm), is transferred from trucks and rail cars to a low-pressure dewar (1.17 atm). This problem can be reduced if transportation of liquid hydrogen is carried out at atmospheric pressures. Furthermore, some of the low-pressure hydrogen can be captured and reliquefied.

Slush Hydrogen

Slush hydrogen is a mixture of liquid and frozen hydrogen in equilibrium with the gas at the triple point, 13.8 K. The density of the ice-like form is about 20% higher than that of the boiling liquid. To obtain the ice-like form, one has to remove the heat content of the liquid at 20.3 K until the triple point is reached and then remove the latent heat of fusion. The "cold content" of the ice-like form of hydrogen is some 25% higher than that of the saturated vapor at 20.3 K. It is of great interest for the space program because of its potential in reducing its physical size and significantly cutting the projected gross liftoff weight (Kandebo 1989). Some of the problems related to liquid hydrogen storage, such as low density, temperature stratification, short holding time due to its low latent heat, hazards associated with high vent rates, and unstable flight conditions caused by sloshing of the liquid in the fuel tank, may be eliminated or reduced if slush hydrogen is used.

Baker and Matsh (1965) gave a tabulated comparison among the properties of liquid and slush hydrogen. According to them, the smaller enthalpy of slush hydrogen reduces the evaporation losses during storage and transport and as a consequence permits longer storage without venting. The increased density permits ground transport equipment to carry 15% greater loads; for space vehicles, a given fuel load can be carried in tankage 13% smaller in volume.

There are basically two methods for slush hydrogen production, namely: freeze-thaw and auger. Sindt (1970) summarized the basic principles of the freeze-thaw method and its feasibility. Two of the methods reported by Sindt involve the formation of solid particles by spraying precooled liquid into an environment well below the triple-point pressure and inducing solid formation by allowing precooled helium gas to flow through a triple-point temperature liquid. The spray technique produced a very low-density solid particle, which had to be partially melted to form a liquid-solid mixture. The technique of allowing helium to flow through the liquid produced solid by reducing the partial pressure of hydrogen below its triple-point pressure which resulted in a clear, rigid tube of solid forming around the helium bubble train. The solid was found not to be convertible to a mixture without further treatment such as crushing. Sindt found that a more desirable preparation method was to evacuate the ullage over the liquid, which created a solid layer at the liquid surface. The solid layer produced had a texture that was very dependent upon the rate of vacuum pumping. Low pumping rates produced a dense, nearly transparent layer of solid. Subsequent breaking of the solid left large rigid pieces that formed inhomogeneous mixtures even with vigorous mixing. Higher pumping rates produced a solid layer that was porous and consisted of agglomerates of loosely attached fine solid particles. The solid layer was found to be easily broken and mixed with the liquid to form a homogeneous liquid-solid mixture that could be defined as slush. Sindt discovered, however, that solids could not be continuously generated at the liquid surface without vigorous mixing. Sindt also reported on a continuous preparation method, which he developed, capable of forming solid layers in cycles by periodic vacuum pumping. During the pumping (pressure reduction) portion of the cycle, the solid layer formed as previously described. When the pumping was stopped, the solid layer melted and settled into liquid, forming slush. Although mixing was not necessarily required, the preparation method was accelerated by some mechanical mixing which helped break the layer of solids. This technique has been called the freeze-thaw process.

Although the freeze-thaw production method was proved to be technically feasible and fully developed, it was shown to have disadvantages. These disadvantages include the fact that the freeze-thaw process is a batch process and that it operates at the triple-point pressure of the cryogen. For hydrogen, this pressure was estimated to be 0.07 bars absolute, a pressure capable of drawing air through inadvertent leaks in the system. The air could be thought of as presenting a potential safety problem in a hydrogen system. Voth (1985) showed that the freeze-thaw process required either costly equipment to recover the generated triple-point vapor or the loss of approximately 16% of the normal boiling-point liquid hydrogen and approximately 24% of the normal boiling-point liquid oxygen if the vapor was discarded.

Voth (1985) reported on a study for producing liquid-solid mixtures of oxygen or hydrogen using an auger. He described an auger used to scrape frozen solid from the inside of a refrigerated brass tube in order to produce slush hydrogen. He showed that slush hydrogen could be continuously produced by this method, and since it could be immersed in liquid, slush was produced at pressures above the triple-point pressure. Voth was also able to produce the increased pressure pneumatically or by generating temperature stratification near the surface of the liquid. He observed that the auger system produced particles in the size realm of the particle produced by the freeze-thaw method so that, like the freeze-thaw-produced slush, the auger-produced slush could be readily transferred and stored. According to Voth (1985), the thermodynamic reversible energy required to produce slush hydrogen was equal to the thermodynamic availability of the slush. Also using normal hydrogen at a temperature of 300 K and a pressure of 1.013 bars as the base fluid, the reversible energy required to produce normal boiling-point para-hydrogen was 3971.4 W.h/kg while the reversible energy required to produce slush with a solid mass fraction of 0.5 was 4372.8 W.h/kg. Voth explained that the energy required by practical systems was higher because of component inefficiencies. The calculated energies for the four cases were based on liquefier and refrigerator efficiencies of 40% of that of Carnot. He assumed that the vacuum pumps required by the freeze-thaw production had an efficiency of 50% of the isothermal efficiency. However, he did not include the increased production energy due to heat leak into the containers and transfer lines because it was difficult to estimate without a firm system definition and because the heat leak would have been nearly equal for all of the systems studied. Because heat leak was not included, the calculated energies were lower than those of an actual system.

It is apparent that the use of slush hydrogen should be considered only for cases in which higher density and greater solid content are really needed. This is mainly because the production costs of slush hydrogen now and in the near future are greater than the liquefaction costs of hydrogen due to the larger energy use involved.

Hydrogen Transport and Distribution

It is envisaged that from the production plants and/or storage, hydrogen will be transmitted to consumers by means of underground pipelines (gaseous hydrogen) and/or supertankers (liquid hydrogen). Presently, hydrogen transportation through pipelines is used either in links between nearby production and utilization sites (up to 10 km) or in more extensive networks (roughly 200 km). Table 7.4 lists the principal existing hydrogen pipelines (Pottier 1995). Future developments will certainly entail greater flow rates and distances. However, it would be possible to use the existing natural gas pipelines with some modifications. For hydrogen pipelines, it is necessary to use steel less prone to embrittlement by hydrogen under pressure (particularly for very pure hydrogen, >99.5% purity). Reciprocating compressors used for natural gas can be used for hydrogen without major design modifications. However, special attention must be given to sealing (to avoid hydrogen leaks) and to materials selection for the parts subject to fatigue stress. Use of centrifugal compressors for hydrogen creates more problems due to hydrogen's exceptional lightness.

TABLE 7.4
Some Major Hydrogen Pipelines

Location	*Years of Operation*	*Diameter, mm*	*Length, km*	*Pressure, MPa*	*Purity*
AGEC, Alberta Canada	Since 1987	273	3.7	3.79	99.9%
Air Liquide, France	Since 1966	various	290	6.48–10	pure and raw
Air Products, Houston, TX	Since 1969	114.324	100	0.35–5.5	pure
Air Products, Louisiana	Since 1990	102–305	48	3.45	
Chemische Werke Huls	Since 1938	168–273	215	to 2.5	raw gas
Cominco, B.C., Canada	Since 1964	5	0.6	>30	62–100%
Gulf Petroleum, Canada		168.3	16		93.5%
Hawkeye Chemical, Iowa	Since 1987	152	3.2	2.75	
ICI Bilingham, UK			15	30	pure
Philips Petroleum	Since 1986	203	20.9	12	

As a rule, hydrogen transmission through pipelines requires larger diameter piping and more compression power than natural gas for the same energy throughput. However, due to lower pressure losses in the case of hydrogen, the recompression stations would need to be spaced twice as far apart. In economic terms, most of the studies found that the cost of large-scale transmission of hydrogen is about 1.5 to 1.8 times that of natural gas transmission. However, transportation of hydrogen over distances greater than 1,000 km is more economical than transmission of electricity (Oney 1991).

To match the consumption demand, hydrogen can be regionally transported and distributed, both as a gas or as a liquid, by pipelines or in special cases in containers by road and rail transportation. Gaseous (and liquid) hydrogen carriage is subject to strict regulations ensuring public safety, which in some countries is very constraining. Transportation of gaseous or liquid hydrogen in a discontinuous mode is currently employed by occasional or low volume users. The cost of discontinuous transport is very high (it can be as high as 2 to 5 times the production cost). In the future energy system, discontinuous transportation of hydrogen would see little use, except for special users (mainly non-energy related). Hydrogen in the gas phase is generally transported in pressurized cylindrical vessels (typically at 200 bar) arranged in frames adapted to road transport. The unit capacity of these frames or skids can be as great as 3,000 m^3. Hydrogen gas distribution companies also install such frames at the user site to serve as a stationary storage.

Hydrogen Conversion Technologies

Hydrogen as an energy carrier can be converted into useful forms of energy in several ways:

- Combustion in internal combustion engines as well as jet and rocket engines
- Combustion with pure oxygen to generate steam
- Catalytic combustion to generate heat
- Electrochemical conversion to electricity
- Metal hydride conversions

Hydrogen Combustion in Internal Combustion, Jet, and Rocket Engines

Hydrogen is a very good fuel for internal combustion engines. Hydrogen powered internal combustion engines are on average about 20% more efficient than comparable gasoline engines. The ideal thermal efficiency of an internal combustion engine is

$$\eta = 1 - \left(\frac{1}{r}\right)^{k-1} \tag{7.5}$$

where r = compression ratio and k = ratio of specific heats (C_p/C_v).

This equation shows that the thermal efficiency can be improved by increasing either the compression ratio or the specific heat ratio. In hydrogen engines both ratios are higher than in a comparable gasoline engine due to hydrogen's lower self-ignition temperature and ability to burn in lean mixtures.

However, the use of hydrogen in internal combustion engines results in loss of power due to lower energy content in a stoichiometric mixture in the engine's cylinder. A stoichiometric mixture of gasoline and air, and gaseous hydrogen and air, premixed externally occupies ~2% and 30% of the cylinder volume, respectively. Under these conditions, the energy of the hydrogen mixture is only 85% of the hydrogen mixture, thus resulting in about 15% reduction in power. Therefore, the same engine running on hydrogen will have ~15% less power than when operated with gasoline. The power output of a hydrogen engine can be improved by using more advanced fuel injection techniques or liquid hydrogen. For example, if liquid hydrogen is premixed with air, the amount of hydrogen that can be introduced in the combustion chamber can be increased by approximately one-third (Norbeck et al. 1996).

One of the most important advantages of hydrogen as a fuel for internal combustion engines is that hydrogen engines emit by far fewer pollutants than comparable gasoline engines. Basically, the only products of hydrogen combustion in air are water vapor and small amounts of nitrogen oxides. Hydrogen has a wide flammability range in air (5–75% vol) and therefore high excess air can be utilized more effectively. The formation of nitrogen oxides in hydrogen/air combustion can be minimized with excess air. NO_x emissions can also be lowered by cooling the combustion environment using techniques such as water injection, exhaust gas recirculation, or using liquid hydrogen. The emissions of NO_x in hydrogen engines are typically one order of magnitude smaller than emissions from comparable gasoline engines. Small amounts of unburned hydrocarbons, CO_2, and CO have been detected in hydrogen engines due to lubrication oil (Norbeck et al. 1996).

The low ignition energy and fast flame propagation of hydrogen has led to problems of pre-ignition and backfire. These problems have been overcome by adding hydrogen to the air mixture at the point where and when the conditions for pre-ignition are less likely, such as delivering the fuel and air separately to the combustion chamber, and/or injecting hydrogen under pressure into the combustion chamber before the piston is at the top dead center and after the intake air valve has been closed. Water injection and exhaust gas recirculation techniques are also used in hydrogen engines to help control premature ignition. It should be noted that most of the research on hydrogen combustion in internal combustion engines has been conducted with modifications of existing engines designed to burn gasoline. Redesign of the combustion chamber and coolant systems to accommodate hydrogen's unique combustion properties could be the most effective method of solving the problems of pre-ignition and knocking (Norbeck et al. 1996).

Hydrogen use in turbines and jet engines is similar to use of conventional jet fuel. The use of hydrogen avoids the problems of sediments and corrosion on turbine blades, which prolongs life and reduces maintenance. Gas inlet temperatures can be pushed beyond normal gas turbine temperatures of 800°C, thus increasing the overall efficiency. The only pollutants from the use of hydrogen in turbines and jet engines are nitrogen oxides.

Steam Generation by Hydrogen/Oxygen Combustion

Hydrogen combusted with pure oxygen results in pure steam, i.e.:

$$2H_2 + O_2 \rightarrow 2H_2O \tag{7.6}$$

The above reaction creates temperatures in the flame zone above 3,000°C, therefore, additional water has to be injected so that the steam temperature can be regulated at a desired level. Both saturated and superheated vapor can be produced.

The German Aerospace Research Establishment (DLR) has developed a compact hydrogen/oxygen steam generator (Sternfeld and Heinrich 1989). The steam generator consists of the ignition, combustion, and evaporation chambers. In the ignition chamber, a combustible mixture of hydrogen and oxygen at a low oxidant/fuel ratio is ignited by means of a spark plug. The rest of the oxygen is added in the combustion chamber to adjust the oxidant/fuel ratio exactly to the stoichiometric one. Water is also injected in the combustion chamber after it has passed through the double walls of the combustion chamber. The evaporation chamber serves to homogenize the steam. The steam's temperature is monitored and controlled. Such a device is close to 100% efficient, since there are no emissions other than steam and little or no thermal losses.

The hydrogen steam generator can be used to generate steam for spinning reserve in power plants, for peak load electricity generation, for industrial steam supply networks, and as a micro steam generator in medical technology and biotechnology (Sternfeld and Heinrich 1989).

Catalytic Combustion of Hydrogen

Hydrogen and oxygen in the presence of a suitable catalyst may be combined at temperatures significantly lower than flame combustion (from ambient to 500°C). This principle can be used to design catalytic burners and heaters. Catalytic burners require considerably more surface area than conventional flame burners. Therefore, the catalyst is typically dispersed in a porous structure. The reaction rate and resulting temperature are easily controlled, by controlling the hydrogen flow rate. The reaction takes place in a reaction zone of the porous catalytic sintered metal cylinders or plates in which hydrogen and oxygen are mixed by diffusion from opposite sides. A combustible mixture is formed only in the reaction zone and assisted with a (platinum) catalyst to burn at low temperatures (Figure 7.2). The only product of catalytic combustion of hydrogen is water vapor. Due to low temperatures no nitrogen oxides are formed. The reaction cannot migrate into the hydrogen supply, since there is no flame and hydrogen concentration is above the higher flammable limit (75%). Possible applications of catalytic burners are in household appliances such as cooking ranges and space heaters. The same principle is also used in hydrogen sensors.

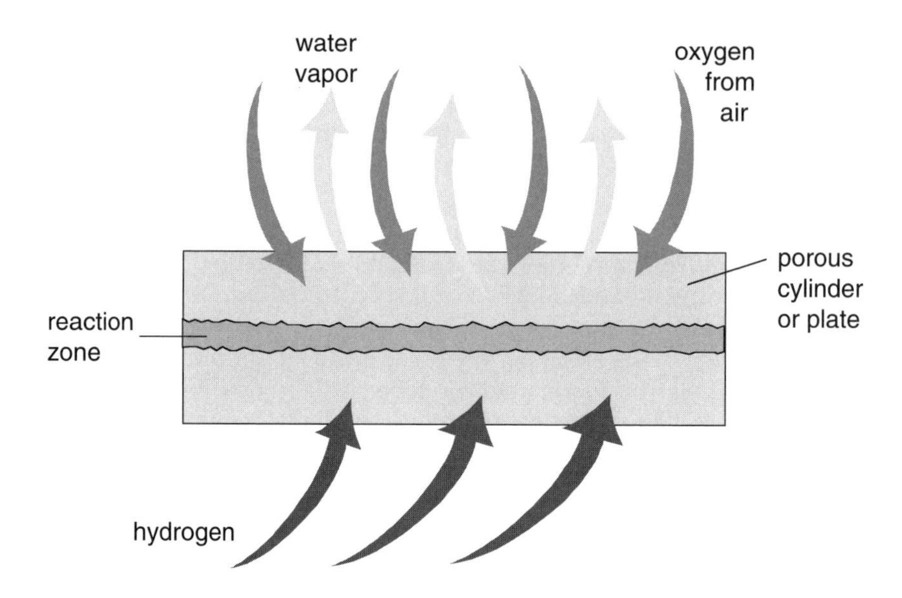

FIGURE 7.2. Schematic representation of catalytic burner.

Electrochemical Conversion (Fuel Cells)

Hydrogen can be combined with oxygen without combustion in an electrochemical reaction (reverse of electrolysis) and produce electricity (DC). The device where such a reaction takes place is called the electrochemical fuel cell or just fuel cell. Depending on the type of the electrolyte used, there are several types of fuel cells

- Alkaline fuel cells (AFC) use concentrated (85 wt%) KOH as the electrolyte for high temperature operation (250°C) and less concentrated (35–50 wt%) for lower temperature operation (<120°C). The electrolyte is retained in a matrix (usually asbestos), and a wide range of electrocatalysts can be used (such as Ni, Ag, metal oxides, and noble metals). This fuel cell is intolerant to CO_2 present in either the fuel or the oxidant (Kinoshita et al. 1988).
- Polymer electrolyte membrane or proton exchange membrane fuel cells (PEMFC) use a thin polymer membrane (such as perfluorosulfonated acid polymer) as the electrolyte. The membranes as thin as 12–20 microns have been developed, which are excellent proton conductors. The catalyst is typically platinum with loadings about $0.3\,mg/cm^2$, or, if the hydrogen feed contains minute amounts of CO, Pt-Ru alloys are used. Operating temperature is usually below 100°C, more typically between 60°–80°C.

- Phosphoric acid fuel cells (PAFC), use concentrated phosphoric acid (~100%) as the electrolyte. The matrix used to retain the acid is usually SiC, and the electrocatalyst in both the anode and cathode is platinum black. Operating temperature is typically between 150°–220°C (Kinoshita et al. 1988; Blumen and Mugerwa 1993).
- Molten carbonate fuel cells (MCFC) have the electrolyte composed of a combination of alkali (Li, Na, K) carbonates, which is retained in a ceramic matrix of $LiAlO_2$. Operating temperatures are between 600°–700°C where the carbonates form a highly conductive molten salt, with carbonate ions providing ionic conduction. At such high operating temperatures, noble metal catalysts are typically not required (Kinoshita et al. 1988; Blumen and Mugerwa 1993).
- Solid oxide fuel cells (SOFC) use a solid, nonporous metal oxide, usually Y_2O_3-stabilized ZrO_2 as the electrolyte. The cell operates at 900°–1000°C where ionic conduction by oxygen ions takes place (Kinoshita et al. 1988; Blumen and Mugerwa 1993).

A typical fuel cell consists of the electrolyte, in contact with porous electrodes, on both sides. A schematic representation of a fuel cell with reactant and product gases, and ions flow directions for the major types of fuel cells are shown in Figure 7.3. The electrochemical reactions occur at the three-phase interface: porous electrode/electrolyte/reactants. The actual electrochemical reactions that occur in the above listed types of fuel cells are different (as shown in Table 7.5) although the overall reaction is the same, i.e., $H_2 + \frac{1}{2}O_2 \rightarrow H_2O$. Low temperature fuel cells (AFC, PEMFC, PAFC) require noble electrocatalysts to achieve practical reaction rates at the anode and cathode. High temperature fuel cells (MCFC and SOFC) can also utilize CO and CH_4 as fuels. The operating temperature is high enough so that CO and CH_4 can be converted into hydrogen through the water-gas shift and steam reforming reactions, respectively.

The electrolyte not only transports dissolved reactants to the electrode, but it also conducts ionic charge between the electrodes and thereby completes the cell electric circuit, as shown in Figure 7.3.

The reversible potential of the above electrochemical reactions is 1.229 V (at standard conditions, i.e., 25°C and atmospheric pressure), and it corresponds to the Gibbs free energy according to the following equation:

$$\Delta G^{o} = nFE^{o} \quad (7.7)$$

where:
ΔG^{o} = Gibbs free energy at 25°C and atmospheric pressure
n = number of electrons involved in the reaction
F = Faraday's constant
E^{o} = reversible potential at 25°C and atmospheric pressure (V)

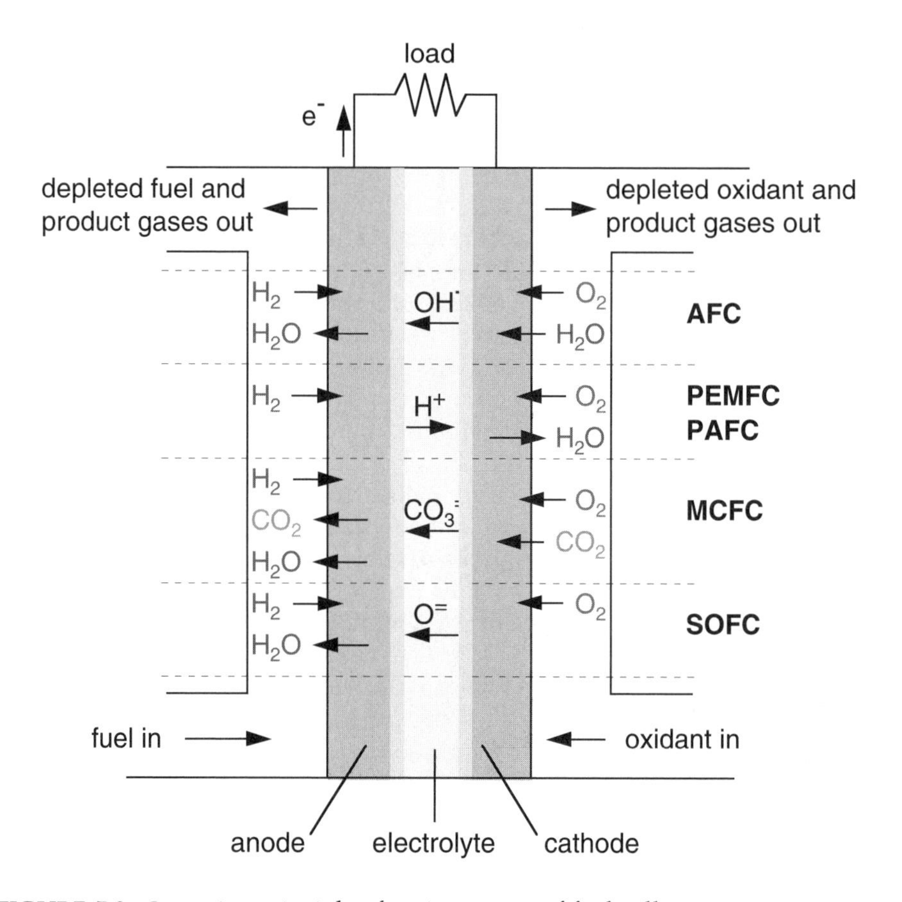

FIGURE 7.3. Operating principle of various types of fuel cells.

TABLE 7.5
Fuel Cell Reactions

Fuel Cell Type	*Anode Reaction*	*Cathode Reaction*
Alkaline	$H_2 + 2OH^- \rightarrow 2H_2O + 2e^-$	$\frac{1}{2}O_2 + H_2O + 2e^- \rightarrow 2OH^-$
Proton exchange	$H_2 \rightarrow 2H^+ + 2e^-$	$\frac{1}{2}O_2 + 2H^+ + 2e^- \rightarrow H_2O$
Phosphoric acid	$H_2 \rightarrow 2H^+ + 2e^-$	$\frac{1}{2}O_2 + 2H^+ + 2e^- \rightarrow H_2O$
Molten carbonate	$H_2 + CO_3^= \rightarrow H_2O + CO_2 + 2e^-$	$\frac{1}{2}O_2 + CO_2 + 2e^- \rightarrow CO_3^=$
Solid oxide	$H_2 + O^= \rightarrow H_2O + 2e^-$	$\frac{1}{2}O_2 + 2e^- \rightarrow O^=$

The reversible potential changes with temperature and pressure; in general it is lower at higher temperatures (reaching ~1.0 V at 1,000 K), and it is higher at higher pressures or higher concentrations of reactants. The actual voltage of an operational fuel cell is always lower than the reversible potential due to various irreversible losses, such as activation polarization, concentration polarization, and ohmic resistance. While ohmic resistance is directly proportional to the current, activation polarization is a logarithmic function of the current, and is thus more pronounced at very low current densities. Concentration polarization is an exponential function of the current and thus becomes a limiting factor at high current densities. Figure 7.4 shows a typical fuel cell polarization curve with pronounced regions of predominant irreversible losses. Figure 7.5 shows actual polarization curves of some representative fuel cells. The fuel cells are typically operated in a range between 0.6 and 0.8 V. The Space Shuttle fuel cell (alkaline) is designed to operate at 0.86 V and 410 mA/cm^2 (Kinoshita et al. 1988). PEM fuel cells have the highest achievable current densities, between 1 and 2 mA/cm^2 at 0.6 V with pressurized hydrogen and air.

The fuel cell efficiency is a function of cell voltage. The theoretical fuel cell efficiency is:

$$\eta_{FC} = \Delta G / \Delta H \tag{7.8}$$

where ΔH is hydrogen's enthalpy or heating value (higher or lower). The theoretical fuel cell efficiency, defined as a ratio between produced elec-

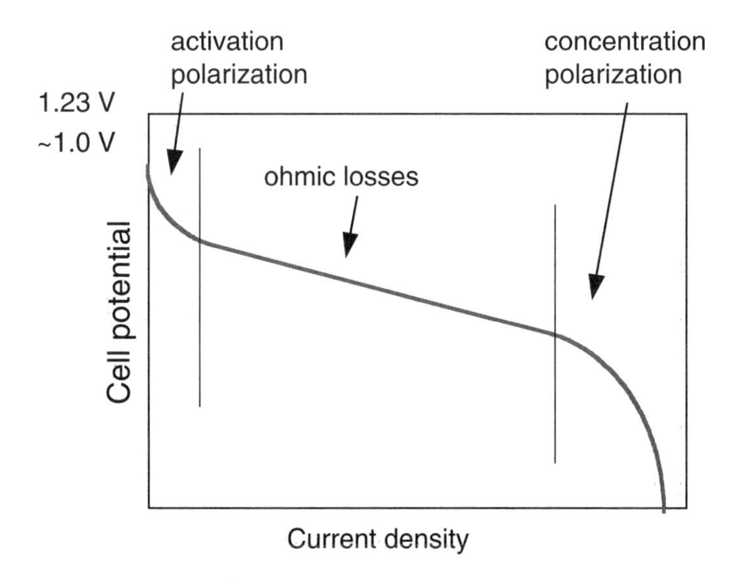

FIGURE 7.4. Typical fuel cell polarization curve.

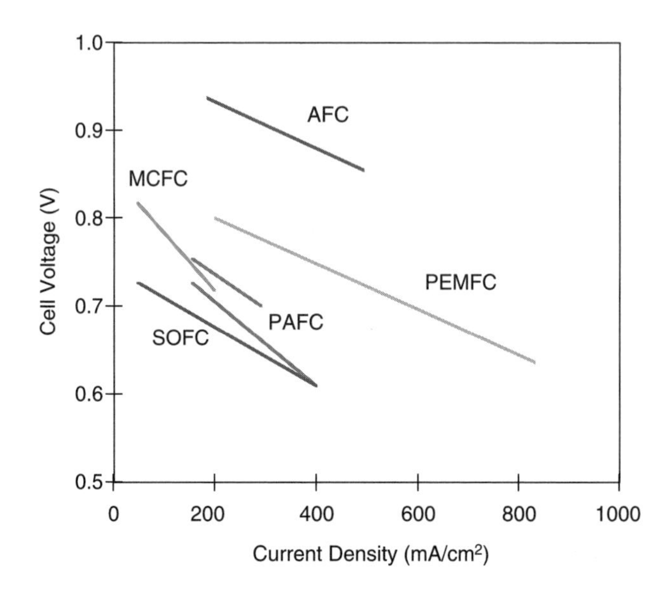

FIGURE 7.5. Polarization curves of some representative fuel cells. AFC, United Technologies' fuel cell operating in Space Shuttle, with H_2 and O_2 at 80°–90°C, 410 kPa.

PEMFC, Ballard's 25 kW fuel cell operating in Daimler-Benz vehicle (NECAR II) with H_2 and air (1996), PAFC-IFCs fuel cell operating in ONSI PC25 200 kW power plant, operating on reformed natural gas and air at atmospheric pressure (1996) (*upper line* is IFCs PC23 operating at 820 kPa); MCFC, ERCs atmospheric pressure fuel cell operating with natural gas (1994); SOFC, Westinghouse fuel cell operating at atmospheric pressure using natural gas (1991).

tricity and higher heating value of hydrogen consumed is therefore 83%. The lower heating value of hydrogen results in an efficiency of 98%. Since the actual voltage of an operational fuel cell is lower than the reversible potential, the fuel cell efficiency is always lower than the theoretical one. Generally, the fuel cell efficiency is a product of several efficiencies:

$$\eta_{FC} = \eta_{Th} \cdot \eta_V \cdot \eta_F \cdot \eta_U \tag{7.9}$$

where:

η_{Th} = thermal efficiency, i.e., ratio between Gibbs free energy of the reaction and heating value of the fuel, $\Delta G_r/\Delta H_{fuel}$ (similar to internal combustion engines, the fuel cell efficiency is often expressed in terms of lower heating value)

η_V = voltage efficiency, defined as a ratio between actual voltage (V) and thermodynamic voltage (E), i.e., V/E

η_F = Faradaic efficiency, or ratio between the actual current and current corresponding to the rate at which the reactant species are consumed, I/nFm, where m is the rate (in moles/s) at which the reactants are consumed.

η_U = fuel utilization, or ratio between the amount of fuel actually consumed in the electrochemical reaction and fuel supplied to the fuel cell.

For a hydrogen/oxygen or hydrogen/air fuel cell operating with 100% fuel utilization, the efficiency is a function of cell voltage only. For such a fuel cell the efficiency in an operating range between 0.6 V and 0.8 V is between 0.48 and 0.64.

In order to get useable voltages (i.e., tens or hundred Volts), the cells are combined in a stack. The cells are physically separated from each other and electrically connected in series by a bipolar separator plate. Figure 7.6 shows a schematic representation of a typical fuel cell stack.

Alkaline fuel cells have been used in the space program (Apollo and Space Shuttle) since the 1960s. Phosphoric acid fuel cells are already

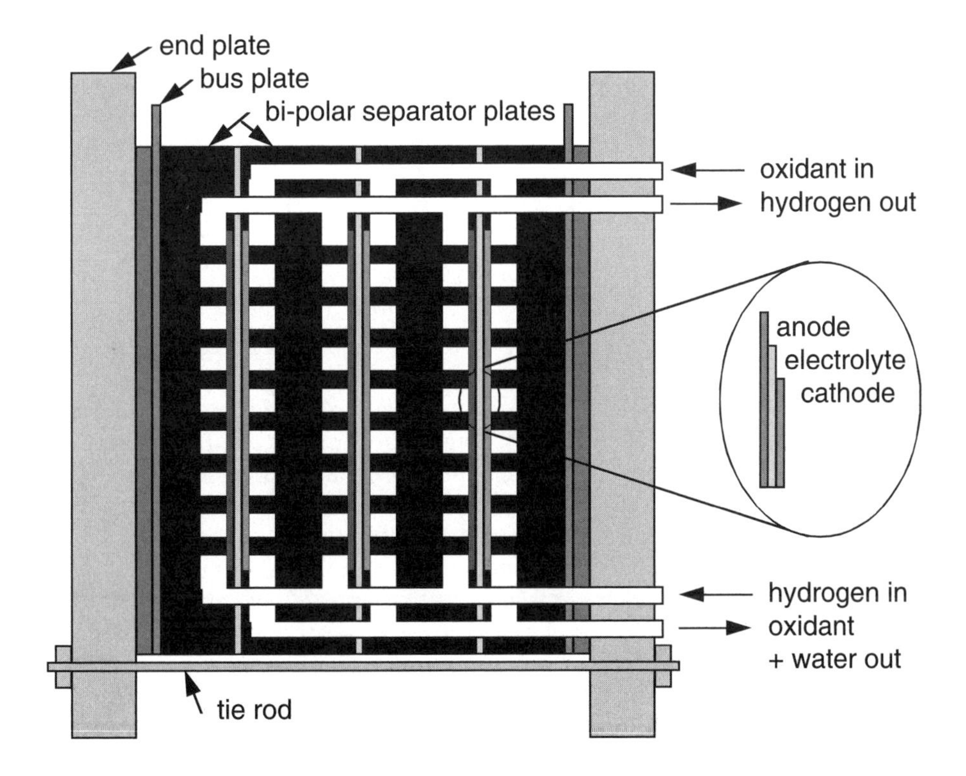

FIGURE 7.6. Schematic representation of a typical fuel cell stack.

commercially available in container packages for stationary electricity generation. PEM fuel cells are a serious candidate for automotive applications, but also for small-scale distributed stationary power generation. High temperature fuel cells, such as molten carbonate and solid oxide fuel cells, have been developed to a pre-commercial/demonstration stage for stationary power generation.

Energy Conversions Involving Metal Hydrides

Hydrogen's property to form metal hydrides may be used not only for hydrogen storage but also for various energy conversions. When a hydride is formed by the chemical combination of hydrogen with a metal, an element or an alloy, heat is generated, i.e., the process is exothermic. Conversely, in order to release hydrogen from a metal hydride, heat must be supplied. These processes can be represented by the following chemical reactions:

$$\text{Charging or absorption: } M + xH_2 \rightarrow MH_{2x} + \text{heat} \qquad (7.10)$$

$$\text{Discharging or desorption: } MH_{2x} + \text{heat} \rightarrow M + xH_2 \qquad (7.11)$$

where M represents the hydriding substance, which could be a metal, an element, or an alloy. The rate of these reactions increases with an increase in the surface area. Therefore, in general, the hydriding substances are used in powdered form to speed up the reactions.

Elements or metals with unfilled shells or subshells are suitable hydriding substances. Metal and hydrogen atoms form chemical compounds by sharing their electrons in the unfilled subshells of the metal atom and the K shells of the hydrogen atoms.

Ideally, for a given temperature, the charging or absorption process and the discharging or desorption process takes place at the same constant pressure. However, actually, there is a hysteresis effect and the pressure is not absolutely constant—for a given temperature charging pressures are higher than the discharging pressures. The heat generated during the charging process and the heat needed for discharging are functions of the hydriding substance, the hydrogen pressure and the temperature at which the heat is supplied or extracted. Using different metals and by forming different alloys, different hydriding characteristics can be obtained. In other words, it is possible to make or find hydriding substances which are more suitable for a given application, such as waste heat storage, electricity generation, pumping, hydrogen purification, and isotope separation.

Hydriding substances can be used for electricity storage in two ways. In one of the methods, electricity (DC) is used to electrolyze the water, and the hydrogen produced is stored in a hydriding substance. When electricity is needed, the hydrogen is released from the hydriding substance by adding heat and used in a fuel cell to produce direct current electricity. Heat from fuel cell can be used to release hydrogen from the metal hydride. In the second method, one electrode is covered with a hydriding substance (e.g., titanium nickel alloy). During the electrolysis of water, the hydriding substance covering the electrode immediately absorbs the hydrogen produced on the surface of the electrode. Then, when electricity is needed, the electrolyzer operates in a reverse mode as a fuel cell producing electricity using the hydrogen released from the metal hydride.

Hydrogen together with hydriding substances can be used for heating or air-conditioning the buildings. Figure 7.7 shows how one of the proposed systems works. The system consists of four hydride tanks, a solar collector (or a heat source) and a number of heat exchangers. Hydride tank 1 is connected to hydride tank 3 with a hydrogen pipe in order to allow the movement of hydrogen from one tank to the other. Similarly, hydride tanks 2 and 4 are connected in the same fashion. Tanks 1 and 2 contain the same hydriding substance ($CaNi_5$), and tanks 3 and 4 contain another hydriding material ($LaNi_5$). Heat exchangers and the hydride tanks are connected by water-carrying pipe circuits or loops, equipped with a set of switches and valves, so that a hydride tank in a given water loop can be replaced by another hydride tank.

When the system works as a heater, the heat from a heat source (e.g., solar collector) is carried to tank 1 at about 100°C. The heat drives the hydrogen from tank 1 to tank 3, where hydrogen is absorbed forming a hydride and heat is released at 40°C. The water loop carries this heat to the building heat exchangers and heats the air in the building. At the same time, water in the other loop absorbs heat from the ambient and carries it to tank 4. This heat drives off the hydrogen from tank 4 to tank 2 where hydride is formed and heat is generated at 40°C. The whole operation of driving hydrogen from tanks 1 and 4 to tanks 3 and 2 takes about two minutes. At the end of this cycle, the hydride tanks are switched from one loop to the other in cycle II (as shown in Figure 7.7). Now the solar and ambient heats are used to drive off the hydrogen in tanks 2 and 3 to tanks 4 and 1, respectively. The heat produced during the absorption processes in tanks 1 and 4 is used for heating the building. After this, the cycles are repeated.

When the system works as an air conditioner, the building heat exchangers are placed in the 8°C water loop, while the outside heat exchangers are placed in the 40°C water loops, and operation proceeds in two cycles as described above.

If passed through a turbine or expansion engine, hydrogen moving from one hydride tank to the other could produce mechanical and electrical energy, as shown in Figure 7.8. The system is somewhat similar to the one proposed for heating and cooling. However, it consists of only three tanks containing the same kind of hydriding substance (in this case $LaNi_5$ alloy). During the first cycle, hydrogen driven off from the desorption tank (tank 1) by means of solar heat (or heat from any other source) passes through the expansion turbine producing, mechanical energy and electricity. It is then

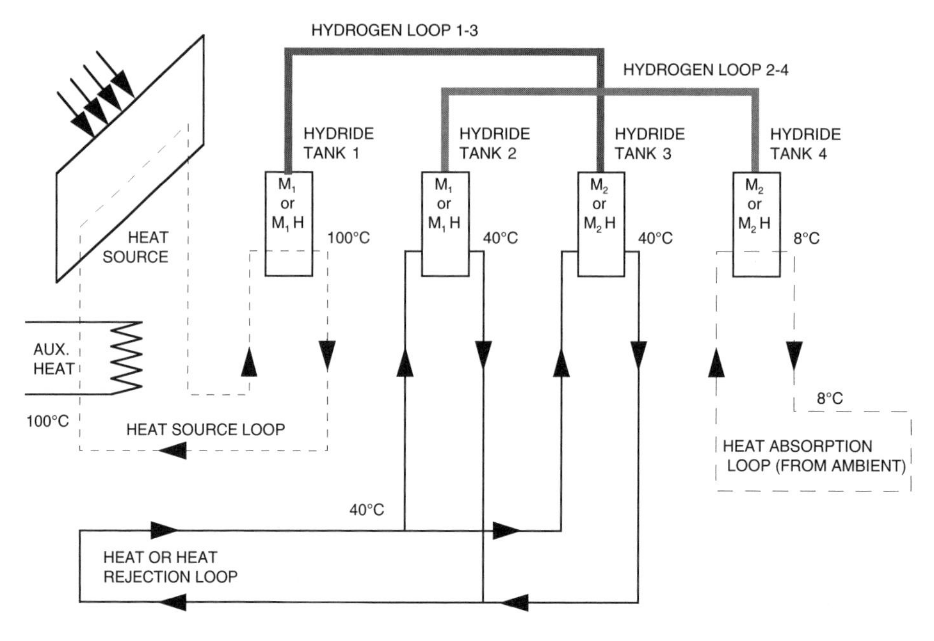

Cycle I

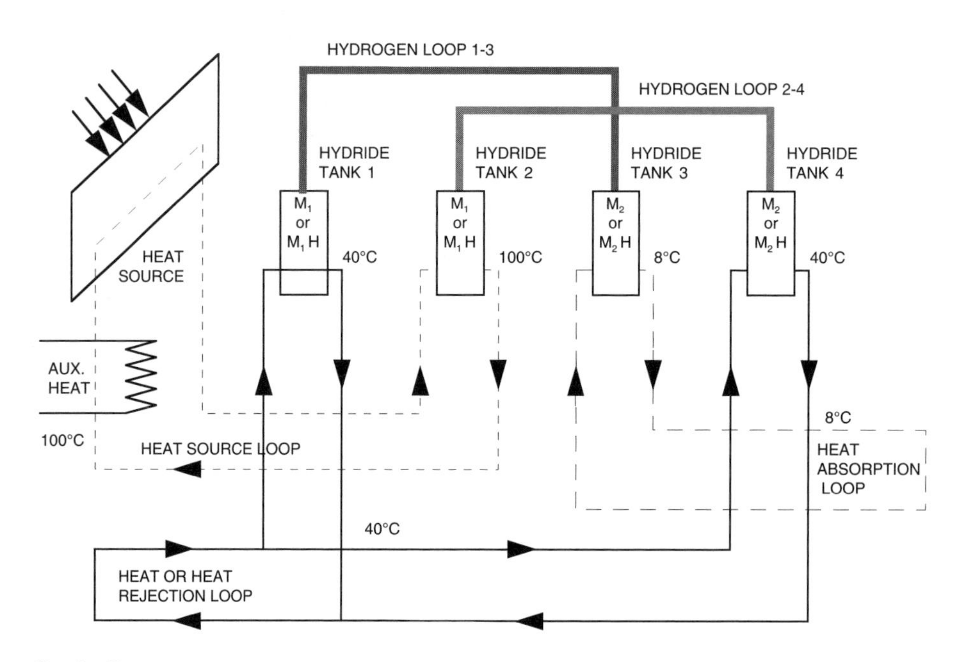

Cycle II

FIGURE 7.7. Hydrogen/hydride heating-cooling system.

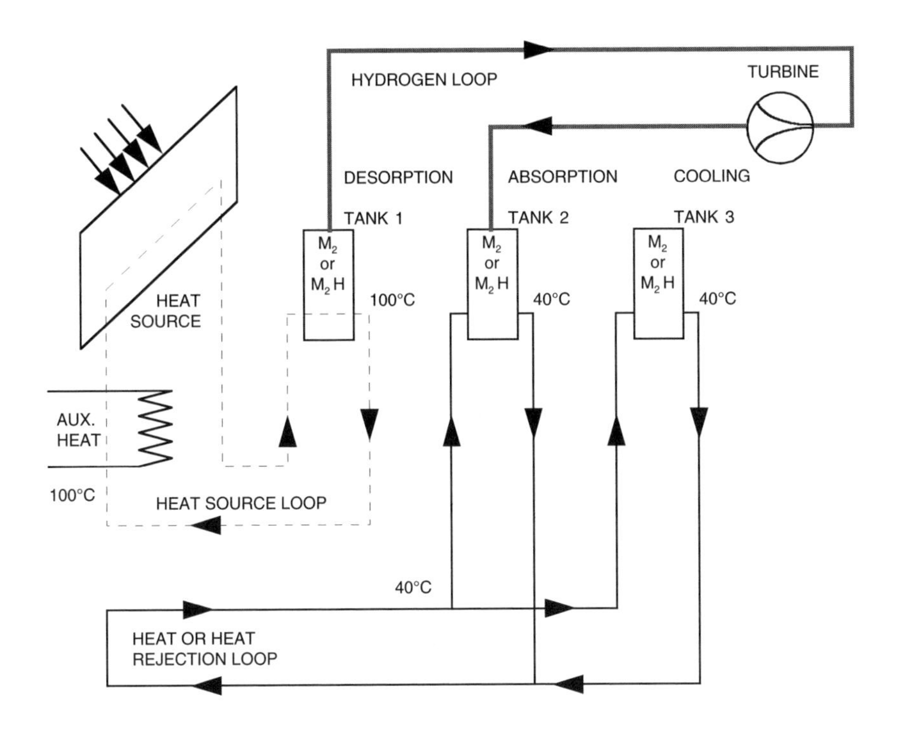

FIGURE 7.8. Electricity generation via hydrogen and hydrides.

absorbed by the hydriding substance in tank 2 (at a lower pressure), thus producing heat at 40°C. In this case, the heat is produced at a lower temperature than the temperature of the desorption since hydrogen is at a lower pressure after passing through the turbine. The heat produced in the absorption tank (tank 2) is rejected to the environment through the water cooling system. The same water cooling system is also used to cool down the cooling tank (tank 3) from 100°C to 40°C, since it has served as the desorption tank in the previous cycle. In the second cycle through a system of switches and valves, the tanks are displaced one step to the right in the diagram, i.e., the cooling tank becomes the absorption tank, the absorption tank becomes the desorption tank, and the desorption tank becomes the cooling tank. Then, the cycles are repeated. Using this method, low quality heat could be converted to electricity.

Hydrogen Safety

Like any other fuel or energy carrier hydrogen poses risks if not properly handled or controlled. The risk of hydrogen, therefore, must be considered relative to the common fuels such as gasoline, propane, or natural gas. The specific physical characteristics of hydrogen are quite different from those common fuels. Some of those properties make hydrogen potentially less hazardous, while other hydrogen characteristics could theoretically make it more dangerous in certain situations.

Since hydrogen has the smallest molecule it has a greater tendency to escape through small openings than other liquid or gaseous fuels. Based on properties of hydrogen such as density, viscosity and diffusion coefficient in air, the propensity of hydrogen to leak through holes or joints of low pressure fuel lines may be only 1.26 to 2.8 times faster than a natural gas leak through the same hole (and not 3.8 times faster as frequently assumed based solely on diffusion coefficients). Experiments have indicated that most leaks from residential natural gas lines are laminar (Thomas 1996). Since natural gas has over 3 times the energy density per unit volume the natural gas leak would result in more energy release than a hydrogen leak.

For very large leaks from high-pressure storage tanks, the leak rate is limited by the sonic speed. Due to higher sonic velocity (1,308 m/s) hydrogen would initially escape much faster than natural gas (sonic velocity of natural gas is 449 m/s). Again, since natural gas has more than 3 times the energy density than hydrogen, a natural gas leak will always contain more energy.

Some high strength steels are prone to hydrogen embrittlement. Prolonged exposure to hydrogen, particularly at high temperatures and pressures, can cause the steel to lose strength, eventually leading to failure. However, most other construction, tank and pipe materials are not prone to hydrogen embrittlement. Therefore, with proper choice of materials, hydrogen embrittlement should not contribute to hydrogen safety risks.

If a leak should occur for whatever reason, hydrogen will disperse much faster than any other fuel, thus reducing the hazard levels. Hydrogen is both more buoyant and more diffusive than gasoline, propane, or natural gas. Table 7.6 compares some properties and leak rates for hydrogen and natural gas.

Hydrogen/air mixture can burn in relatively wide volume ratios, between 4% and 75% of hydrogen in air. The other fuels have much lower flammability ranges, viz., natural gas 5.3–15%, propane 2.1–10%, and gasoline 1–7.8%. However, the range has a little practical value. In many actual leak situations the key parameter that determines if a leak would ignite is the lower flammability limit, and hydrogen's lower flammability limit is 4 times higher than that of gasoline, 1.9 times higher than that of propane and slightly lower than that of natural gas.

Hydrogen has a very low ignition energy (0.02 mJ), about one order of magnitude lower than other fuels. The ignition energy is a function of fuel/air ratio, and for hydrogen it reaches a minimum at about 25–30%. At the lower flammability limit, hydrogen ignition energy is comparable with that of natural gas (Swain and Swain 1992). Hydrogen has a flame velocity

TABLE 7.6
Properties and Leak Rates of Hydrogen and Natural Gas

Flow parameters	*Hydrogen*	*Natural gas*
Diffusion coefficient (cm^2/s)	0.61	0.16
Viscosity (μ-poise)	87.5	100
Density (kg/m^3)	0.0838	0.651
Sonic velocity(m/s)	1308	449
Relative leak rates		
Diffusion	3.80	1
Laminar flow	1.23	1
Turbulent flow	2.83	1
Sonic flow	2.91	1

7 times faster than that of natural gas or gasoline. A hydrogen flame would therefore be more likely to progress to a deflagration or even a detonation than other fuels. However, the likelihood of a detonation depends in a complex manner on the exact fuel/air ratio, the temperature and particularly the geometry of the confined space. Hydrogen detonation in the open atmosphere is highly unlikely.

The lower detonability fuel/air ratio for hydrogen is 13% to 18%, which is 2 times higher than that of natural gas and 12 times higher than that of gasoline. Since the lower flammability limit is 4% an explosion is possible only under the most unusual scenarios, e.g., hydrogen would first have to accumulate and reach 13% concentration in a closed space without ignition, and only then an ignition source would have to be triggered. Should an explosion occur, hydrogen has the lowest explosive energy per unit stored energy in the fuel, and a given volume of hydrogen would have 22 times less explosive energy than the same volume filled with gasoline vapor.

Hydrogen flame is nearly invisible, which may be dangerous, because people in the vicinity of a hydrogen flame may not even know there is a fire. This may be remedied by adding some chemicals that will provide the necessary luminosity. The low emissivity of hydrogen flames means that near-by materials and people will be much less likely to ignite and/or get hurt by radiant heat transfer. The fumes and soot from a gasoline fire pose a risk to anyone inhaling the smoke, while hydrogen fires produce only water vapor (unless secondary materials begin to burn).

Liquid hydrogen presents another set of safety issues, such as risk of cold burns, and the increased duration of leaked cryogenic fuel. A large spill of liquid hydrogen has some characteristics of a gasoline spill, however, it will dissipate much faster. Another potential danger is a violent explosion of a boiling liquid expanding vapor in case of a pressure relief valve failure.

In conclusion, hydrogen appears to pose risks of the same order of magnitude as other fuels. In spite of public perception, in many aspects hydrogen is actually a safer fuel than gasoline and natural gas. As a matter of fact, hydrogen has a very good safety record, as a constituent of the "town gas" widely used in Europe and United States in the 19th and early 20th century, as a commercially used industrial gas, and as a fuel in space programs. There have been accidents, but nothing that would characterize hydrogen as more dangerous than other fuels.

One of the most remembered accidents involving hydrogen is the Hindenburg dirigible disaster in 1937. However, hydrogen did not cause that accident and hydrogen fire did not directly cause any casualties. The accident appears to be caused by static electricity discharge, and it was the balloon's lining that caught fire first (DiChristina 1997). Once hydrogen that the balloon was filled with for buoyancy (instead of helium as it was originally designed to be filled with) was ignited it burned (as any fuel is supposed to). However, hydrogen fire went straight up and it did not radiate heat so people in the gondola underneath the balloon were not burned or suffocated. As a matter of fact 56 survivors walked out of the gondola once it landed after all the hydrogen and balloon structure burned out. Therefore, even in a worst case scenario accident, hydrogen proved to be a safer fuel.

Review Questions

1. What is the difference between an energy source and an energy carrier?
2. What are some of the issues hindering the use of direct thermal decomposition of water as a means of hydrogen production?
3. What is the mechanism for using polyhydride complexes for hydrogen storage and what are the advantages of using this storage method?
4. What are some of the problems involved in liquid hydrogen production and storage?
5. What is the difference between the freeze-thaw and the auger methods for the production of slush hydrogen?

References

Baker, C. R., L. C. Matsch. 1965. Production and distribution of liquid hydrogen. *Advances in Petroleum Chemistry and Refining* 10:37–81.

Baker, C. R., R. L. Shaner. 1978. A study of the efficiency of hydrogen liquefaction. *Int. J. Hydrogen Energy* 3:321–334.

Barclay, J. A. 1982. Use of a ferrofluid as the heat-exchange fluid in a magnetic refrigerator. *J. Applied Physics* 53(4):2887.

Barclay, J. A. 1983. The theory of an active magnetic regenerative refrigerator. NASA-R-2287.
Barclay, J. A. 1984. Magnetic refrigeration for low-temperature applications. *Proceedings of the Cryocooler Conference*, Colorado, September.
Barclay, J. A., S. Sarangi. 1984. Selection of regenerator geometry for magnetic refrigerator applications. *Proceedings of the ASME Winter Annual Meeting*, New Orleans, LA, December.
Barclay, J. A., W. F. Stewart. 1982. The effect of parasitic refrigeration on the efficiency of magnetic liquefiers. *Proceedings IECEC '82: 17th Intersociety Energy Conversion Engineering Conference*, Los Angeles, CA, August, 1166.
Barclay, J. A., W. A. Steyert. 1982. Materials for magnetic refrigeration between 2K and 20K. *Cryogenics* February 22:3–79.
Barclay, J. A., W. C. Overton, Jr., W. F. Stewart. 1983. Phase I final report: Magnetic reliquefication of LH_2 storage tank boil-off. Kennedy Space Center, Ref. PT-SPD/6037/2511 C/000000/04/82. NASA—Defense Purchase Req. CC—22163B.
Barclay, J. A., W. F. Stewart, W. C. Overton, R. C. Candler, O. D. Harkleroad. 1985. Experimental results on a low-temperature magnetic refrigerator. *Cryogenic Engineering Conference*, Boston, MA, August.
Barron, R. F. 1985. *Cryogenic Systems*. Oxford, UK: Oxford University Press.
Baykara, S. Z., E. Bilgen. 1989. An overall assessment of hydrogen production by solar water thermolysis. *Int. J. Hydrogen Energy* 14(12):881–889.
Blank, H., A. Szyszka. 1992. Solar hydrogen demonstration plant in Neunburg vorm Wald. In T. N. Veziroglu, C. Derive, J. Pottier, eds. *Hydrogen Energy Progress IX*, 2. Paris: M.C.I., 677–686.
Blumen, L. J. M. J., M. N. Mugerwa, eds. 1993. *Fuel Cell Systems*. New York: Plenum Press.
Bockris, J.O'M. 1975. *Energy: The Solar—Hydrogen Alternative*. New York: Halsted Press.
Bockris, J.O'M., T. N. Veziroglu. 1985. A solar—hydrogen energy system for environmental compatibility, Environmental Conservation, 12(2):105–118.
Bockris, J.O'M., T. N. Veziroglu, D. Smith. 1991. *Solar Hydrogen Energy: The Power to Save the Earth*. London: Optima.
Bonner, M., T. Botts, J. McBreen, et al. 1984. Status of advanced electrolytic hydrogen production in the United States and abroad. *Int. J. Hydrogen Energy* 9(4):269–275.
Brewer, G. D. 1982. The prospects for liquid hydrogen fueled aircraft. *Int. J. Hydrogen Energy* 7:21–41.
Brown, N. S. 1986. Advanced long term cryogenic storage systems. NASA Marshall Space Flight Center, Huntsville, AL. *NASA Conference Publication 2465P*, Washington, DC, 7–16.
Bull, S. R. 1988. Hydrogen production by photoprocesses. *Proc. Int. Renewable Energy Conf.*, Honolulu, HI, 413–426.
Carpetis, C. 1984. An assessment of electrolytic hydrogen production by means of photovoltaic energy conversion. *Int. J. Hydrogen Energy* 9(2):969–992.

Carpetis, C. 1988. Storage, transport and distribution of hydrogen. In C.-J. Winter, J. Nitsch, eds. *Hydrogen as an Energy Carrier*. Berlin/Heidelberg: Springer-Verlag, 249–289.

Chambers, A., C. Park, R. T. K. Baker, N. M. Rodriguez. 1998. Hydrogen storage in graphite nano-fibers. *J. Phys. Chem* B, 102(22):4253–4259.

DiChristina, M. 1997. What really downed the Hindenburg. *Popular Science* November, Vol. 251, No. 5, pp. 70–77.

Dillon, A. C., K. M. Jones, M. J. Heben. 1996. Carbon nanotube materials for hydrogen storage. *Proc. U.S. DOE Hydrogen Program Review* II:747–763, Golden, CO: National Renewable Energy Laboratory.

Dutta, S. 1990. Technology assessment of advanced electrolytic hydrogen production. *Int. J. Hydrogen Energy* 15(6):379–386.

Engels, H., J. E. Funk, K. Hesselmann, K. F. Knoche, 1987. Thermochemical hydrogen production. *Int. J. Hydrogen Energy* 12(5):291–295.

Ewe, H. H., H. J. Selbach. 1987. The storage of hydrogen. In W. E. Justi, ed. *A Solar Hydrogen Energy System*. London: Plenum Press.

Garcia-Conde, A. G., F. Rosa. 1992. Solar hydrogen production: A Spanish experience. In T. N. Veziroglu, C. Derive, J. Pottier, eds. *Hydrogen Energy Progress IX*, 2. Paris: M.C.I., 723–732.

Goebel, E.D., R. M. Coveney, Jr., E. E. Angino, E. J. Zeller, G. A. M. Dreschhoff. 1984. Geology, composition, isotopes of naturally occurring H_2/N_2 rich gas from wells near Junction City, Kansas. *The Oil and Gas Journal* Vol. 82, May, pp. 215–221.

Grasse, W., F. Oster, H. Aba-Oud. 1992. HYSOLAR: The German—Saudi Arabian program on solar hydrogen—5 years of experience. *Int. J. Hydrogen Energy* 17(1):1–8.

Hancock, Jr., O. G. 1984. A photovoltaic-powered water electrolyzer: Its performance and economics. In T. N. Veziroglu, J. B. Taylor, eds. *Hydrogen Energy Progress V.*, New York: Pergamon Press, 335–344.

Hands, B. A. 1986. *Cryogenic Engineering*. New York: Academic Press.

Heydorn, B. 1998. *SRI Consulting Chemical Economics Handbook*. Menlo Park, CA: Stanford Research Institute.

Huston, E. L. 1984. Liquid and solid storage of hydrogen. *Hydrogen Energy Progress V*. T. N. Veziroglu (Ed.), Pergamon Press, New York.

Jensen, C. 1996. Hydrogen storage via polyhydride complexes. *Proc. U.S. DOE Hydrogen Program Review*, II:787–794, Golden, CO: National Renewable Energy Laboratory.

Kandebo, S. W. 1989. Researchers explore slush hydrogen as fuel for national aerospace plane. *Aviation Week & Space Technology* June 26:37–38.

Kinoshita, K., F. R. McLarnon, E. J. Cairns. 1988. Fuel cells: A handbook. DOE/METC88/6069. Morgantown, WV: U.S. Department of Energy.

Lehman, P., C. E. Chamberlain. 1991. Design of a Photovoltaic—Hydrogen—Fuel Cell Energy System. *Int. J. Hydrogen Energy* 16(5):349–352.

Liepa, M. A., A. Borhan. 1986. High-temperature steam electrolysis: Technical and economic evaluation of alternative process designs. *Int. J. Hydrogen Energy* 11(7):435–442.

Lund, P. D. 1991. Optimization of stand alone photovoltaic system with hydrogen storage for total energy self-sufficiency. *Int. J. Hydrogen Energy* 16(11):735–740.

Mitlitsky, F. 1996. Development of an advanced, composite, lightweight, high pressure storage tank for on-board storage of compressed hydrogen. *Proc. Fuel Cells for Transportation TOPTEC: Addressing the Fuel Infrastructure Issue.* Alexandria, VA, SAE, Warrendale, PA.

National Hydrogen Association. 1991. The hydrogen technology assessment, phase I. A Report for NASA. Washington, DC: Author.

Newton, C. L. 1967a. Hydrogen production, liquefaction and use. *Cryogenic Engineering News* Part I, 8:50–60.

Newton, C. L. 1967b. Hydrogen production, liquefaction and use. *Cryogenic Engineering News* Part II, 9:24–29.

Norbeck, J. M., J. W. Heffel, T. D. Durbin, et al. 1996. Hydrogen fuel for surface transportation. Warrendale, PA: SAE Publishing.

Oney, F. 1991. The comparison of pipelines transportation of hydrogen and natural gas. M.S. Thesis, University of Miami.

Pottier, J. D. 1995. Hydrogen transmission for future energy systems. In Y. Yurum, ed. *Hydrogen Energy System, Utilization of Hydrogen and Future Aspects.* NATO ASI Series E-295. Dordrecht, Netherlands: Kluwer Academic Publishers, 181–194.

Pottier, J. D., E. Blondin. 1995. Mass storage of hydrogen. In Y. Yurum, ed. *Hydrogen Energy System, Utilization of Hydrogen and Future Aspects.* NATO ASI Series E-295. Dordrecht, Netherlands: Kluwer Academic Publishers, 167–180.

Rambach, G., C. Hendricks. 1996. Hydrogen transport and storage in engineered glass microspheres. *Proc. U.S. DOE Hydrogen Program Review* II:765–772, Golden, CO, National Renewable Energy Laboratory.

Schwartz, J. A., K. A. G. Amankwah. 1993. Hydrogen storage systems. In D. G. Howell, ed. *The Future of Energy Gases.* U.S. Geological Survey Professional Paper 1570. Washington, DC: U.S. Government Printing Office, 725–736.

Scott, R. B. 1962. *Cryogenic Engineering.* Princeton, NJ: Van Nostrand Company, Inc.

Sherif, S. A., M. Lordgooei, M. T. Syed. 1989. Hydrogen liquefaction. In T. N. Veziroglu, ed. *Solar Hydrogen Energy System.* Final Technical Report, Clean Energy Research Institute, University of Miami, Coral Gables, FL, August: C1–C199.

Siegel, A., T. Schott. 1988. Optimization of photovoltaic hydrogen production. *Int. J. Hydrogen Energy* 13(11):659–678.

Sindt, C. F. 1970. A summary of the characterization study of slush hydrogen. *Cryogenics* October 10:372–380.

Steeb, H., A. Brinner, H. Bubmann, W. Seeger. 1990. Operation experience of a 10 kW PV-electrolysis system in different power matching modes. In T. N. Veziroglu, P. K. Takahashi, eds. *Hydrogen Energy Progress VIII*, 2. New York: Pergamon Press, 691–700.

Sternfeld, H. J., P. Heinrich. 1989. A demonstration plant for the hydrogen/oxygen spinning reserve. *Int. J. Hydrogen Energy* 14:703–716.

Swain, M. R., M. N. Swain. 1992. A comparison of H_2, CH_4, and C_3H_8 fuel leakage in residential settings. *Int. J. Hydrogen Energy* 17(10):807–815.

Taylor, J. B., J. E. A. Alderson, K. M. Kalyanam, A. B. Lyle, L. A. Phillips. 1986. Technical and economic assessment of methods for the storage of large quantities of hydrogen. *Int. J. Hydrogen Energy* 11(1):5–22.

Thomas, C. E. 1996. Preliminary hydrogen vehicle safety report. The Ford Motor Company, Contract, DE-AC02-94CE50389. U.S. Department of Energy.

Veziroglu, T. N. 1987. Hydrogen technology for energy needs of human settlements. *Int. J. Hydrogen Energy* 12(2):99–129.

Veziroglu, T. N., F. Barbir. 1992. Hydrogen: The wonder fuel. *Int. J. Hydrogen Energy* 17(6):391–404.

Voth, R. O. 1985. Producing liquid—solid mixtures (slushes) of oxygen or hydrogen using an auger. *Cryogenics* September 25:511–517.

Wendt, H. 1988. Water splitting methods. In C.-J. Winter, J. Nitsch, eds. *Hydrogen as an Energy Carrier*. Berlin Heidelberg: Springer-Verlag, 166–238.

Willner, I., B. Steinberger-Willner. 1988. Solar hydrogen production through photobiological, photochemical and photoelectrochemical assemblies. *Int. J. Hydrogen Energy* 13(10):593–604.

Winter, C. J., J. Nitsch. 1988. *Hydrogen as an Energy Carrier*. Berlin: Springer-Verlag.

Yalcin, S. 1989. A review of nuclear hydrogen production. *Int. J. Hydrogen Energy* 14(8):551–561.

CHAPTER 8

Chemical Solutions

Patrick Sullivan

Introduction

Throughout history, water resources have always been one of the primary receptacles for man's waste. When coupled with an ever-expanding population, it was only a matter of time before the beneficial uses of our water resources (e.g., drinking water, aquatic habitat, recreational use, agricultural use, industrial use) would be impaired, and in some cases impaired irreversibly. Today, the pollution of potential drinking water resources has led the American Water Works Association to conclude that "the boundaries between water and wastewater are already beginning to fade" (Maxwell 2001). This result should not be unexpected since the guiding principle for protecting the environment allows man-made chemicals to be intentionally discharged into our water resources as long as established chemical-specific standards are not exceeded. This principle is based on the United States Environmental Protection Agency (EPA) establishment of water quality standards. These standards, once established, identify individual compounds as regulated chemicals.[1] In other words, as long as the concentration of a regulated chemical is not exceeded in either a wastewater discharged to a water resource or exceeded in the receiving water itself, then by definition there is (presumably) no harm to human health or the environment.

Our confident reliance on a regulatory system that gives each discharger a "license to pollute" is contingent on not only knowing the potential aquatic impairment or human health risk associated with each regulated chemical, but believing that (1) methods of estimating the risk of individually regulated chemicals are both accurate and valid; (2) mixtures

1. This same set of compounds can be regulated at the state level with additional compounds being selected for regulation by individual states. For the purpose of this chapter, state water quality standards will not be addressed.

of regulated chemicals have no risk, and (3) nonregulated chemicals have no risk either individually or in a mixture (i.e., because they are nonregulated). In our chemically dependent society, any regulatory system that is based on establishing or identifying "regulated" chemicals as a means of protecting the environment is clearly out of date. The National Research Council estimates that there are approximately 72,000 organic chemicals in commerce within the United States, with nearly 2,000 new chemicals being added each year. Given this number of chemicals, it is impossible for the EPA to identify those chemicals that pose a hazard to the environment. For example, under the Toxic Substances Control Act (TSCA) of 1979, the EPA has conducted an assessment program to determine which new chemicals present an unreasonable risk to human health or the environment. Since 1979, the EPA has only reviewed some 32,000 new chemical substances National Research Council 1999).

Of these tens of thousands of chemicals, the list of regulated compounds is very short. Those compounds with EPA recommended water quality criteria (U.S. EPA 2002) are given in Table 8.1. These compounds tend to be the industrial and pesticide pollutants of the past. The compounds in Table 8.1 can be contrasted to the chemicals now being found in our water resources today (see Table 8.2).

In addition to the specific water quality criteria given in Table 8.1 that are used to protect the environment, the Clean Water Act (CWA) requires that industries pretreat any wastewater prior to its discharge into surface water or into a publicly owned treatment works (POTW). In other words, effluent limitations are placed on a discharge to regulate the concentration of selected chemicals into a receiving water or POTW. Thus, a water resource is presumably protected by limiting the discharge concentration of selected chemicals. These selected regulated chemicals can be any of the compounds given in Table 8.1 or chemicals that are industry specific and pose a threat to the environment (i.e., as determined by the EPA). Therefore effluent limitations for industry specific compounds have been developed (e.g., aquaculture; waste treatment; mining; feedlots; laundries; iron and steel manufacturing; landfills; leather tanning and finishing; meat products; metal products and machinery; oil and gas extraction; pesticide formulation and packaging; pharmaceuticals manufacture; pulp and paper; transportation equipment cleaning). As an example, the effluent guidelines (U.S. EPA 1998a) for the pharmaceutical manufacturing point source category required that water soluble compounds that will pass through a POTW be regulated. These effluent limitations listed in Table 8.3 are a combination of industry specific chemicals and compounds having established water quality criteria. However, the use of effluent limitations has its own problems.

A review of the effluent limitation process conducted by the General Accounting Office (1994), found that 236 facilities from three industrial sectors (pulp and paper, pharmaceutical, and pesticides manufacturing) did

TABLE 8.1
Compounds with U.S. EPA Recommended Water Quality Criteria

Priority Pollutants	*Non-Priority Pollutants*	*Organoleptic Pollutants*[1]
Antimony	Alkalinity	Acenaphthene
Arsenic	Aluminum	Monochlorobenzene
Beryllium	Aesthetic Qualities	3-Chlorophenol
Cadmium	Bacteria	4-Chlorophenol
Chromium (III & VI)	Barium	2,3-Dichlorophenol
Copper	Boron	2,5-Dichlorophenol
Lead	Chloride	2,6-Dichlorophenol
Mercury	Chlorine	3,4-dichlorophenol
Methylmercury	2,4,5-TP	2,4,5-Trichlorophenol
Nickel	2,4-D	2,4,6-Trichlorophenol
Selenium	Chloropyrifos	2,3,4,6-Tetrachlorophenol
Silver	Color	2-Methyl-4-chlorophenol
Thallium	Demeton	3-Methyl-4-chlorophenol
Zinc	Ether, Bis (Chloromethyl)	3-Methyl-6-chlorophenol
Cyanide	Gases, total dissolved	2-Chlorophenol
Asbestos	Guthion	Copper
2,3,7,8-TCDD	Hardness	2,4-Dichloropohenol
Acrolein	Hexachlorocyclo-hexane	2,4-Dimethylphenol
Acrylonitrile	Iron	Hexachlorocyclopentadiene
Benzene	Malathion	Nitrobenzene
Bromoform	Manganese	Pentachlorophenol
Carbon tetrachloride	Methoxychlor	Phenol
Chlorodibromomethane	Mirex	Zinc
Chloroethane	Nitrates	
2-Chloroethylvinyl ether	Nitrosamines	
Chloroform	Dinitrophenols	
Dichlorobromomethane	Nitrosodibutylamine, N	
1,1-Dichloroethane	Nitrosodiethylamine, N	
1,2-Dichloroethane	Nitrosopyrrolidine, N	
1,1-Dichloroethylene	Oil and Grease	
1,2-Dichloropropane	Oxygen, dissolved	
1,3-Dichloropropene	Parathion	
Ethylbenzene	Pentachlorobenzene	
Methyl bromide	pH	
Methyl chloride	Phosphorus (elemental)	
Methylene chloride	Nutrients	
1,1,2,2-Tetrachloroethane	Solids dissolved and salinity	
Tetrachloroethylene	Solids suspended and turbidity	
Toluene	Sulfide-hydrogen sulfide	
1,2-trans-Dichloroethylene	Tainting substances	
1,1,1-Trichloroethane	Temperature	

TABLE 8.1 *(continued)*

Priority Pollutants	*Non-Priority Pollutants*	*Organoleptic Pollutants*[1]
1,1,2-Trichloroethane	Tetrachlorobenzene, 1,2,4,5-	
Trichloroethylene	Tributyltin	
Vinyl Chloride	Trichlorophenol, 2,4,5-	
2-Chlorophenol		
2,4-Dichlorophenol		
2-Nitrophenol		
4-Nitrophenol		
3-Methyl-4-chlorophenol		
Pentachlorophenol		
Phenol		
2,4,6-Trichlorophenol		
Acenaphthene		
Acenaphthylene		
Anthracene		
Benzidine		
Benzo(a)anthracene		
Benzo(a)pyrene		
Benzo(b)fluoranthene		
Benzo(ghi)perylene		
Benzo(k)fluoranthene		
Bis(2-chloroethoxy)methane		
Bis(2-Chloroethyl)ether		
Bis(2-Chloroisopropyl)ether		
Bis(2-ethylhexyl)phthalate		
4-Bromophenyl phenyl ether		
Butylbenzyl phthalate		
2-Chloronaphthalene		
4-Chlorophyenyl phenyl ether		
Chrysene		
Dibenzo(a,h)anthracene		
1,2-Dichlorobanzene		
1,3-Dichlorobenzene		
1,4-Dichlorobenzene		
3,3′-Dichlorobenzidine		
Diethyl phthalate		
Dimethyl phthalate		
Di-n-butyl phthalate		
2,4-Dinitrotoluene		
2,6-Dinitrotoluene		
Di-n-octyl phthalate		
1,2-Diphenylhydrazine		
Fluoranthene		
Hexachlorobenzene		

TABLE 8.1 *(continued)*

Priority Pollutants	*Non-Priority Pollutants*	*Organoleptic Pollutants*[1]
Hexachlorobutadiene		
Hexachlorocyclopentadiene		
Hexachloroethane		
Indeno(1,2,3-cd)pyrene		
Isophorone		
Naphthalene		
Nitrobenzene		
N-Nitrosodimethylamine		
N-Nitrosodi-n-propylamine		
N-Nitrosodiphenylamine		
Phenanthrene		
Pyrene		
1,2,4-Trichlorobenzene		
Aldrin		
alpha-BHC		
beta-BHC		
gamma-BHC (lindane)		
delta-BHC		
Chlordane		
4,4′-DDT		
4,4′-DDE		
4,4′-DDD		
Dieldrin		
alpha-Endosulfan		
beta-Endosulfan		
Endosulfan sulfate		
Endirn		
Endrin aldehyde		
Heptachlor		
Heptachlor epoxide		
Polychlorinated biphenyls		
Toxaphene		

[1] Organoleptic effects (e.g., taste and odor) are those that would make water and edible aquatic life unpalatable but not toxic to humans.

not control 77% of the chemicals discharged to the environment. According to the GAO, "the current permit process does not limit the vast majority of toxics being discharged from the nation's factories and sewage treatment plants [i.e., POTWs]. Although most of these toxicants are 'nonpriority' pollutants [see Table 8.1], they do pose risks to both human health and aquatic life. GAO tried to examine the implications of uncontrolled pollution cases identified in the facility sample population, but the

TABLE 8.2
Chemicals In the Water Environment (Primarily Non-Regulated)

Study or Report	*Pollutants*
Mississippi River Study[1]	Methylene-blue substances from synthetic and natural anionic surfactants (e.g., detergents)
	Linear alkylbenzenesulfonates, a complex mixture of anionic surfactant compounds used in soak and detergent products
	Nonionic surfactants such as nonylphenol and polyethylene glycol
	Adsorbable halogen-containing organic compounds including solvents and pesticides
	Polynuclear aromatic hydrocarbons from the combustion of fuels.
	Caffeine from beverages, food products and medications.
	Ethylenediaminetetraacetic acid (EDTA), a widely used synthetic chemical for complexing metals.
	Volatile organic compounds including chlorinated solvents and aromatic hydrocarbons.
	Semivolatile organic compounds including priority pollutants and compounds such as trimethyltriazinetrione (a by-product of methylisocyanate) and trihaloalkylphosphates (a flame-retardant).
Wastewater from a Publicly Owned Treatment Works[2]	Benzaldehyde
	1,2-Benzene dicarboxylic acid
	Benzene, 1-methyl-4-2(methyl propyl)
	Benzophenone
	Benzothiazole, 2,2-(methylthio)
	Bromodichloromethane
	Butyl 2-methylpropyl ester
	Cholestanol
	3-Chloro-2-butanol
	1,3,5-Cycloheptatriene
	Cyclohexanone
	Cyclohexanone, 4-(1,1-dimethyethyl)
	Cyclopentanol 1,2-dimethyl-3-(methylethenyl)
	Decahydro naphthalene
	Decanal
	Diacetate, 1,2-ethanediol
	Dibromochloromethane
	N,N-Diethyl-3-methyl benzamide
	2,5-Dimethyl 3-hexanol
	2,2-Dimethyl 3-pentanol
	Ethanol, 2-butoxy-phosphate
	Ethanone 1-(2-naphthalenyl)
	Ethyl citrate
	Heptanal
	2-Heptanone,3 hydroxy-3 methyl
	Hexadecanoic acid
	Hexanal
	3-Hexanol
	3-Methoxy-3methyl-hexane

TABLE 8.2 *(continued)*

Study or Report	*Pollutants*		
	2-Methoxy-1-propanol		
	Nonanal		
	Octandecanoic acid		
	Octadiene,4,5 dimethyl-3,6 dimethyl		
	Octanal		
	Phenol 2,4 (bis(1,1-dimethylethyl))		
	Phenol 4,4(1,2-diethyl-1,2-ethanediyl)bisphenol nonyl		
	1-Phenyl ethanone		
	Propanic acid 2 methyl-2,2-dimethyl-1-(2hr . . .)		
	1-Propanol, 2-(2-hydroxypropoxy)		
	2-Propanone, 1-(1-cyclohexen-1-yl)		
	Tetradecanal		
	Tetradecanoic acid		
Pharmaceuticals found in water resources[3]	*Type of Drug*	*Human Drug*	*Veterinary Drug*
	Analgesic	Aspirin	
		Dextropropoxyphene	
		Dichlorfenac	
		Ibuprofen	
		Indometacin	
	Antibiotic	Erythromycin	Oxytetracycline
		Phenicilloyl groups	
		Sulphamethoxazote	
		Tetracycline	
	Antiparacitic		Ivermactin
	Anxiolytic	Diazepam	
	Cancer treatment	Bleomycin	
		Cyclophosphamide	
		Ifosfamide	
		Methotrexate	
	Hormone	Estrogen	Estrogen
		Estradiol	Testroesterone
		Estrone	
		Ethinylestradiol	
		Norethisterone	
		Oral Contraceptives	
		Testosterone	
	Cholesterol-lowering	Clofibrate	
		Clofibric acid	
	Narcotic	Morphinan-structure	
	Psychomotor	Caffeine	
		Theophylline	

[1] United States Geological Survey, 1995 "Organic Contamination of the Mississippi River from Municipal and Industrial Wastewater," *Circular* 1133.
[2] Levine, B. B., et al., "Treatment of Trace Organic Compounds by Ozone-Biological Activated Carbon for Wastewater Reuse: The Lake Arrowhead Pilot Plant," *Water Environmental Research*, 72(4), July/August 2000.
[3] Hunt, T., "Water quality—Studies Indicate Drugs in Water May Come from Effluent Discharges," *Water Environment & Technology*, (July 1998).

TABLE 8.3
Effluent Limitations for Pharmaceutical Manufacturing

Acetone
Ammonia
n-Amyl acetate
Benzene*
n-Butyl acetate
Chloroform*
Dichlorobenzene*
Dichloroethane*
Diethylamine
Ethyl acetate
n-Heptane
n-Hexane
Isobutyraldehyde
Isopropyl acetate
Isopropyl ether
Methylene Chloride*
Methyl cellosolve
Methyl formate
MIBK
Monochlorobenzene*
Tetrahydrofuran
Toluene*
Triethylamine
Xylenes*

* Established U.S. EPA Water Quality Criteria (see Table 8.1).

majority of cases could not be evaluated because of a lack of criteria for assessing the health risks posed by the discharges."

When considering the explosive growth of new man-made organic chemicals in the environment, all these data strongly suggest that neither water quality criteria nor effluent limitations are totally capable of protecting human health and the environment. There are simply too many existing and new chemicals that occur in our water resources with little or no information on their potential effects. The use of water quality criteria and effluent limitations based on individual organic chemical compounds is seriously limited and out of touch with the realities of our time. Fortunately, advances in technology finally allow an alternative to chemical specific standards. The implementation of a standardless approach can be achieved by uniting the appropriate combination of water treatment and chemical fingerprinting technologies to render chemical specific standards obsolete. Thus, the innovative application of technological solutions will advance the protection of our water resources.

Protecting the Environment with Technology

The cornerstone of a chemical control program that can be independent of chemical standards must be based on water quality monitoring using existing commercially available state-of-the-art analytical instrumentation to fingerprint wastewater, raw drinking water resources, and treated water products on either a batch or continuous basis. By obtaining the chemical fingerprint of a wastewater or raw drinking water, the pattern of the pollutants can be identified and selectively treated until the fingerprint is eliminated or minimized to the point where the application of sequential treatments (i.e., same or different appropriate technology) yields no appreciable change. For example, no further treatment would be required if the total area of the fingerprint cannot be reduced by more than 10% and representative chemical surrogates are removed below 1 part per billion.

Such an approach would allow the application of treatment technologies that will remove all pollutants and not just those small numbers of compounds that have gone through the political/regulatory process of being defined as a hazard. Therefore, the purpose of this paper is to outline the chemical treatment and monitoring technologies that, if applied in a comprehensive program, could offer a solution to the chemical pollution of our water resources irrespective of an artificial risk management process.

Waste Treatment Solutions

Both inorganic and organic pollutants are distributed throughout the environment. A wide range of these chemical pollutants already exist in surface and groundwater resources as a consequence of past waste disposal practices or current sources of point and nonpoint sources of pollution. Therefore, protecting both human health and the environment necessitates that the pollution load to receiving bodies of water be minimized while ensuring that drinking water treatment systems maximize the removal of pollutants within the reality of budgetary constraints.

The optimization of water treatment systems is no simple task since each wastewater stream and drinking water resource is chemically unique. A given water may require treatment to remove a singular combination of inorganic and/or organic compounds. Therefore, it is important to address each group of chemicals individually.

Inorganic Chemical Solutions

The occurrence of inorganic compounds in water is the natural result of their contact with soil and rock. Therefore, the chemical elements that will be naturally dissolved in water are based on their relative abundance in minerals. For example, the average abundance of the most common chemical elements in the earth's crust (in decreasing order of abundance, see Table 8.4) are: oxygen, silica, aluminum, iron, calcium, sodium, magnesium, and

TABLE 8.4
Average Elemental Abundance in the Earth's Crust

Element	*Concentration (mg)*[1]	*Percentage*
Oxygen	464,000	46.16
Silica	282,000	28.05
Aluminum	82,000	8.16
Iron	56,000	5.57
Calcium	41,000	4.08
Sodium	24,000	2.39
Magnesium	23,000	2.39
Potassium	21,000	2.01
Titanium	5,700	0.56
Hydrogen	1,400	0.14
Phosphorous	1,050	0.11
Manganese	950	0.10
Fluorine	625	0.062
Barium	425	0.042
Strontium	375	0.037
Sulfur	260	0.025
Carbon	200	0.020
Zirconium	165	0.016
Vanadium	135	0.013
Chlorine	130	0.012
		99.95%
Chromium	100	
Rubidium	90	
Nickel	75	
Zinc	70	
Cerium	67	
Copper	55	
Yttrium	33	
Neodymium	28	
Lanthanum	25	
Cobalt	25	
Scandium	22	
Lithium	20	
Nitrogen	20	
Niobium	20	
Gallium	15	
Lead	12.5	
Boron	10	
Thorium	9.6	
Samarium	7.3	
Gadolinium	7.3	
Praseodymium	6.5	
Dysprosium	5.2	
Ytterbium	3	

TABLE 8.4 *(continued)*

Element	*Concentration (mg)*[1]	*Percentage*
Hafnium	3	
Cesium	3	
Beryllium	2.8	
Erbium	2.8	
Uranium	2.7	
Bromine	2.5	
Tin	2	
Arsenic	1.8	
Germanium	1.5	
Molybdenum	1.5	
Tungsten	1.5	
Holmium	1.5	
Europium	1.2	
Terbium	1.1	
Lutetium	0.8	
Iodine	0.5	
Thallium	0.45	
Thulium	0.25	
Cadmium	0.2	
Antimony	0.2	
Bismuth	0.17	
Indium	0.1	
Mercury	0.08	
Silver	0.07	
Selenium	0.05	
Gold	<0.05	
Platinum	<0.05	
Tellurium	<0.05	
Rhenium	<0.05	

[1] Average elemental concentration in 1,000 grams of rock.
Adapted from Krauskopf, 1967.

potassium. These first eight elements are the building blocks of the most common minerals that make up soil and rock. The next 12 most common elements are: titanium, hydrogen, phosphorous, manganese, fluorine, barium, strontium, sulfur, carbon, zirconium, vanadium, and chlorine. Combined, these 28 elements make up 99.95% of the earth's crust.

Therefore, it would be anticipated that natural water would contain some or most of these dissolved elements. Based up the natural properties of water (H_2O) and the properties of each chemical element and mineral combination, the actual occurrence and concentration of an element in water can vary widely. When minerals dissolve in water, the chemical elements are usually ionized. These ions occur as either cations (a positively charged ion) or as anions (a negatively charged ion). The major cations and anions that occur in both surface water and ground water are: Na^+ (sodium) > K^+ (potassium) > Ca^{2+} (calcium) > Mg^{2+} (magnesium) > Fe (iron) for the cations and HCO_3^- (bicarbonate) > SO_4^{2-} (sulfate) and SiO_4^{4-} (silicate and as aqueous SiO_2) > Cl^- (chloride) > PO_4^{-3} (phosphate), and NO_3^- (nitrate; from both natural organic and rock sources) for the anions. In general, the most common naturally occurring trace elements in our water resources are aluminum, arsenic, boron, copper, cadmium, manganese, nickel, lead, selenium, and zinc.

The most common trace elements from industrial sources are arsenic, cadmium, chromium, copper, lead, mercury, silver, zinc, while huge volumes of anions (primarily chloride and sulfate) are discharged from industrial sources. Obviously, a water resource can contain a combination of both natural or industrial sources of inorganic compounds. Because the inorganic compounds occur naturally, pollution of water resources must be defined on the basis of background inorganic compound concentrations. Therefore, when inorganic compounds in wastewater exceed the receiving stream background concentrations, water treatment technologies should be employed to ensure that effluent concentrations do not exceed receiving stream background concentrations.

On the other hand, when treating raw drinking water sources, there should be no detectable concentrations of known toxic inorganic compounds such as antimony, arsenic, barium, beryllium, bromine, cadmium, chromium, copper, fluoride, lead, mercury, nitrogen (as nitrate and nitrite), selenium, silver, and thallium. Furthermore, the concentration of total dissolved solids should not exceed 250 mg/L which is used to define natural "mineral water."

The ability to meet all of these water quality objectives is well within the capabilities of existing water treatment technologies. Depending on the source water to be treated, the use of any combination of chemical precipitation (e.g., pH modification or use of sulfides precipitation), flocculation (e.g., with iron, aluminum, or synthetic organic polymers), specific ion exchange resins, or microfiltration/reverse osmosis technologies can be used. Furthermore, the ability to monitor the quality of the treated water for inorganic compounds is also straight forward since extremely sensitive analytical methods for all the compounds of concern are readily available, such as inductively coupled plasma mass spectrometry and ion chromatography, and selective ion electrodes. All of these technologies can also be configured to provide on-line analysis of water sources.

Organic Chemical Solutions

The occurrence of man-made organic compounds in our water resources is already a national epidemic. This condition is illustrated using surface water data collected around the United States (U.S. Geological Survey 1998) (Table 8.5). Furthermore, the vast majority of these known toxic compounds

TABLE 8.5
Summary of Surface Water Data for Pesticides

Compound	*Frequency of Detection (%)*
Acetochlor	9.80
Acifluorfen	2.74
Alachlor	39.02
Aldicarb	0.32
Aldicarb sulfoxide	0.11
Atrazine	88.62
Atrazine, deethyl	62.20
Azinphos-methyl	1.22
Benfluralin	2.61
Bentazon	8.68
Bromacil	0.74
Bromoxynil	0.64
Butylate	8.16
Carbaryl	21.14
Carbofuran	11.99
Chlorothalonil	0.32
Chlorpyrifos	40.67
Cyanazine	45.93
2,4-D	13.50
Dacthal mono-acid	0.43
2,4-DB	0.32
DCPA	30.20
DDE	6.30
Diazinon	74.85
Dicamba	1.83
Dichlobenil	1.59
Dichlorprop	0.96
Dieldrin	6.90
2,4-Diethylaniline	4.90
Dinoseb	0.11
Disulfoton	0.92
Diuron	13.02
DNOC	0.32
EPTC	25.13
Ethalfluralin	3.30
Ethiprop	3.40
Fenuron	0.45
Fluometuron	2.86
Fonofos	10.20
HCH, alpha	0.41
HCH, gamma	2.86
Linuron	3.40
Malathion	19.57
MCPA	4.82
Methiocarb	0.45

TABLE 8.5 *(continued)*

Compound	*Frequency of Detection (%)*
Methomyl	0.96
Methyl parathion	0.90
Metolachlor	81.30
Metribuzin	14.29
Molinate	4.90
Napropamide	7.59
Neburon	0.32
Norflurazon	0.64
Oryzalin	3.81
Parathion	0.31
Pebulate	4.08
Pendimethalin	19.57
Permethrin, cis	0.82
Phorate	0.41
Prometon	83.79
Pronamide	2.85
Propachlor	3.66
Propanil	2.75
Propargite	5.30
Propham	0.63
Propoxur	0.33
Simazine	87.77
Tebuthiuron	34.15
Terbacil	4.64
Terbufos	0.41
Thiobencarb	3.10
Triallate	8.50
Triclopyr	2.25
Trifluralin	17.50

are not regulated. When considering the contributions of synthetic industrial chemicals and pharmaceuticals to the mix in addition to disinfection byproducts from the chlorination of natural humic substances, there is truly a witch's brew of man-made chemicals in our water resources. Fortunately, a wide range of treatment technologies are available to remove organic compounds from water.

Historically, the use of biological treatment systems have usually been the first method employed to reduce the concentration of dissolved organic compounds in waste and wastewater. With the advent of more chemically resistant organic compounds (i.e., high chemical oxygen demand with little or no biological oxygen demand), biological treatment systems have been found to be less efficient. This has led to the need for nonbiological treatment methods. For example, activated carbon has been used to remove organic compounds from water since the turn of the 20th century and, today, granular-activated carbon is for many applications the preferred choice for removing organic compounds directly from wastewater and drinking water resources. For example, commercially available treatment systems around the United States that use granular activated carbon consistently report removal efficiencies from 92% to 99% depending on chemical specific characteristics.

Over the last several decades, there have been major advances in the development of new commercially available treatment methods to remove organic compounds for water. Since the 1980s, wet oxidation and supercritical wet oxidation have been used to treat wastewater (i.e., the combustion of organic compounds that are dissolved or suspended in water). In general, these technologies report reductions in chemical oxygen demand (COD) in the range of 70–99%, and 98% reduction in a wide range of aliphatic and aromatic hydrocarbons as well as pesticides. In the past 10 years there have been major advances in the development of effective advanced photochemical oxidation processes. Commercially available systems are based on ultraviolet (UV)/oxidation or photo-Fenton processes.

The UV/oxidation process generates a very reactive hydroxyl radical ($^{\bullet}OH$) through UV photolysis of either hydrogen peroxide (H_2O_2) or ozone (O_3). In other words, a water is injected with H_2O_2 or O_3 and exposed to a UV light source. The photo-Fenton process generates $^{\bullet}OH$ by reacting ferrous iron (Fe^{2+}) with H_2O_2 in a source water. In both cases, the reactive $^{\bullet}OH$ will oxidize dissolved organic compounds in water. The treatment efficiencies compiled by the EPA (U.S. EPA 1998b) for these processes are given in Table 8.6. Clearly, a wide range of treatment technologies are available to remove organic compounds from water. The only remaining technical issue is the ability to monitor for all organic compounds in a water to ensure that the treatment process has been effective.

Because of the vast number of organic compounds that can be dissolved in water, it is beyond the ability of chemical specific monitoring (i.e., quantification of individual compounds) to adequately characterize a water resource. This means that nonspecific chemical fingerprinting methods should be utilized to characterize the complete range of organic compounds that can be found in a given water resource or wastewater. Given this need, the only commercially available technology with adequate sensitivity is high-performance liquid chromatography (HPLC) coupled with a mass spectrometer (MS) or HPLC/MS. For example, pesticide studies conducted by the United States Geological Survey (Lee and Strahan 2003) showed that the estimated mean method detection limit (MDL) for all of the compounds and their degradation products ranged from 0.004 to 0.051 g/L. A list of the MDLs for each pesticide is given in Table 8.7. An example of a total ion chromatogram of a 1.0 g/L standard in a buffered reagent-water sample is given in Figure 8.1. It should be noted that in this sample (i.e., Figure 8.1), the herbicide 2,4-D is used as an internal standard.

TABLE 8.6
Treatment Efficiency Examples of Advanced Photochemical Oxidation Processes

Process	*Organic Compounds (μg/L)*	*Removal(%)*
UV/H_2O_2	Benzene @ 52	>96
	Chloroform @ 240	93.6
	Methylene Chloride @ 8	>86
	NDMA @ 20	>99.9
	PAH @ 2,000	>99.9
	PCE @ 150	>98.7
	PCP @ 15,000	99.3
	Phenol @ 2,000	>99.9
	TCE @ 1,300	>99.9
	TCA @ 130	92.9
	Vinyl Chloride @ 1,200	>95.8
UV/O_3	Benzene @ 310	93
	DCE @ 250	>99.9
	Ethyl Benzene @ 41	92
	PCE @ 11	>98
	TCE @ 1,800	99.9
	Vinyl Chloride @ 34	86
	Pesticides (alachlor, atrazine, bentaxon, metolachlor, metribuzin trifluraline, carbofuran, malathion)	
	@ 10,000	>99.9
	@ 1,000,000	75 to 85
Photo-Fenton	COD @ 3,000,000	>98.3

TABLE 8.7
Method Detection Limits for Selected Pesticides

Compound	*Mean Concentration (μg/L)*	*Estimated Method Detection Limit (μg/L)*
Acetochlor	0.027	0.021
Alachlor	0.021	0.019
Dimethenamid	0.024	0.018
Flufenacet	0.023	0.011
Metolachlor	0.023	0.004
Propachlor	0.022	0.008

This chromatogram illustrates the ability for HPLC/MS to provide a chemical fingerprint. However, even with a chemical fingerprint of a wastewater or drinking water source, it is important to spike samples prior to treatment with known chemical surrogates (i.e., like the 2,4-D internal standard). Selected chemical surrogates (i.e., compounds representative of

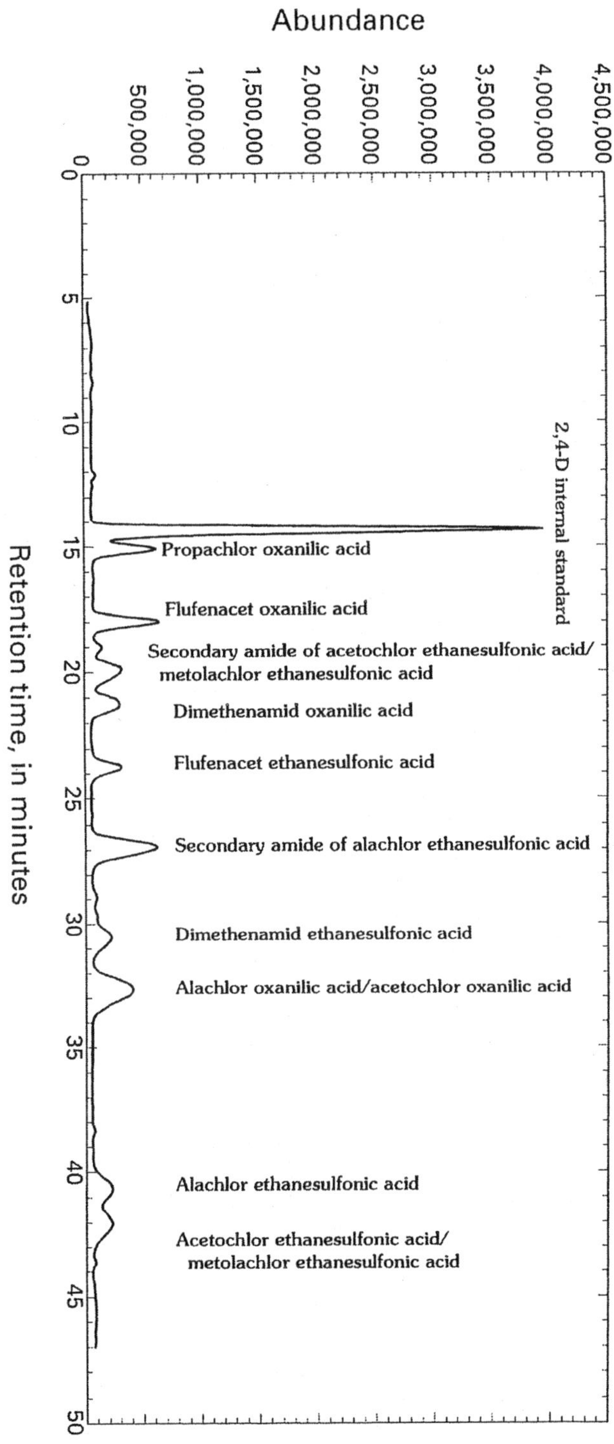

FIGURE 8.1. HPLC/MS fingerprint example.

various organic chemical classes) can help define the chemical characteristics of a fingerprint and serve as an indicator of treatment efficiency. HPLC is not the only monitoring technology that could be used for fingerprinting polluted water. Raman spectroscopy may also be appropriate for online line monitoring applications.

Although there are no commercially available Raman spectroscopy units with the current sensitivity as HPLC, the technology has been used to characterized dissolved organic compounds in groundwater in the part-per-billion range. Raman spectroscopy provides chemical information about molecular vibrations that can be used for sample identification and quantization. The technique involves shining a laser, at a selected wavelength, on a sample and detecting the scattered light. The majority of the reflected light will be the same frequency as the laser. However, a very small amount of light interacts with the molecules of a compound so that the original wavelength emitted by the laser is shifted. This shifted light produces a Raman spectra or chemical fingerprint. For example, studies (Premasiri et al. 2001) on monitoring cyanide in waste water using surface enhanced Raman spectroscopy found that cyanide in water could be detected down to 10g/L. An example of the cyanide spectra is given in Figure 8.2.

Applying Technological Solutions

Each wastewater and raw drinking water resource is chemically unique. Thus, any water subjected to chemical treatment needs to be characterized using standard analytical methods (i.e., ASTM and EPA-SW846) and then fingerprinted. Once characterized, a water can then be sequentially treated with those technologies which are appropriate to removing the identified fingerprint and chemical specific compounds.

With the selection of the proper treatment technology (or sequence of technologies) that is optimized to remove the maximum amount of pollutants, the water treatment process can be continually monitored by comparing the source water fingerprint with the post-treatment fingerprint. When all of these technologies, both treatment and monitoring, are combined into a water treatment system, the protection of water quality will no longer be dependent on arbitrary definitions of pollution and environmental harm.

Summary

Our water resources are polluted with a vast array of man-made chemicals. Furthermore, reports of the prevalence of this pollution are becoming more

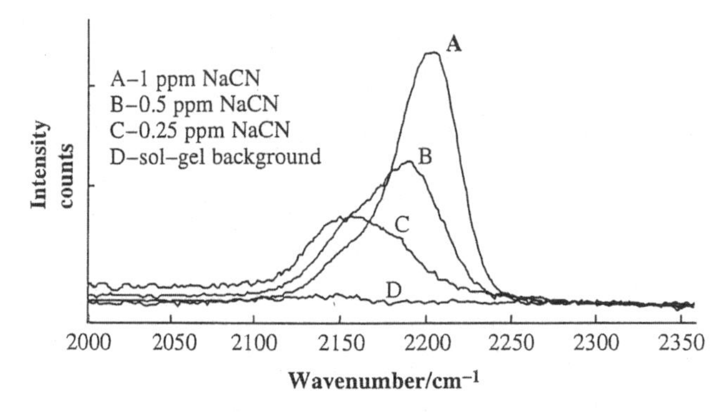

Surface enhanced Raman spectra of NaCN (with different concentrations) on gold sol–gel.

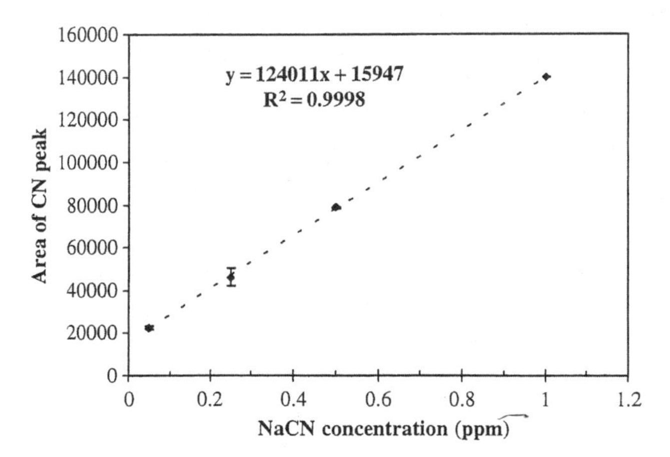

Calibration curve for NaCN in water on gold sol–gel.

FIGURE 8.2. Raman spectra fingerprint example.

and more frequent. For example, a 1999–2000 United States Geological Survey study of U.S. rivers and streams revealed that almost every surface water body is polluted with industrial chemicals, pesticides, pharmaceuticals, hormones, and home care product chemicals; while the United States Environmental Protection Agency's 2001 National Drinking Water Contaminant Occurrence Database shows that approximately 90% of the public water systems have found chemical pollution in their water supplies.

For the majority of these chemicals, the potential health effects to humans from consuming low levels of these compounds is unknown. This is critical because the federal government will never be able to provide water quality standards in "real time." Therefore, it is necessary that our current standards-based approach to managing water quality be changed to a technology-based approach.

Review Questions

1. Is the selection of specific water quality standards based on economics, science, and/or politics?
2. Why aren't the existing chemicals listed in the National Water Quality Criteria the same as current EPA Primary and Secondary Drinking Water Standards?
3. The time period between the introduction of a chemical into the environment and when it is recognized as being toxic to humans or animals often spans decades. How can this problem be solved?
4. If unregulated chemicals are found in drinking water resources, should the community be notified? Why or why not?
5. If the cost to treat drinking water to remove the maximum amount of chemical pollutants would only increase a community's water bill by 20%, how could this program be implemented and still be economically viable?

References

General Accounting Office. 1994. Water pollution: Poor quality assurance and limited pollutant coverage undermine EPA's control of toxic substances. Chapter Report, February 17, GAO/PEMD-94-9.

Lee, E. A., A. P. Strahan. 2003. Methods of analysis by the U.S. Geological Survey Organic Geochemistry Research Group—Determination of acetamide herbicides and their degradation products in water using online solid-phase extraction and liquid chromatography/mass spectrometry. Open-File Report 03-173.

Maxwell, S. 2001. Ten key trends and developments in the water industry. *Journal of the American Water Works Association* 93(4).

National Research Council. 1999. Identifying future drinking water contaminants. Washington, DC: National Academy Press.

Premasiri, W. R., R. H. Clark, S. Londhe, M. E. Wombie. 2001. Determination of cyanide in waste water by low-resolution surface enhanced Raman spectroscopy on sol-gel substrates. *Journal of Raman Spectroscopy* 32:919–922.

U.S. EPA. 1998a. *Development Document for Final Effluent Limitations Guidelines and Standards for the Pharmaceutical Manufacturing Point Source Category.* EPA-821-R-98-005, September.

U.S. EPA. 1998b. *Advanced Photochemical Oxidation Processes, Handbook.* EPA/625R-98/004.

U.S. EPA. 2002. *National Recommended Water Quality Criteria: 2002*. EPA-822-R-02-047, November.

U.S. Geological Survey. 1998, July 22. Pesticides in surface and ground water of the United States: Summary of results of the National Water Quality Assessment Program.

CHAPTER 9

Electrical and Thermal Solutions

John B. Wilcox

Part 1

Introduction

When the avoidance or reduction of pollutant or waste generation by means discussed in Chapters 5 through 8 is not feasible, "end-of-pipe" control is often the only remaining option. The subjects of electrical and thermal treatment technologies in this chapter are recognized end-of-pipe control options that offer a wide range of flexibility, performance, and cost-effectiveness in the treatment of a variety of contaminated media and wastes. Selecting and applying a treatment or control technology for a given application involves a process that includes defining the needs and constraints of the application, identifying technologies that may satisfy those needs and constraints, and determining which of those technologies are economically feasible compared with other options.

The goal of Part I of this chapter is to introduce the reader to the basic steps in the process of identifying and selecting treatment and disposal technologies for specific applications and provide an overview of a variety of commercially available thermal and electrical treatment technologies. The reader may find the information in this chapter useful for developing a general understanding of these technologies and their applications, or may use it as a starting point in the process of planning and conducting a comprehensive technical and economic feasibility assessment and technology selection for a specific application. Details of operating theory and design are not covered in this chapter. The reader is directed to other texts for this level of detail.

Part I of this chapter is organized into four major sections. The first section discusses the general mechanisms of thermal and electrical treatment technologies and their applicability in the treatment of various types of media. Information in this section will help the reader determine if electrical or thermal technologies may be applicable to a specific application of interest.

The second section discusses the type of information about a specific application that is typically necessary to compile before proceeding with assessment of the technical and economic feasibility of potentially applicable technologies.

The third section discusses the basics of how the individual components of the technical and economic feasibility assessment are prepared and used.

The final section presents summaries of information for a variety of thermal and electrical technologies, such as operating principals, physical characteristics, operational constraints, cost factors, and so forth. The type of information provided in this section should acquaint the reader with these technologies and assist the reader in developing preliminary assessments of the applicability of a variety of technologies to a specific application.

Applicability of Thermal and Electrical Technologies

Thermal or electrical technologies can eliminate or reduce the undesirable characteristics (e.g., toxicity, flammability, reactivity, infectivity) associated with organic, inorganic and biological contaminants in a wide range of solid, liquid, or gaseous media or wastes. The types of treatment achievable with these technologies range from the removal of relatively low concentrations of contaminants from media such as air, water, or soil, to the destruction or necessary pretreatment prior to disposal of bulk waste materials.

The necessity for these technologies springs from their unique ability to do things such as the following:

- Destroy or encapsulate toxic or infectious materials
- Prevent reuse of products and materials for safety or other reasons
- Allow disposal of materials in a landfill that would otherwise not have been allowed
- Conserve landfill space by reducing the volume of wastes
- Recover energy
- Avoid or reduce disposal costs

The general applicability of a given electrical or thermal technology to the treatment of a specific contaminated medium or disposal of a specific waste stream is ascertained by matching the basic chemical and physical nature of the contaminants and medium to the ability of the mechanism of the given treatment technology to achieve the desired action (e.g., destruction, capture, etc.) on that type of contaminant.

Characteristics of contaminated media and wastes that are key to the selection of treatment technologies are discussed in the following section.

The specific applicability of and actual selection of a treatment or disposal technology for a given contaminated medium or waste involves numerous additional technical, economic, and regulatory considerations.

Categories of Wastes and Contaminated Media

For the purposes of determining the applicability of specific thermal and electrical technologies, contaminated media and wastes may be grouped based on whether they are primarily organic or inorganic, and further by physical state (i.e., solid, liquid, or gas, including particulate matter). Table 9.1 provides examples of the major categories of wastes and contaminated media that are treatable with thermal or electrical technologies. The table includes the type (organic, inorganic, or infectious) of contaminant, the phase or phases of the waste category, and examples of specific media or wastes fitting each category. Following the table is a discussion of the characteristics of the contaminant categories.

Organic Contaminants and Wastes

Organic contaminants and wastes are compounds primarily composed of carbon, hydrogen, and potentially other elements. Gas or liquid-phase organic contaminants may include volatile organic compounds (VOCs) in the form of solvents, chemical precursors and intermediates, petroleum compounds, etc. Solid organic compounds may include sludges, still bottoms, resins, chemicals, waxes, paper, plastic, wood, foodstuffs, etc. Because the undesirable properties of organic compounds (e.g., toxicity, flammability, etc.) almost always arise from the structure of the compounds rather than the basic elements comprising them, organic compounds may be eliminated from contaminated media and destroyed in bulk by mechanisms that alter their structure such as oxidation or thermal decomposition. Under the right conditions, gas phase organic compounds may also be removed from gaseous media by condensation at reduced temperature or collection by electrostatic filtration.

Inorganic Contaminants and Wastes

Inorganic compounds may consist of heavy metals and toxic elements (e.g., lead, mercury, chromium, arsenic, etc.) in pure form or combined with other elements. These compounds exist primarily in the solid phase, but can also exist at ambient temperatures in the gas phase if finely divided or have a high vapor pressure (e.g., mercury, hydrogen sulfide) or the liquid phase if water soluble. In general the undesirable (e.g., toxic or hazardous) nature of these compounds stems from the element(s) they contain rather than their structure. Therefore, the undesirable in most inorganic compounds cannot be eliminated by oxidation or thermal decomposition of those compounds. Notable exceptions include compounds such as hydrogen sulfide (H_2S) and ammonia (NH_3), which when oxidized become far less toxic.

TABLE 9.1
Examples of Media and Wastes Amenable to Treatment with Thermal and Electrical Technologies

Waste or Contaminated Media	*Contaminants*	*Phase*	*Examples*
Municipal solid waste (MSW)	Organic Inorganic	S	Trash (paper, plastics, metals, glass, food wastes, etc.) from residences and commercial establishments.
Industrial wastes (nonhazardous)	Organic Inorganic	S, L	Process wastes (e.g., plastic, sawdust, wood, contaminated raw materials, etc.)
Medical wastes	Infectious	S, L	Redbag wastes, pathological wastes, anatomical remains, etc.
Special wastes	Organic Inorganic	S, L	Off-specification and returned products, foods, pharmaceuticals, chemotherapy wastes, agricultural wastes, sewage sludge, tires, contraband, currency, etc.
Hazardous wastes (as defined under the Resource Conservation and Recovery Act)	Organic Inorganic	S, L	Toxic, ignitable and listed organic and metals-bearing solids and sludges, discarded chemicals, propellants explosives and pyrotechnics, etc.
Contaminated environmental media	Organic Inorganic	S	Soils, debris and residues from remedial actions contaminated with a range of petroleum, toxic, ignitable or listed materials
Other solid media	Infectious	S, L	Mail, equipment, supplies, food, water, livestock etc.
Gaseous wastes and contaminated gaseous media	Organic Inorganic Infectious	G	Vent streams from painting and coating operations, reactors, tanks, ovens, pumps, condensers, water treatment systems, odors from sewage collection and treatment systems, etc. Exhaust streams from materials handling operations, boilers, dryers, incinerators, etc.

The treatment of solid inorganic compounds by electrical or thermal means is most often accomplished by binding hazardous inorganic constituents into a noncrystalline, glass-like product that is resistant to leaching. Under the right conditions, inorganic particulate matter may also be removed from gaseous media by electrostatic filtration. Aqueous solutions of inorganic compounds may be treated to reduce their volume and weight by evaporation.

Infectious Contaminants and Wastes
Infectious contaminants may be living or dormant forms of bacteria, viruses, or other organisms and biological agents capable of causing illness and death in plants, animals, and humans. Such contaminants may arise from the process of diagnosis and treatment of humans and animals, outbreaks in livestock and food crops, the use of biological weapons, etc. Most organisms are relatively intolerant to heat, and may be deactivated or destroyed by exposing them to elevated temperatures for an appropriate period of time. Alternatively, organisms may also be deactivated (rendered unable to reproduce or properly function) in shorter periods of time by exposing them to high fluxes of ionizing radiation. This approach may often be conducted with no appreciable effect to the media being decontaminated, making it useful for treatment of mail, foodstuffs, etc.

Treatment and Disposal Requirements

Another essential consideration in determining the applicability of a given treatment or disposal technology to a specific contaminated medium or waste is the required or desired performance of the technology. In this sense, performance means the degree to which contaminants have been removed from media or rendered unleachable, waste materials have been destroyed, the population of infectious organisms has been reduced, etc. In many cases the performance a technology must achieve is directly tied to federal, state, or local environmental regulations that apply to the treatment or disposal of the contaminated media or wastes.

In other cases the treatment of contaminants or disposal of wastes does not involve regulatory performance requirements, but is driven by considerations including economics, trade secrets, and safety.

In the United States, performance is often driven by the following regulations:

- The Clean Air Act (CAA)
- The Resource Conservation and Recovery Act (RCRA)
- The Toxic Substances Control Act (TSCA)
- The Comprehensive Environmental Response, Compensation, and Liability Act (CERCLA)

- State/local rules pertaining to:
  - Municipal solid waste (MSW)
  - Industrial wastes (nonhazardous)
  - Medical wastes
  - Special wastes

In order to establish the regulatory treatment and disposal requirements for the contaminant or waste stream, it is necessary to characterize the materials involved and research the potentially applicable regulations to determine which, if any, include standards that must be followed for the project being undertaken. At a minimum, this task requires knowledge of the following for the contaminant or waste stream:

- Identities of the contaminants and wastes to be treated
- Composition (identities of components of a mixture, chemical composition, metals content, flammability, etc.)
- Physical state (solid, liquid or gas)
- Source of media (e.g., what industrial process it originates from, etc.)
- Location of source

The above information must be compared with standards and requirements in the potentially applicable regulations with due consideration to the regulatory status of the location where the treatment or disposal process is to be installed. Consideration of the regulatory status of the location in which the treatment or disposal process is proposed to be installed is important because the nature of the existing facility and its environmental permits can affect the applicability of regulatory requirements on new treatment of disposal processes.

Treatment and Disposal Technologies

Technologies for treatment of the contaminated media and disposal of wastes considered in this chapter rely on the direct application of thermal or electrical energy (including electrically generated ionizing radiation) to remove inorganic and organic compounds from contaminated media through destruction or collection, alter the physical properties of waste through stabilization to prevent the leaching of pollutants, dispose of wastes by physical or chemical change through a variety of mechanisms, or kill or deactivate infectious organisms using thermal energy or ionizing radiation.

Treatment and disposal mechanisms include electrostatic capture or solidification of inorganic pollutants, destruction of organic compounds by oxidation or pyrolysis, collection of organic compounds by condensation, and deactivation or destruction of infectious organisms with heat or ionizing radiation, etc. Examples of these technologies are presented in Table 9.2

TABLE 9.2
Applicability of Electrical and Thermal Treatment Technologies

Technology	*Method*	*Type*		*Contaminants*				*Phase*	
		Contaminated Media	*Wastes*	*Organic*	*Inorganic*	*Infectious*	*Solid*	*Liquid*	*Gas*
Oxidation thermal	Destruction	x	x	x		x	x	x	x
Regenerative thermal Oxidation	Destruction	x		x					x
Catalytic oxidation	Destruction	x		x					x
Pyrolysis or reforming	Destruction	x	x	x		x	x		
Thermal desorption	Collection	x		x			x		
Autoclaving	Deactivation	x	x			x	x	x	
Microwave irradiation	Deactivation, destruction	x		x		x	x	x	x
Plasma	Destruction, stabilization		x	x	x	x	x	x	
Glass furnace	Destruction, stabilization	x	x		x		x		
Evaporation	Concentration	x			x			x	
Condensation	Collection	x		x					x
Electrostatic precipitation	Collection	x		x	x				x
E Beam irradiation	Deactivation, destruction	x		x		x	x	x	x
UV irradiation, photo oxidation	Deactivation, Destruction	x		x		x		x	x
X ray irradiation	Deactivation	x	x			x	x	x	x

along with the types of media, contaminants, and phases to which they are generally applicable.

Selection of a Treatment and Disposal Technology for a Specific Application

Once potentially applicable treatment or control technologies have been identified for a particular application, the technical and economic feasibility of each technology identified must be assessed and compared before a final selection can be made.

Development of a process flow diagram (PFD) for the pollutant or waste streams associated with the application is essential in order to identify and perform the detailed evaluation of technical feasibility of commercially available equipment embodying the technology of interest. Application data typically incorporated into a PFD include the following (minimum, maximum and average):

- Flow rate of the compounds comprising the media to be treated
- Temperature and pressure
- Heating value, moisture content
- Duty cycle of the process generating the media
- Trace contaminants of potential concern

The following information about the plant or location where the technology is to be installed or operated is also essential to the assessment of the technical and economic feasibility of specific commercially-available equipment:

- Available utilities (gas, electricity, water, wastewater treatment, compressed air, steam, etc.)
- Type and cost of available solid and hazardous waste disposal options
- Space, shape, weight or other physical constraints for installed equipment
- Ability of existing workforce to support the operation and maintenance
- The type and extent of siting studies required, if any
- The type and extent of zoning and other local approvals and permits required
- The types and extents of environmental permits (or permit modifications) required

The above process and plant data are typically provided to equipment manufacturers in discussions of the applicability and selection or design of available equipment and ultimately for solicitation of bids for treatment or disposal equipment. Once the equipment vendors have provided preliminary cost estimates and specifications for their equipment, the process of estimating and comparing capital and operating costs can commence.

Cost Estimation Methodology

Cost is typically the factor that ultimately determines which specific technology or commercially available equipment is selected. Costs of the many of the technologies discussed in Part I of this chapter may be quantified and compared using the general method described in EPA's OAQPS Control Cost Manual (EPA/452/B-02-001). The method in this reference involves reducing estimated total capital investment and direct operating costs to a total annual cost, which facilitates making cost comparisons among technologies.

Review Questions

1. Name at least four fundamental reasons why the decision to select an electrical or thermal means of treatment or disposal may be made.
2. What basic information must be compiled about a specific application in order to conduct a technical and economic feasibility study of electrical or thermal technologies?
3. How does the chemical composition of a waste or contaminant affect the applicability of the various electrical and thermal technologies discussed?
4. What are the primary environmental regulations applicable to the technologies discussed?
5. Name the four basic categories of costs that must be estimated in order to compare the economics of treatment and disposal technologies, and give an example of each.

Alternative Energy Solutions

Nelson Leonard Nemerow

Part 2

Introduction

Since the current use of fossil fuels (coal and oil) and nuclear fuels are either too costly or arise from nonrenewable fuel resources or result in too great an adverse environmental impact, the search goes on for renewable, non-polluting and economical alternative energy sources. In this chapter I present potential alternate solutions to the use of fossil and nuclear fuels and the adverse environmental effects they create. The reader is urged to attempt to use any of these suggested alternatives in applications suitable to their particular situations.

Alternative Energy Sources

The following six fuel sources are suggested as alternatives to fossil and nuclear fuels for producing electricity with little or no adverse environmental impacts:

- Hydrogen fuels
- Wind energy
- Solar energy
- Wave energy
- Geothermal energy
- Other electrical energy sources

Each of the above six sources of energy should be considered in solving environmental problems when it is desired to conserve and diminish the polluting effects of other nonrenewable fuels.

Hydrogen Fuels

Although these fuels are sufficiently important and currently in use to warrant a complete chapter (Chapter 7), they are mentioned briefly here to make certain that the reader considers their potential to the fullest. Hydrogen is produced today primarily from the electrolysis of water into its separate constituents of hydrogen and oxygen. In this case some conventional electric fossil fuel power is required. But theoretically hydrogen can also be produced in a number of other ways (see Chapter 7). It can also be produced biologically from anaerobic decomposition of various agricultural wastes (see below).

The production of hydrogen by electrical disassociation of water is shown in Figure 9.1 (directly from the presentation by Sherif, Veziroglu, and Barbir in Chapter 7).

The most recent use of hydrogen fuel as an alternative to fossil fuel (gasoline) is in automobiles. The hybrid cars—partly fueled by hydrogen—have been quite successful. Automobiles completely powered by hydrogen are also forecasted for the near future—within the next few years. One maker is even proposing a hydrogen fueled auto with the hydrogen fuel tank installed in the rear seat area of the automobile to avoid the fuel delivery and loading problem. This would represent a highly desirable alternative solution to the combustion gas, air pollution problem caused by gasoline fueled autos.

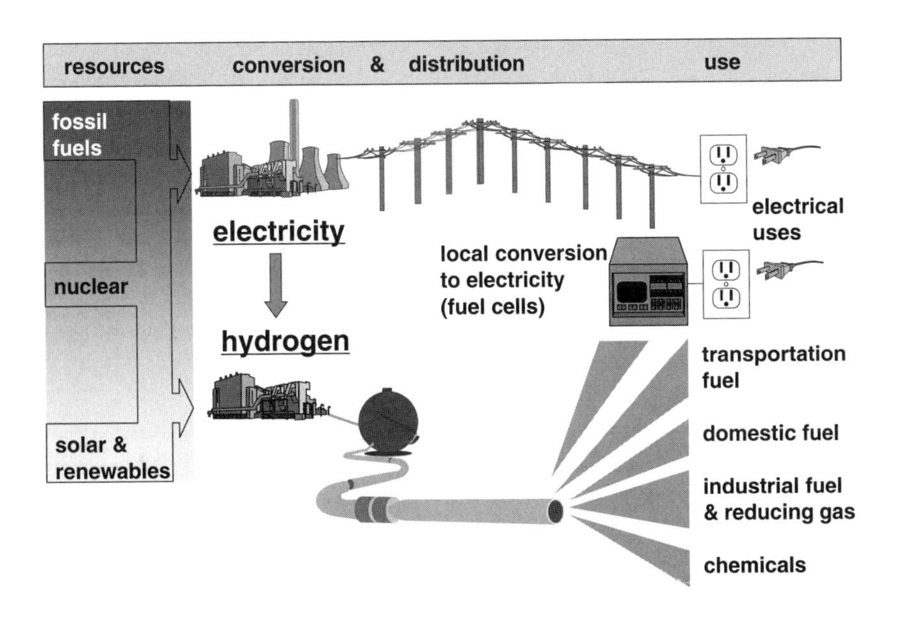

FIGURE 9.1. Hydrogen Production.

Wind Generated Power

By the year 2004 more than 13,000 megawatts of wind power had been installed in the world. California alone had 1,600 megawatts of wind power in use to provide enough electricity for over 750,000 homes. Windfarms—a collection of individual windmills at one location—have been increasing to a point where the U.S. Department of Energy predicts that wind power costs will drop to 2 cents per kilowatt-hour from the current value of 3 to 6 cents.

Wind Production

Wind farms use large blades to catch the wind and turn rotors to produce electricity. A modern wind farm may contain as many as 500 wind turbines connected to a transmission grid. They produce electricity much the same as steam engines use steam to turn rotors of generators to produce electricity (except that in this case wind instead of steam does the work).

A wind speed production of at least 12 miles per hour is usually required to produce electricity. When the wind is not blowing at this speed, no electricity is produced, and the main transmission line is not being supplemented with wind-generated electricity. However, as soon as the wind speed picks up sufficiently, the wind electricity begins again to supplement that in the main transmission line. This creates no problems, since wind energy usually represents a small (2–4%) of the total power being transmitted.

Environmental Impacts of Wind Power

Wind plants create no air pollution, nor use or waste any water, nor despoil the land, but they may produce some environmental effects on vision, audio, and wildlife.

Visual

Since wind farms contain so many turbines which are mounted on top of tall towers (some 350 feet), they often are visible for a very great distance away from the farm. Some people object to the sight of large wind farms just as some people object to crowded buildings in a city. However, as these farms become more prevalent and as people get used to seeing them they may become more acceptable to the majority of people in the area. In fact, some people may even claim that their appearance is desirable from an artistic standpoint. Modern planners and architects face the challenge of designing these farms so as to make them desirable for the surrounding humans.

Audio

Other people may be concerned about the noise created by the wind farms. That noise may be mechanical or aerodynamic in nature.

Mechanical noise is produced by parts rubbing against or hitting other parts and has virually disappeared in the newer designed rotors. The aerodynamic noise which is that swishing sound emitted as the blades pass the tower can be masked by proper use of sound barrier construction. Some people even claim that the swishing sound is rather soothing, similar to that generated by ocean waves hitting the shore.

Wildlife Effects

Bird populations can be threatened by wind farms. The ground below the mills is disturbed during construction—which in turn attracts mice, prairie dogs, and burrowing animals. These in turn attract raptors such as hawks and eagles that prey on them. These birds may perch on top of the wind generators for reasons of hunting for prey and often get caught in the spinning blades. Lately the farms are designed to contain tubular towers to prevent the birds from perching on them. In addition, they turn more slowly than those of earlier design. When compared to other industries such as mining and coal combustion, the environmental effects on birds are much less.

The use of wind farms is greatly enhanced by the realization of farmers of the value of their land. They have found that by leasing the land for the wind farms, they can overcome the increasing costs of fuel for cultivating crops on the land.

Solar Energy Power

The science of converting the suns' energy into electricity is often referred to as photovoltaic science. Although the concept was known since Edmond Becquerel discovered it in 1839, it was not until 1954 when the first photovoltaic cell was created at Bell Labs. These cells are semiconductor devices that convert light directly into electricity. They are generally made of silicon with traces of other elements. Although photovolaic (PV) cells are quite sophisticated in design, they are very simple to use. The PV cells are mainly low voltage DC devices with no moving or wearing parts. Once they are installed no maintenance outside of an occasional cleaning is required. However, most of these systems do include battery (storage) and need some water just like automobile batteries.

PV Cells

These cells (solar) consist of layers of semiconductor materials with various electronic properties. The bulk of the cell is usually silicon based along with a minor amount of boron to lend a positive electrical charge to it. A thin layer on the front of the cell is painted with phosphorous to render the other part of the cell negatively. The union between the two cell layers will then contain an electric field at the junction. When the photons of daylight hit the solar cell, some of these photons are absorbed at the junction which then free the electrons of the silicon material. If and when the photons possess enough energy, the electrons will be able to move through the silicon and into an external circuit. They give up their energy as they flow through the external circuit and result in producing electricity to power all kinds of small and often larger electronic devices before returning to the solar cell. The entire process is solid state with no moving parts and no materials released or consumed in the process. Despite what many people believe these solar cells work better in colder weather than in hot—mainly because they produce electricity from light not heat. As long as sunlight is reduced no more that 20% of full sun, sufficient photons will be released to operate under these partly cloudy conditions.

Benefits of Solar Powered Electricity

Some of the many benefits of using solar energy to replace fossil or nuclear-fueled sources are:

- The fixed costs of operation remain constant for the life of the system
- The solar system is independent from any other source of fuel
- Several solar cells can be joined to increase the output capacity of the system.
- No noise occurs from operating the system
- No so-called greenhouse gases are evolved as is the case with fossil fuels
- The solar system possesses a long operating life
- Cost of operation and maintenance are competitive with other energy types
- Because of its nonpolluting nature solar cells are widely known as "clean and green"

Some Environmentally Related Uses of Solar Cells

Because so many people in developing countries have little or no access to electricity, these solar cells are predicted to compete more favorably with conventional sources of power. Some of the uses which normally you might not consider for solar-derived electricity are the following, which also are environmentally enhancing.

- Powering weather stations to provide dependable, economical electricity
- Powering pumps to transmit water from remote reservoirs and lakes or from rivers which are not readily accessible
- Telecommunication of river water stages and even magnetic detection of earthquakes causing tsunami effects in oceans
- Electricity to remote homes such as vacation or seldom used buildings
- Nuclear power radiation detection systems
- Portable light and electric systems

- Remote lagoon aeration systems for waste treatment
- Disaster and all types of civil defense warning systems
- Corrosion systems for pipes (cathodic protection)
- Remote charging of batteries in hybrid and other autos
- Powering difficult to access air pollution sampling and analysis stations

The reader is urged to think of his/her own uses of solar generated electricity that may also benefit the environment and serve as an alternative to fossil-fueled electricity.

Geothermal Energy

Geothermal energy is obtained from heated water, steam, or soil which is derived from deep in certain land masses. There are two main uses for this energy: (1) hot water is used to create electricity or to provide hot water heating or warming; (2) the thermal mass of the soil or groundwater is used to drive heat pumps which provide either heating or cooling. The first use is more widely known and used and is obtained from geothermal geysers that find their way to the earth's crust.

The above uses are not really from renewable resources, however, with properly calculated use they can almost approach the "renewable" classification. The heated water, steam, or soil will gradually be depleted if overdrawn from the ground. This valuable ground resource will slowly regenerate itself over time so that if the withdrawal at the surface is timed to match that regeneration rate, the resource will be considered renewable. In any event, it will not deplete itself as fast as fossil or oil fuels are depleted by normal mining techniques. In addition, heat reservoirs are considered immense in magnitude compared to its current or even projected use, thus rendering it practically renewable.

In the United States the production of electricity from the geothermal energy of the earth's interior heat is centered in northern California. Here these geothermal sources provided just over 7% of California's electricity in the 15-year period ending the 20th century. However, the geyser production has decreased from supplying about 2,000 megawatts in 1989 to 1,100 megawatts near the turn of the century. Unfortunately, because of the specific location of these geothermal fields most individual households cannot make use of this energy. However, direct use of the heated water can save establishments as much as 80% in their fuel bills.

Geothermal Ground Source Heat Pumps for Residential Use

Heat pumps can reduce both air conditioning peak loads as well as winter heating loads, In addition, they are normally used to heat water (or as hot water) in households and buildings.

Economics of Geothermal Energy

Geothermal electricity can be produced practically and economically for about 5 cents per kilowatt-hour—slightly higher than wind or solar energy. This higher cost is largely due to the fact that it is necessary to drill deeper today to produce a given amount of power than in earlier years. It has been suggested (and even used) that the economics of geothermal power can be improved through co-production of other goods from high-temperature brine extracted from the depths of the ground. While geothermal power applications require more advances in exploration and drilling, heat pump, direct uses require that the engineer and the consumer understand the technology. It may be more expensive to install geothermal energy systems at the start, but over the long term the benefits may make it economically and environmentally worthwhile.

Effects on the Environment

Air pollution relative to conventional fossil fuel energy production will be minimized when selecting geothermal energy instead. It produces only about one-sixth of the CO_2 and none of the NOxs or sulfur gases that fossil fuel plants emit. For these reasons alone this method of energy production can be a very environmentally friendly alternative to fossil fuel energy.

Amount of This Energy Already Being Produced

In 1998 geothermal energy provided 0.4% of the electricity generated in the United States. This amounted to 14.3 billion kilowatts of electricity to over 1,400,000 homes. At that time it was growing at a rate of slightly less than 3% over an 8-year period. Worldwide, geothermal energy was slightly more than 8 million kilowatts or about 3% of the 3,180 kilowatts used.

Some Examples of Geothermal Energy Uses

The Oregon Institute of Technology has been heated by the direct geothermal energy since 1964. In Iceland geothermal energy is used to provide the majority of households with residential heat. Tax neutrality, continued and increased federal funding, continued and expanded production tax credits, resource identification, renewable portfolio standards, contractor education, and the issuing of air emission standards have been and are being used to encourage the continued use of geothermal energy.

The reader of this section of the chapter is urged to consult the U.S. Department of Energy's Web site for more information of geothermal energy. In addition, the Renewable Energy Policy Project maintains a rather detailed bibliography of the uses of this form of energy. It is located at 1612 K Street, N.W., Suite 202, Washington, DC, 20006.

Ocean Wave Energy

How Wave Electric Energy Is Created

The entire earths' surface including the ocean is heated by the sun. This creates wind that pushes against the surface of the ocean and forms waves. Waves can travel hundreds and thousands of miles from the beginning of their propagation. They are being continuously supplemented by new winds. These waves keep their energy long after the winds that created them have abated. These same waves represent one of the most concentrated and consistent sources of renewable energy. When compared to conventional fossil fuel generation, wave energy provides the increased advantage of a limitless free supply of energy along with a total lack of environmentally polluting emissions.

However, even today there appears to be no agreement among professionals as to the most efficient technological approach to the use of wave energy.

Types of Wave Energy Conversion Systems

The kinetic energy of waves may be converted into electrical energy mainly by four different systems.

1. **Tapered channel systems** that funnel incoming waves into shoreline reservoirs that raise the water above sea level. The head of water then is directed down through a turbine which then drives a generator producing electric energy.
2. **Float systems** consisting of buoys which sit out on the ocean's surface. As the ocean rises and falls, the relative motion between the float and the ocean floor drives hydraulic pumps or pistons. This kinetic energy is also used to drive a turbine and a generator producing electricity.
3. **Oscillating water column systems** are fixed in place and are devices in which waves enter the column and force air up past a turbine. As the wave retreats, the air pressure drops resulting in the turning of a turbine which once again drives a generator and produces electric energy. The first of this type of system was produced in Japan to power a light on a buoy used for navigation.
4. **Underwater turbines** collect and contain the movement of the ocean's currents and utilize this energy to drive slow-moving blades. These in turn drive a generator directly—similar to an above ground windmill—to produce electricity.

Recent Advances

Many of today's professionals as well as equipment manufacturers feel that the time is now here for the era of wave energy usage to accelerate. Recent technological advances have progressed sufficiently to make this form of energy cost-effective when compared to fossil fueled power. These advances include those of marine engineering that have come from the offshore drilling industry, which provide ocean-tested "off the shelf" components at reasonable prices. Also the cost of electronic control devices that optimize the efficiency of the technology has been reduced.

Environmental Considerations

Wave generated electrical energy is a source of clean, renewable energy and does not produce any objectionable greenhouse gases. When selecting this method to produce electrical energy, one must also be aware of certain disadvantages, which may hamper their acceptance. First, the sea is unpredictable at best—and devastating at worst (such as when the tsunami hit the Indian Ocean and South Pacific in 2004). Under these latter situations the facilities must be designed to be able to withstand pressures many times the normal wave pressures. Second, these wave generating systems may cause alterations to shore lines and local ecosystems. And third, the electricity produced will vary because of the variability of the waves.

Generally, the average wave power level should be above 15 kw per meter in order to generate wave energy at competitive prices.

Other Electric Generated Systems

Agricultural residues, farm animal wastes, human sewage sludges and other biomasses can be fermented to produce combustible gases such as methane. The gas can be burned directly in a boiler to convert water to steam which then can drive a turbine connected to a generator to produce electrical energy. It is not the intent to describe these systems at this point in this chapter. However, the reader is encouraged to refer to scientific descriptions presented in Chapter 6 by Dr. Oether and the practice of using these systems presented in Chapter 11 by this author.

Environmental Considerations

Wherever one has a situation with an ample supply of these biomasses available, it is desirable to consider them as an alternate energy source. Not only does one produce a valuable energy resource at a competitive cost, but also one rids the environment of a source of waste causing adverse environmental effects. In these systems, bacteria do the work required in digesting the organic matter of these wastes to free CH_4 (methane). Bacteria require no compensation for this work, but they do require proper design and operation of equipment. Some of the latter may be obtained by referring to Chapter 5 by Dr. Tewari.

Review Questions

1. List and describe six alternate energy systems.
2. Under what circumstances would you recommend the use of wind generated energy, solar, geothermal, and wave energies.
3. What common characteristic is inherent in all of these alternate systems that make them desirable as energy sources?
4. What is needed to make biomass conversion into useful energy a practical solution for the environmental dilemma?
5. What practical considerations have hindered the acceptance of the use of wave energy systems; of solar energy systems?

CHAPTER 10

Environmental Health Solutions

Edwin M. Kilbourne and Henry Falk

Definitions and Conceptual Approach

Definitions

The term ***environmental health*** refers to the **science, techniques, and professional practice involved in protecting people from adverse health effects caused by exposure to their surroundings**. In this chapter, we refer to the pathogenic or noxious features of the environment as ***environmental agents***.

Because it is an activity oriented toward the health of groups of people or of whole populations, ***environmental health*** is a discipline of ***public health***. Accordingly, we frequently use the term ***environmental public health*** as a synonym for ***environmental health***.

Scope

The scope of "environment" applied to human beings is enormous, since all aspects of a person's surroundings may be considered. "Environment" is somewhat more limited when applied to human health, because it comprises only environmental influences that ***limit or potentially limit the length or quality of life*** or that ***cause or might cause actual clinical disease***. Risk factors and early indicators of illness are also part of the broad field of environmental health. Large numbers of known and potential environmental agents cause myriad adverse effects on health. Prevention and regulation are correspondingly complex.

The field of environmental health is oriented toward prevention. For example, there is intensive current research in the area of ***biomarkers*** which allow the detection of early biological effects of exposure prior to occurrence of actual disease. In fact, the potential health effects of exposures that have not yet even been associated with human disease receive substantial scrutiny in environmental health research and practice. Because of this conservative, preventive approach, major epidemics involving acute clinical disease due to environmental agents have been relatively infrequent in the United States in recent decades.

Due to its enormous scope, environmental health is frequently subdivided in various ways. Concentrating on one or a few aspects of the field gives greater focus to such activities as training, scientific investigation, medical practice, and public health interventions. Dividing environmental health can be done usefully in at least four ways.

By agent—Groups of similar or related environmental agents
By environmental medium—Air, water, soil, food, etc.
By type of health impact—Categories linking similar diseases or other adverse health effects
Place—Locations defined by social context rather than by geography in which certain hazardous exposures may occur

Etiologic Agents

These four modes of conceptual subdivision (agent, medium, outcome, and place) have differing utilities depending on (1) the specific environmental issue under consideration and (2) the orientations and expertise of the groups addressing the issue. Scientists and technicians who evaluate exposures necessarily focus on the agent, and their fields of training and specialized knowledge may be exclusively dedicated to a particular class of agent. Environmental agents can be classified as biological, chemical, or physical. (An additional category of "complex" agents may also be required.) A biological agent (for example, a bacterium) may need evaluation by a microbiologist, and a potentially harmful chemical will more likely require the attention of a chemist or toxicologist. In contrast, an exposure involving a radionuclide may involve specialists in health physics and/or radiobiology.

Biological environmental agents are living organisms that cause disease by infection. That is, they grow and multiply within the human body. These organisms include pathogenic microorganisms (for example, viruses and bacteria), as well as protozoan and metazoan parasites. Other species that might injure people (e.g., feral predators) are generally excluded from consideration, and we shall not consider them here.

Vectors are not themselves biological environmental agents. Rather, they are the fauna and flora that may transmit the pathogenic agents described above. Nevertheless, vectors require both study and attention, since vector control may be the key intervention for managing (diminishing) the risk from biological agents. Vectors include mosquitoes (that may carry the malaria parasite or encephalitis-causing arboviruses), other insects and arthropods (e.g., *Ixodes* spp ticks that may carry the spirochete responsible for Lyme disease, birds (a reservoir for West Nile virus), monkeys (a reservoir for the flavivirus that causes yellow fever), and snails (the host for one stage in the multistage life cycle of schistosome parasites that also infect humans).

Chemical environmental agents may cause gross structural damage to body components (as do strong acids and alkalis) or may adversely affect vital body functions at the molecular level by interfering with important physiological or metabolic processes. Taken as a whole, chemical agents are sometimes referred to as **xenobiotics** (literally, that which is alien to the body), but xenobiotics are frequently also divided into groups according to their origins.

Toxins are chemical agents that arise from living organisms. We classify toxins as chemical (rather than biological) agents because they cause adverse effects even if there is no infection by the responsible organism. On the other hand, **toxicants** (e.g., arsenic, lead) are not elaborated by living organisms. They may be natural substances (for example, lead mine tailings) brought into contact with people by human activities. Alternatively, they may be chemical compounds that are not found (or not found in abundance) in nature and which arise from such human activities as industrial-scale chemical synthesis (e.g., petrochemical production).

At one time or another, many readers of this book have probably suffered from toxin-mediated disease because of exposure to ***staphylococcal enterotoxin***. When certain strains of staphylococci are introduced into poorly refrigerated or lukewarm food, they grow and produce this toxin. When the food is eaten, nausea, vomiting, and gastrointestinal distress occur, and the symptoms are frequently severe. This condition was formerly known as "ptomaine."

Paralytic shellfish poisoning (PSP) is another example of a toxin-mediated disease. When certain dinoflagellates (especially *Gymnodinium* spp) "bloom" (occur in large numbers), a so-called "red tide" results. Bivalve mollusks filter the dinoflagellates and concentrate various toxins, collectively referred to as ***saxitoxin***. If these contaminated shellfish are ingested, they cause PSP. In its mildest form PSP causes tingling and numbness around the mouth and in the extremities and may be accompanied by nausea and vomiting. In its full-blown form, PSP causes complete muscle paralysis. Without intensive care, victims may die from the inability to breathe due to weakness or paralysis of the respiratory muscles.

Cyanide (as inhaled hydrogen cyanide gas or as an ingested cyanide salt) is a well-known and common toxicant with powerful acute effects. Relatively small amounts rapidly diffuse into cells, interfering with the biochemical reactions by which the body combines oxygen with its cellular fuels to create energy. As a result, cells depend on glycolysis as the major pathway for energy production, which results in production of lactic acid and thus, systemic acidosis and other serious metabolic derangements.

Cyanide exposure may be rapidly lethal, unless the dose is very small or unless an appropriate antidote is administered within minutes of intoxication.

Physical environmental agents are a highly diverse group. Any physical property or manifestation of matter and energy may be a physical "agent," if it has an impact on health. Among physical agents, ionizing radiation is of major concern. Large doses of ionizing radiation cause acute radiation sickness, but low doses may also cause serious illnesses, such as cancer, many months or (more often) years after an exposure. Although ionizing radiation is a major concern, any manifestation of energy, force, or any physical quality of matter may act as an agent. Microwaves, ultraviolet radiation, visible light, sound, heat, and cold may all act as environmental agents of disease. Other physical agents, including electrical and magnetic fields and infrasonic vibration, are suspected of causing adverse health effects.

Recent developments in environmental health suggest the need for an additional agent class that we shall call "***complex agent***." An example of such an agent is the human-made or "built" environment. Such a term is inadequately characterized by the terms biological, chemical, and physical. Evaluation of the built environment has revealed the potential adverse health impact of urban and suburban sprawl. This problem is attributable to more than the simple physical form, shape, or functional characteristics of inhabited spaces. Those raising this concern have argued that the make-up of one's living area, particularly the area outdoors, in addition to the commuting lifestyle associated with sprawl, probably have an impact in determining certain behaviors, some of which may influence health. These behaviors, in turn, affect social context and other related habits (Frumkin et al. 2004).

Examples of specific biological, chemical, and physical environmental agents of major concern are listed in Table 10.1.

Environmental Media

Virtually all agents of concern in environmental health are associated with one or more environmental media. Understandably, the media of importance are those with which people are most frequently in contact. They are the air, water (both potable and nonpotable), and various solids. However, some environmental agents do not occur in environmental media. The various forms of energy (e.g., ionizing radiation and environmental heat) that may adversely influence health are not necessarily associated with an environmental medium. Thus, classification of environmental health issues by medium is useful, but it is not a comprehensive means of subdividing environmental health activities.

Most would include food as a part of the human environment. Food, however, is a special case, because of the potentially very strong health

TABLE 10.1
Classes and Examples of Environmental Agents of Major Concern

Biological
- Vector borne infectious agents
- Enteric agents

Chemical
- Allergens & sensitizers
  - Chemical sensitizers (organic and inorganic)
    - Beryllium
    - Toluene di-isocyanate
  - Grain dusts
- Toxins
  - Enterotoxins
  - Plant toxins
  - Marine toxins
  - Mycotoxins
  - Reptile venoms
- Toxicants
  - Heavy metals
  - Persistent organic pollutants
  - Volatile liquids and solvents (halogenated and non-halogenated)
  - Oxides of Nitrogen and SO2
  - Fumigants and other toxic gases
  - Pesticides
  - Asphyxiants
  - Polycyclic Aromatics
  - Oxidizers, strong acids, strong bases
  - Other caustics

Physical
- Ionizing radiation
- Ultraviolet radiation
- Electro-magnetic fields
- Environmental temperature (heat, cold)
- Vibration and sound

impact of food exposures. Accordingly, federal agencies in the United States, including agencies that do not have responsibilities for other environmental media (the Food and Drug Administration [FDA] and the Department of Agriculture [USDA]), implement protective regulations for food. Both food and drinking water are of particular concern because the act of ingesting them brings them directly into contact with the gastrointestinal tract. Most gastrointestinal surfaces are lined with cells specifically designed to absorb the chemical components of what is eaten. Contaminated food and drinking water thus represent particularly high-risk items.

There are, however, important differences requiring different approaches to drinking water and foods. Potable water is (ideally) a pure compound with a benign toxicological profile. Foods, however, are complex mixtures. Many foods, both "organic" and processed, include substances that are toxic if eaten in large quantities. These substances include methanol (so-called "wood alcohol"), solanine (a glycoalkaloid that occurs naturally in potatoes), essential oils (which have numerous toxic components), cyanogenic glycosides (which release cyanide when digested) and many other toxic compounds. All of these substances are native to common foods (that is, they are not contaminants), but they would be highly toxic if ingested in sufficient quantities.

In addition to these naturally occurring substances, many foods have extrinsic contamination that may be impractical to prevent or completely remove. For example, although good management of harvested grains minimizes the problem, some level of contamination with aflatoxin (a carcinogenic hepatotoxin) in corn and peanuts is difficult to prevent. Although the extent of organic mercury contamination of fish varies by size and species and depends on the environment from which the fish are taken, virtually all fish available for human consumption contain quantifiable levels of methyl mercury, a potent neurotoxin that is particularly dangerous to the fetus *in utero*. Nevertheless, because of the health benefits of eating fish, public health interventions are meant to prevent consumption of only the most contaminated fish. Pregnant women, however, should be wary of excessive fish consumption. Mercury in fish is a good example of the need to take into account complex risk and benefit trade-offs when applying regulations to the environment.

Certain practical considerations support classification of the environment by medium. U.S. federal environmental laws tend to support such distinctions. Several laws carry names indicating the environmental medium that they address. For example, we have the Clean Air Act, the Clean Water Act, and the Safe Drinking Water Act. Other major environmental laws are also oriented toward one medium or another. The Comprehensive Environmental Response, Compensation, and Liability Act (CERCLA or "Superfund") and the Resource Conservation and Recovery Act (RCRA) largely address issues of solid wastes, although neither is entirely limited to that medium.

Protecting health requires monitoring of the environment, an activity that is medium-specific. For example, there are enormous differences in methods for field collection of water and air specimens. Water specimens are typically taken for environmental health monitoring as a simple specimen of the liquid at the point of interest. On the other hand, the ways of collecting air specimens are many and varied. The air may be drawn through a filter designed to capture particulates or designed to react with or adsorb the analyte(s) of interest. Other approaches involve simply drawing the air through an electrochemical detector. In the 2001 anthrax attacks, one of the most sensitive methods for detecting spores in air was by use of an "impactor." The Anderson Impactor was used at several sites and draws air against a blood agar plate or other culture medium in such a way that spores hit that medium and remain adherent. Quantitative or semi-quantitative measurements are possible based on the numbers and morphology of the colonies that result after the culture medium is incubated (Weis et al. 2002).

Adverse Health Effects

Although useful in focusing the training of specialists in exposure, classes of health outcomes do not correspond neatly to specific agent categories. Thus, the study of outcomes requires a different conceptual approach. Cancer, for example, may have infectious (biological), toxic (chemical), or radioactive (physical) causes. Nevertheless, cancer is best dealt with as a public health issue by practitioners who are among a defined group of clinical and public health specialists. The clinical specialists are both surgical and medical oncologists. The public health specialists who study and evaluate cancer issues are environmental, cancer, and chronic diseases epidemiologists.

The environmental health practitioners may focus their professional activities on the causes of cancer, but in so doing, they must have basic knowledge of the wide array of potentially etiological environmental exposures. This requires training and expertise that is based, not just on the exposures of concern, but also the molecular events they generate in vivo. Similarly broad thinking should be applied to disease-specific research problems and issues other than cancer.

Even if limited to chemical substances, the scope of adverse health effects possibly occurring from environmental exposures is extremely broad. Environmental health professionals spend the most time on issues involving health outcomes that are particularly severe, common, or frightening to the public at large. In the authors' experience, the public is particularly concerned about (1) cancer; (2) genetic damage (whether or not it is associated with clinical illness); (3) birth defects, neurobehavioral effects, and developmental disabilities; and (4) degenerative neurologic disease. Consequently, the public is especially concerned about exposures that might lead to any of these outcomes.

Occupational exposures tend to be more intense than exposure occurring outside the workplace. Thus, the most severe and clinically significant environmental illnesses are frequently the ones resulting from workplace exposures. As a result, previously unrecognized health effects of environmental chemicals are often first recognized in groups who are occupationally cxposcd.

The hazards of asbestos were recognized and described in workers from such facilities as shipyards, asbestos mines, and insulation and brake lining factories. ***Asbestosis*** (asbestos-caused lung disease) is progressive and often severe. Following an initial latent period measured in years, patients develop diminished lung function with consequently reduced oxygen transfer to the blood. The shortness of breath (dyspnea) that results may be incapacitating. Ultimately, the condition is lethal in many patients. In addition, patients with asbestos-related lung disease are at greatly increased risk of lung cancer and an otherwise rare malignant tumor called ***mesothelioma***. Lung cancer and mesothelioma that occur because of asbestos exposure are extremely lethal. Except in the few cases that may be cured surgically, existing treatments are not very effective (ATSDR 2001).

Part of the syndrome of asbestosis is proliferation (abnormal growth) of cells in the membrane or ***pleura*** that lines the outer surface of the lung and the inner surface of the chest cavity. The pleural membranes are made of a tissue called ***mesothelium***. The proliferating mesothelium is visible on chest radiographs as ***pleural plaques*** that are highly characteristic of the pneumoconiosis caused by asbestos (ATSDR 2001).

Asbestosis is one of a class of diseases labeled ***pneumoconiosis***. Pneumoconioses are caused by exposure to airborne particles of a variety of types. Significant occupational illness has resulted from breathing any of a wide variety of inorganic and organic dusts or particulates. Many pneumoconioses are not as severe as asbestosis. In addition, not all such illnesses are associated with an increased risk of cancer.

Among the most interesting non-workplace intoxication epidemics are those that have arisen in attempts to bypass legally restricted access to euphoriants. During the prohibition era of the 1930s, one of the few commercially available sources of potable ethanol was in plant extracts sold for use as flavorings. Ginger extract from Jamaica, known popularly as "Jake," was consumed particularly frequently. Much of it contained the organophosphorus compound tri-ortho-cresyl phosphate, which can cause irreversible nervous system damage. As a result, there were thousands of cases of muscle pain and weakness of both upper and lower extremities. The victims of this epidemic had a characteristic and recognizable gait due to extremity weakness, which was known popularly as "Jake Leg" or "Jake Walk" (Morgan and Tulloss 1976).

In the early 1980s, an attempt to create a compound with effects similar to those of heroin resulted in a substance containing 1-methyl-4-phenyl-1,2,5,6-tetrahydropyridine (MPTP). Persons who injected this preparation intravenously developed a syndrome virtually identical to that of Parkinson's disease. Further work demonstrated the selective toxicity of MPTP for cells in the nigrostriatal pathway, corresponding to the cells affected in idiopathic Parkinson's disease. Although unfortunate for those affected, the explanation of this epidemic yielded an experimental model of parkinsonism that has been of great value and led to the development of pharmacologic interventions that slow the progression of Parkinson's disease (Langston et al. 1983; Burns et al. 1983; Rost 2004).

"Place" and Environmental Health Solutions

The first three frameworks, *agent*, *medium*, and *effect*, constitute categories of discourse among environmental health professionals. They also determine the pathways for training and sometimes for career advancement. For example, one may be said to be a "heavy metal toxicologist" (expertise categorized by exposure), a "water quality expert" (expert characterized by an environmental medium), or a "specialist in occupational lung diseases" (practitioner categorized by outcome). The classifications by agent and health outcome enhance communication among environmental health professionals and other technical professionals.

"Place," our fourth conceptual approach, is not yet a well-developed organizing principle in environmental health. Nevertheless, it is a concept that provides a means of focusing and enhancing both communication about environmental health problems and the interventions needed to solve them. Because interventions, or "environmental solutions," are the theme of this book, we give "place" special attention in this chapter.

One should note the very clear distinction between "place" and "location." Geographic *location* identifies the specific position of an event of interest in environmental health. However, the "location" is defined by latitude, longitude, and elevation and is independent of the social context of that location. A location is a location whether the "place" happens to be the region, a residence, a place of employment, or any other kind of place. This is not to say that location itself is not of interest. Simultaneous proximity of the host (person exposed) and an environmental agent is required for any exposure to take place. Thus, all events of importance in environmental health have an associated location, and the concept of location is certainly useful in evaluating the risk of an etiologic exposure.

Place, on the other hand, includes the social and other context of a given location. For instance, a school can be said to be located at a specific address. Nevertheless, the public health significance of the location lies not in the location itself but in the kind of place it is. The importance of the school as a kind of place lies in its generic characteristics as a school—the fact that it has students, contains certain kinds of furniture, and is affected by certain regulations (such as rules affecting asbestos remediation). Although no standard categorization yet exists, we suggest the labels in Table 10.2 because they identify intuitively understood places around which environmental health communications, regulations, prevention, and practice can be organized.

By highlighting place as an organizing principle, we hope to facilitate communications and other interactions among environmental health scientists, regulators, and the public at large. In particular, place offers promise as a means of focusing sets of prevention activities and regulations (for example, programs for healthful homes or for occupational health) that can be efficiently implemented in a coordinated manner. Ideally the same

TABLE 10.2
Short Operational Definitions of Some Environmental Places

Label(s)	*Operational Definition*
Home/Residential	The immediate area within and around a residence or dwelling over which the occupants have direct influence or control.
Community	An area involving multiple residences but small enough for individuals to know each other within which a single (or small group of) environmental health practitioners can have meaningful impact at a personal level
Regional	Area larger than the "community" and generally smaller than "national" within whch occupants have a common environmental exposure. Rarely a regional environmental problem may cross national boundaries.
National	An area over which the highest level of national sovereignty can impose environmental programs and controls
Global/International	Area of concern to multiple countries
Travel/Transportation	"Place" in which environmental health problems related to the movements of people or substances arise
Occupational	Where "place" is the "workplace"
Disaster	Environmental conditions (either natural or human-made) far more malignant than those usually experienced

people and organizations would promote programs that would crosscut all of the other ways of subdividing environmental health. Thus, a program targeting the "home" might address both lead paint and environmental tobacco smoke, agents that are often addressed in separate programs. This program would also crosscut environmental media (both solids and air) and health (both neurological and respiratory).

Place-specific programs might involve unaccustomed coordination among experts with different backgrounds, but communicating with the public about them would be simplified. Kaleidoscopically confusing public service messages about diverse hazards could be consolidated, and prevention programs could reinforce each other.

Historical Background

Biological Environment

The remarkable efforts of Roman engineers to bring potable water via aqueducts over long distances into cities date back at least two millennia (Hodge

2002). However, centralized systems for the comprehensive management of sewage and wastewater did not become widespread in Europe and North America until the 19th century. Construction of such systems resulted in a tremendous decrease in the rates of enteric illness, and there was a corresponding increase in life expectancy, mostly because of a decrease in infant mortality (Mullan 2000). In parts of the developing world, where effective management of wastes and wastewater does not yet exist, infant mortality has been diminished by modern oral rehydration therapy (Guerrant et al. 2003). Tremendous morbidity still occurs from dehydration caused by gastrointestinal infection and diarrhea. Thus, early, "classical" environmental health focused virtually entirely on infectious diseases (biological agents).

The Physical Environment

The predominance of biological agents as the principal focus of environmental health at the regional and national level in the United States has diminished greatly over time, particularly over the second half of the 20th century. Although many physical hazards cause environmental health problems affect modern life, ionizing radiation is the hazard on which environmental health specialists are called to spend the greatest amount of time.

There was great enthusiasm for the widespread use of x-rays and ionizing radiation from natural radioactivity during the early 20th century. The use and overuse of radiation was rampant as seen in the widespread and uncontrolled utilization of radiation for common consumer applications. For example, from the mid 1920s to the 1950s in Europe and North America, fluoroscopes were used commonly in shoe stores for the simple purpose of evaluating fit. The device permitted the salesperson and customer to visualize the bones and soft tissues of the customer's foot within the shoe as he or she tried it on (Duffin and Hayter 2000). The cancers of bone (osteosarcoma) that afflicted watch dial painters who used radium-based paint to produce luminous numbers is a particularly famous example of the health consequences of the overuse of ionizing radiation (Woodard and Higinbotham 1962).

By the mid 20th century, however, the harmful potential or ionizing radiation was more clearly recognized. The increased risk of both leukemia and solid tumors of various types has been documented (Muirhead 2003). The powerful destructive force of the nuclear fission bombs used in Hiroshima and Nagasaki, Japan, were particularly sobering for the U.S. public. Later, the apparent absence of full operator control over the nuclear reactors at Three Mile Island and (more significantly) at Chernobyl has helped maintain the public's attitude of wariness toward things radioactive.

Nevertheless, ionizing radiation is a pervasive fact of modern life. Despite concerns of consumers, nuclear power plants are still widespread and produce about 20% of electrical energy used in the United States. Radionuclides and radiation sources, are used extensively for both diagnostics and therapeutics in medicine, in geology, archeology, and many other scientific and technical disciplines. Because of heightened concerns about terrorism in the United States, x-ray machines to examine luggage, briefcases, and all manner of items people carry with them are present in numerous places, including airports, courthouses, law enforcement facilities, government and public buildings, and any other place considered to be a potential target. Low intensity radiation sources are incorporated into various consumer items, such as smoke detectors.

With respect to ionizing radiation, the role of environmental health agencies and practitioners depends upon their precise duties. However, the overall goal of the environmental health community is to assure that rules, systems, and procedures are in place that will ensure that radionuclides and devices that emit ionizing radiation do not do so to an extent that will substantially endanger the public.

The Chemical Environment

The need for munitions, synthetic rubber, and many other products required to fight World War II spurred remarkable growth in the in the U.S. chemical industry. That growth continued during the post-war era, based on the enormous success of the rapidly developing technology underlying polymer and synthetic chemistry in the manufacture of consumer products. Currently, new substances are identified (literally) by the minute. The centralized CAS registry of chemical compounds is updated daily, and as of this writing (March 4, 2005), there are 25,434,241 individual organic and inorganic substances listed. (Of course, if all of the known macromolecular sequences are counted, there are many more.) (ACS 2005a)

Outside of the synthetic chemical laboratory, human beings are likely to encounter only a relatively small proportion of CAS-registered compounds. Nevertheless, because the total number is so large, the absolute number of compounds to which humans are likely to be exposed is still considerable. Under the Toxic Substances Control Act (TSCA), the U.S. Environmental Protection Agency (EPA) tracks and maintains an "inventory" of potentially toxic substances that are imported, used in commerce or are present in substantial quantities in U.S. commercial establishments. The current inventory includes some 80,000 substances (ACS 2005b).

Solutions: Environmental Health Infrastructure

U.S. Federal Government

Through the creation of certain agencies and reorganization of others, U.S. federal capability in the protection of environmental quality has been greatly enhanced during the past 40 years. These actions allowed the government to begin to meet, in a substantial and meaningful way, the challenges to public health posed by new hazards in the chemical and physical environments.

The U.S. Environmental Protection Agency (EPA) was created in 1970 by Executive Order of President Richard M. Nixon. Creation of the EPA helped better accomplish the goals expressed by Congress in the earlier National Environmental Policy Act of 1969 (NEPA). Specific authorities were given to the agency in the Clean Air Act (1970), the Clean Water Act (1972), and the Safe Drinking Water Act (1974). Control of solids was authorized under the Resource Conservation and Recovery Act (RCRA, passed in 1976). The Toxic Substances Control Act (TSCA, passed in 1976) provided the agency with authority to monitor and "inventory" chemicals to which people might be exposed. Other authorizing legislation for the EPA is available from its Internet site (EPA 2005a).

In 1978, the environmental contamination near residents at "Love Canal" by Niagara Falls, New York, achieved national notoriety, dramatically highlighting the need for public health intervention to protect people from discarded hazardous wastes. William T. Love began a canal to connect the upper and lower Niagara River to generate energy. After his project was discontinued in 1910 for economic reasons, Hooker Chemical used the three-block tract of land as a place for disposal of chemical waste. Subsequently, the site was rehabilitated (albeit inadequately) by covering it with uncontaminated earth. Houses and a school were built on and around the canal (Beck 1979).

Homeowners were mostly unaware of the site's previous use as a chemical dump until obvious signs of the contaminants began to surface. As the extent of contamination became obvious, residents also complained of a variety of health effects. A confusing array of claims and counterclaims emerged from a variety of research studies attempting to evaluate the relationship of health effects to the chemical contamination. To this day, controversy exists about the extent of health effects suffered by residents (Bown and Clapp 2002). What is clear is that residents were potentially exposed to highly toxic chemical residues in and around their homes, representing an unacceptable public health risk (Beck 1979).

On August 7, 1978, New York Governor Hugh Carey announced to residents that the state government would purchase the homes affected by toxicants. That same day, President Jimmy Carter, announced the first-ever emergency financial aid for an event other than a natural disaster (Beck 1979). "Love Canal" rapidly became a household expression and came to symbolize a fear shared by many Americans: they were at risk because environmental safeguards then in place were inadequate.

The events of the Love Canal episode provided a major impetus for passage of the Comprehensive Environmental Response, Compensation, and Liability Act (CERCLA). CERCLA supported a federal "Superfund" to clean up uncontrolled or abandoned hazardous waste and to take appropriate actions to protect the public against the hazards of accidents, spills, and other emergencies created by the release of pollutants into the environment. Moreover, CERCLA authorized EPA to seek out the parties responsible for any release and assure their cooperation in the cleanup. Also, EPA recovers its costs from financially viable individuals and companies once a response action has been completed (EPA 2005b).

Laboratory Infrastructure

The development of specialized laboratory analyses was required for environmental and biological monitoring required to maintain health. In order to support the requirements of the new environmental legislation of the 1970s and 1980s, analytical chemists and analytical toxicologists, who identify and quantify toxicants in the laboratory, had a great deal of new work to do. Proper monitoring efforts required new and reliable techniques for measuring an unfamiliar set of analytes in what (for many) were novel analytical matrices, both environmental and biological. For example, to develop a remediation plan for a contaminated area it is necessary to know the extent of contamination. Establishing the boundaries of contamination may require a large number of soil samples to be measured. If the agent is a potent one, then the analysis must be especially sensitive, since the presence of small amounts of the contaminant should not be overlooked. On the other hand, because large amounts of money may be expended in cleanup, it is also important that analyses be specific. One would not want to spend scarce health resources to clean-up an area that tested falsely positive.

Workforce Development

Along with new federal programs, came the need for "environmental" professionals to do the jobs that were either (1) required under the new legislation or (2) developed as a logical consequence of the new laws and new interest in environmental health. The new activities offered opportunities even for a variety of professions. Not only were health personnel required to think in new ways, the outlook also changed for professions that assist public health organizations in the performance of their duties. For example, lawyers with environmental and environmental law expertise were required. Because the OSHA (Occupational Safety and Health Administration) and EPA work by means of implementing regulations and there are definite costs for industry and other affected entities to comply with the regulations, it was inevitable that legal actions and court cases would result. Moreover, CERCLA mandated recovery of the costs of clean up from the original polluters, where they could be found. Those responsible were supposed to reimburse the government for its expenses in remediation of contaminated sites to protect the public. Accordingly, environmental law has grown, and like other areas of public health law, it plays an important role in protecting the public.

Specialized lawyers, engineers, chemists, physicists, and other professionals all play key roles in environmental health. However, surprisingly few professionals from more traditional health professions choose to specialize in environmental health. In particular, a shortage of physicians has been described (Rosenstock et al. 1991).

Hard data are lacking on why few physicians work in environmental health. We speculate that both compensation and professional traditions may be at issue. Relatively few physicians choose careers in fields that are largely preventive in character, such as environmental health. Physicians in these jobs may see few patients individually, because of the need to approach environmental health at a population level. Moreover, some positions may be regarded as requiring long hours without commensurate compensation.

Physicians may also feel unfamiliar with the subject matter. Although medical training includes the basics of toxicology, this discipline is taught mostly as a part of pharmacology and as instruction in adverse pharmaceutical effects. The complex exposures encountered in the environment are infrequently addressed in medical school. In addition, except for their use in diagnostics and radiotherapy, not much is taught about ionizing radiation and radionuclides. Physicians may be reluctant to enter a field in which they have little prior experience or training without a strong economic incentive to do so.

The apparent shortage of physicians in environmental health is unfortunate. Because of the developing nature of the field, the fact that not all environmental hazards have been identified or adequately characterized, and the many ways in which familiar agents pose new problems, the situations with which the environmental professional must deal are frequently ambiguous. Of all the professions contributing to environmental health, physicians are probably the best trained and most experienced in dealing with ambiguous situations in which health is at issue, since such situations occur so frequently among human patients. Physicians' abilities in the handling of ambiguity and their experience in making judgments in spite of inadequate or contradictory data can be usefully applied to environmental health.

The reluctance of physicians to undertake careers in environmental health may soon change. Recently, the largely preventive field of environmental toxicology has drawn the interest of clinical specialists in medical toxicology, who usually deal one-on-one with poisoned patients in the clinical encounter. The core content of this specialty has recently been rewritten, and now includes specific mention of public health and preventive approaches to toxicological problems (Wax et al. 2003).

Solutions: Acute Health Effects

Field Epidemiology and Outbreak Control

Whenever actual illness occurs and a chemical, physical, or other environmental agent is suspected as the cause, an epidemiological investigation may be required. The nature of the investigation depends on both the illness and the circumstances surrounding its occurrence. A clear-cut outbreak in which large numbers of cases occur over a relatively short period is the simplest situation to detect and investigate. The most important steps involved are:

1. **Case finding**—Hospitals, clinics, physician's offices and/or other healthcare entities may be contacted to determine whether cases that are unusual, either in nature or in number, have been seen during the epidemic period. Arrangements are made to obtain the relevant clinical data. Other sources of information, such as school or work absences, numbers of prescriptions written at pharmacies, and other sources may be consulted to ensure that virtually all cases are detected.
2. **Case definition**—After casting the net broadly to determine the scope of the outbreak and thus acquiring clinical information on a great number of cases, the key components of the clinical picture will begin to come into focus. At this point, in order to be able to further study the epidemic in a way that yields interpretable and reproducible results, the investigator will decide on the clinical features required for an illness to be considered an epidemic "case." Where definitive diagnostic data are available on only a small portion of patients making up the outbreak, it is common practice to define both "definite" and "probable" cases.
3. **Descriptive review**—A systematic summary compilation of findings on key variables is performed. Important variables to summarize include the time of onset, location, age, gender, occupation, and other characteristics. This summary information frequently generates clues regarding the cause of the epidemic.
4. **Analytical study**—Further study to verify etiological hypotheses (or to generate such hypotheses if they are lacking) may be indicated. Most commonly, the characteristics and circumstances of affected persons are compared with data derived identically from unaffected but otherwise similar persons. Logically, the differences between ill and well may bear on the etiology of the illness.

The above steps are best illustrated with an example.

In 1989, clinicians in New Mexico noted the unusual occurrence of three patients with severe muscle pain, a striking increase in eosinophils (a particular kind of white blood cell) and other peculiar features. The patients were taking the amino acid L-tryptophan (LT) as a nutritional supplement. They wondered whether the peculiar syndrome and the food supplement might be related (Kilbourne 1992).

Following recognition of these cases, federal and state public health authorities began case finding. Ultimately, over 1500 cases of the syndrome were recognized in the U.S. Even prior to the completion of case finding, a case definition for surveillance (further case finding) was established. Disabling muscle pain and an eosinophils level of >1000 per mm^3 (normal <350 per mm^3) were required. A descriptive review showed the typical patient to be a middle aged white female who took nutritional supplements. Although the national summary data showed virtually all cases occurring in persons taking LT, that fact was difficult to interpret since the rates of LT use in the groups most affected were unknown.

Case control analyses were performed rapidly. In a study performed in New Mexico, all ill persons and very few of their well neighbors of the same age and gender were taking LT. The difference was highly statistically significant (Kilbourne 1992). Because studies done in various populations in somewhat different ways all yielded similar findings, it was possible to conclude that the L-tryptophan (or much more likely, a contaminant in certain lots of it) caused the disease, which is now know as the eosinophilia-myalgia syndrome. Establishing the causal link provided a basis for the FDA to take action to remove products associated with illness from the market. The episode has been reviewed in detail elsewhere (Kilbourne 1992).

Poison Control

As chemicals of all sorts became ubiquitous following World War II, the impact on children was severe, and by the early 1950s, poisoning was the most common type of injury in young children. Because the number of poisons affecting children was large and because new substances were (and are) constantly coming into use, physicians were not always familiar with them. Accordingly, the first poison control center was set up in Chicago in 1952.

Although there once were over 700 poison control centers in the United States, there are now only 62. Working under the medical direction of a physician who is board-certified in medical toxicology, each poison center answers questions from the public and from medical practitioners. Each year, U.S. poison control centers handle well in excess of two million cases of actual human exposure to potentially toxic substances.

Under a program supported by the Health Resources and Services Administration (HRSA) and the Centers for Disease Control and Prevention (CDC), dialing 1-800-222-1222 anywhere in the United States yields a toll-free connection with the regional poison center corresponding to the region from which the call was made. If caller is a layperson and the exposure is not serious enough to require direct medical attention, the center so advises the caller. Where direct intervention is required, instructions are provided for obtaining appropriate care. If the caller is a healthcare provider, expert advice regarding managing the patient is provided.

Poison control centers do not necessarily come to mind when a practitioner of an environmental health discusses his or her field. Nor do poison center directors and staff often think of themselves as dealing with "environmental health." This split has arisen as the result of the tradition of considering acute medical effects as something different from the low-dose long-term exposure scenarios so frequent in so many of the rest of environmental public health activities.

Closer integration with poison control centers by environmental public health practitioners is desirable and appears to be occurring. Certainly in environmental emergencies (for example, fires at chemical manufacturing plants or releases of toxicants from these sites), poison centers inevitably become involved, because of the need of first responders and clinical practitioners for the poison center's advice in the handling of any casualties that occur. Because of their visibility, availability, reliability, and frequent contacts with the public, most regional poison control centers have become a trusted source of information in their respective regions. Involving the poison center to help communicate the scientific basis for public health decisions about environmental hazards offers the possibility of synergy between poison center operations and those of other aspects of environmental public health.

Solutions: Cancer And Other Chronic Diseases

Cancer Clusters

Acute environmental illnesses constitute a minority of all health problems handled by specialists in environmental health. More frequently, chronic diseases are at issue. A particularly common perception or complaint coming to the attention of environmental health practitioners is the concern that a given community has an excessive number of cancer cases. Although the steps required have been clarified in recent years, much time and effort is required to address effectively these concerns (CDC 2005).

The first step in evaluating such a complaint is to verify that the numbers of cases actually justify concern. The investigator first establishes the geographical and temporal boundaries (place and time) of concern about increased risk. The cancer cases that occur (observed cases) are compared with the expected numbers calculated by applying tumor type-, age-, and gender-specific rates from population-based cancer registries to the numbers of persons in the same age and sex groups in the population at risk. Many times, the difference between observed and expected is not statistically significant, despite the community's perception of a problem.

Even where statistics indicate an actual increase in the numbers of cancers over the number expected after factoring in age and other nonenvironmental variable, identifying an environmental cause is difficult. Carcinogens typically have a latent period that may be as short as about 5 years but more often is measured in decades. In the highly mobile population of the United States, persons developing cancer in the concerned community may not even have lived there at the time a carcinogenic exposure occurred. A diligent search for etiological factors is required, and any explanation offered must take into account the time lag between cancer-causing exposure and its ultimate health outcome.

A cancer cluster involving leukemia at Fallon, Nevada has been studied especially intensively. Among many exposures evaluated, evidence emerged implicating arsenic and the metal tungsten. Despite the intensive study and apparently positive findings, it is still uncertain whether either of these exposures actually played a role in the etiology of leukemia at Fallon (CDC 2005b). Clearly, determining the relationship of exposures to chronic disease outcomes is a complex task, especially if such factors as latent periods need to be taken into account. In evaluating whether environmental exposure causes a chronic disease, frequent weaknesses of epidemiological studies include inadequate (or missing) exposure data, confounding among exposures, and study population that is not sufficiently large to detect a low-level effect with reasonable statistical power (Rachamin 2001).

Animal Studies

Where epidemiological (human observational) studies are absent, data from animal studies must be used. Because cancer is such a prominent concern in environmental public health, an important goal of animal experimentation is to assess the carcinogenicity of chemicals. The National Toxicology Program, a multi-agency program led by the National Institute of Environmental Health Sciences (NIEHS) of the National Institutes of Health (NIH), funds and guides much of the animal toxicology testing done in the United States. A complete work-up of a new chemical substance involves various types of exposure of animals to the compound, followed by intensive veterinary pathological study.

Most frequently, carcinogenicity of chemical compounds is evaluated in test systems with mice or rats given the maximum tolerated dose (MTD) of the chemical, a level of exposure associated with a weight of test animals that is no less than 90% of unexposed controls. Along with a control group (zero exposure), a group of animals given the MTD and one or two groups given lower levels of the test exposure are studied in parallel over the approximately 2-year life span of the animals. The animals are sacrificed at the end of the study and are evaluated for evidence of cancer and other chronic health effects.

The animal studies just described can be done with a great deal of precision. However, the interpretation of the significance of the animal findings for human health is inevitably less precise. Problems addressed in the interpretation of animal studies include:

- The shorter life span of rodents as compared with human beings
- The relatively high levels of exposure that test animals receive as compared with the relatively low doses to which humans are exposed in the environment
- Between-species differences in the metabolism of xenobiotics
- Between-species differences in organs and tissues (for example, there is no human forestomach, making interpretation of the human significance of pathology in this organ difficult to evaluate)

Standard procedures and formulae have been developed for dealing with many of these issues of interpretation, allowing derivation of estimates of human equivalent risks based on animal findings. The use of animal studies is justified by fact that there is a high rate of concordance of human and animal data for many exposures. In general, exposures known to cause cancer in humans also cause cancer in animals. Thus, most environmental health scientists feel that the converse is likely to be true, that animal carcinogens can reasonably be expected to be human carcinogens (Rachamin 2001).

Solutions: Occupational Environmental Health

The challenge of the occupational environment also received significant new federal attention in the early 1970s. Congress passed the Occupational Safety and Health Act in 1970. This bill created the Occupational Safety and Health Administration (OSHA) within the U.S. Department of Labor. To provide a scientific basis for the regulatory work of OSHA, the OSH Act also called for the creation of the National Institute for Occupational Safety and Health (NIOSH). NIOSH currently operates as a unit of the CDC, within the U.S. Department of Health and Human Services. Although NIOSH is not a regulatory agency, its scientists enhance worker safety and health by conducting industry-wide studies to assess workplace hazards; going into the field to investigate directly any specific hazards reported to it by workers, unions, or management; and conducting relevant basic scientific ("bench") research.

Many of the relationships of human diseases to environmental exposures have been discovered among workers. Degeneration of the jaw was recognized to be a consequence of exposure to white phosphorus in the match industry (ATSDR 1997). The peculiar behavior of hatters (hat makers) was eventually understood to arise from exposure to mercury and consequent neurotoxicity (Wedeen 1989). Scrotal cancer was recognized as an occupational risk of chimney sweepers (Cherniack 1992).

Clinically important pneumoconiosis occurred in different forms among a variety of different occupational groups. Pulmonary fibrosis (growth of fibrous tissue within the lung) occurs in anthrasilicosis (black lung) due to coal dust exposure, particularly among miners. Silicosis involves a somewhat different pattern of fibrosis and develops in persons exposed occupationally to silica dust. Tuberculosis frequently complicates silicosis because of increased susceptibility to this infection (Carta et al. 2001). Byssinosis (so-called brown lung) involves constriction of the bronchi and results from excessive exposure to cotton dust. Asbestosis and related diseases have already been described (see above). A pernicious form of pneumoconiosis called berylliosis develops after exposure to beryllium dust. However, only a genetically predisposed subset of persons with significant exposure to inhaled beryllium becomes ill (Tinkle et al. 1996).

Future Solutions in Environmental Health

Sprawl

The evolving approaches to land use, transportation, and community design in many parts of the United States have all resulted in a condition called "sprawl" (or sometimes, "urban sprawl" or "suburban sprawl"). Although few hard data exist, a compelling case can be made for the idea that the factors determining sprawl may also influence the health of people in sprawling communities. Certainly, the quality of life is adversely affected where the presence of large amounts of motor vehicle traffic, the absence of sidewalks, and the distant locations of stores and other community amenities all combine to diminish the desirability of obtaining exercise by walking. When we drive (not walk) by our neighbors, there are fewer chances to get to know them, leading to a sense of anomie and a level of social integration and mutual support that is less than would otherwise exist. Civic involvement suffers when people are isolated from each other. Moreover, in comparison to more resident-friendly communities, sprawled communities are frequently less attractive.

In a sprawled environment, the default lifestyle is sedentary. The locale has insufficient attraction—and there are insufficient sidewalks, safety measures, and other infrastructure—to support a lifestyle of bicycling or walking. Moreover the extent of land required to generate square mile upon square mile of well-spaced, single family dwellings essentially guarantees that some points of interest are sufficiently distant from people to limit the practicality of walking and bicycling. The ubiquitous automobile generates pollution even as its use diminishes our cardiopulmonary reserves.

Although sprawl seems accidental—or at least incidental—to other land use goals, it is actually supported by the current crop of land use plans, tax codes, and other public policies. Such forces as urban pollution, the noxious proximity of heavy industry, high rates of urban crime, and other centrifugal demographic forces led to sprawl. These have diminished, as have other concerns about the urban environment, including poor sanitation and the presence of serious diseases (such as polio, now eliminated from the western hemisphere). Accordingly, the current era may be a propitious time to raise the debate about sprawl, allowing due consideration of new and creative public policies that could diminish sprawl and its adverse impact.

New Paths to Solutions in Environmental Health

Federal programs supporting environmental health in the United States are largely based on legislation of the 1970s and 1980s, which specifically adopted a regulatory approach. Regulations are specific legal requirements and thus involve enforcement. Such approaches reflected the realities of the time. Substantial capital investment and major corporate interests were deeply rooted in energy-producing and manufacturing infrastructure that allowed release of high levels of pollutants to the environment. Adding to the problem, the environmental disinterest of urban industries resonated in rural areas. There, the efficacy of new pesticides that amplified agricultural productivity blinded many to the possible adverse effects these chemicals. The fact that such large-scale releases of poisonous substances to air, earth, and water were not sustainable over the long term was only just then emerging into society's collective awareness. Given the strength of the economic interests allied in opposition to changes that would protect the environment (and the health of the people living in it), the full weight of the federal legal system seemed the only approach with any hope of success.

Judicious national environmental regulation has yielded some important successes. Although full compliance with existing legislation is by no means universal, environmental media, particularly air and water, are noticeably cleaner than a few decades ago. Illness and diminished quality of life still do arise from noxious environmental exposures, but the outlook has considerably improved.

Regulations have been the major environmental interventions to date, but they cannot continue to do the whole job. By our categorization of "place," the place in which regulations are promulgated is the nation. Of necessity, national regulations are written as one size fits all. They are conceived in broad strokes and are meant to address the bulk of a problem nationwide. But the scope of regulation is limited. Regulations may guide major industries but are unlikely to have the same impact in places controlled by individual members of the public.

In maintaining their residences, individual homeowners, do not typically monitor the federal register, implementing all new environmental regulations published. Indeed, home is the most obvious example of an environmental place where regulations will have relatively little impact. This is not to say, however, that there are no means of influencing people to take the steps necessary to protect the health of the inhabitants of their homes. Knowledge that people gain from the media, from authoritative professionals (for example, from physicians and public health authorities), from the school classroom, and from their friends helps to guide their behavior.

Public health professionals have limited access to people in their homes. Interventions requiring a home visit should optimally address multiple home-based environmental health problems. More typically, a home visit represents a single health program, and the opportunity to use the visit for other purposes is missed.

The multi-faceted problem of lead exposure, already introduced above, provides a good example of how this scheme for prevention might work. Currently, the home environment is the greatest contributor to unduly high lead levels in American children. The largest contribution comes from lead paint residue in neighborhoods with houses built prior to limitations on the use of lead in paint. Young children are a concern, because the combination of increased mobility, mouthing behaviors, proximity to the floor, and the lack of fully developed personal hygiene all combine at that age to enhance ingestion of whatever lead may be in the environment.

Nevertheless, until the mid 1970s the major determinant of people's lead levels was the lead breathed in air because of the use of tetraethyl lead use in gasoline. Restriction of most uses of leaded fuels in the United States in the 1970s as well as limitations on other uses lead resulted in a dramatic diminution in the mean blood levels of lead in people around the country.

The falling lead levels were well documented in the U.S. National Health and Nutrition Examination Survey (NHANES, values graphed as Fig. 10.1). Despite the dramatic decrease in children's blood lead levels in recent years, some 2.2% of U.S. children between the ages of 1 and 5 years still have a blood lead level ≥10 mcg/dL (CDC 2005c). Thus, the task of lead poisoning remains incomplete.

The regulatory approach has done its part. Methods of intervention other than (or in addition to) national regulation are needed. Areas with old housing stock are being targeted. In many urban areas with lead problems, the discovery of a child with a high lead level triggers an inspection of the home for the source. If necessary, remediation of the home environment occurs.

While regulations remain essential, we look toward a new era where environmental interventions are more diverse in nature. Clearly, there will be a mix of regulatory and non-regulatory approaches. The active

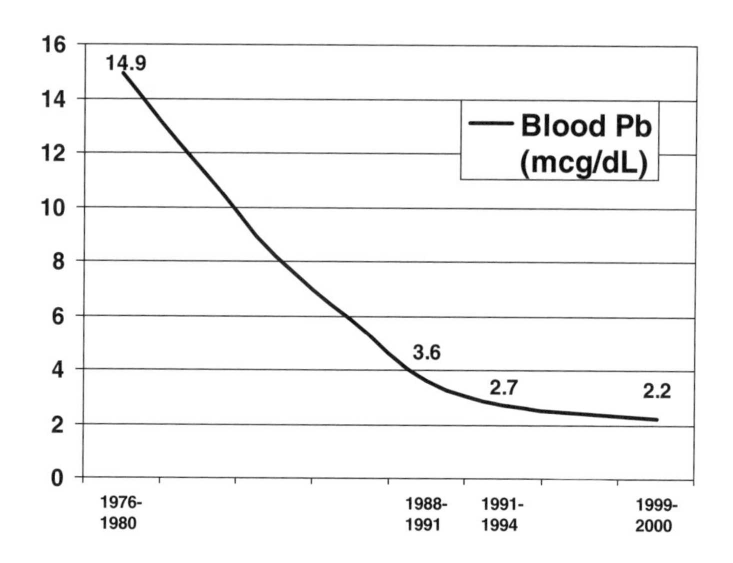

FIGURE 10.1. Geometric mean blood lead levels, Children Aged 1-5 Years by year, National Health and Nutrition Examination Survey (NHANES), United States. (Source: U.S. Centers for Disease Control and Prevention, Atlanta, Georgia, U.S.A.)

participation of the public needs to be solicited, and the many different programmatic activities in environmental health and other healthcare and public health endeavors should be coordinated. Such coordination will best be target at particular places. We envision the specific social context in each type of place ultimately including behavioral cues and prompts. All of these interventions would be targeted toward mitigation of the environmental health problems of importance at that particular place.

Review Questions

The student may benefit from using the following questions as a framework for study.

1. Identify the various ways in which the field of environmental health may be subdivided. Which activities or groups benefit from each of the different ways of subdividing the field?
2. Identify differences and similarities between acute and long-term environmental concerns. How would you study an acute epidemic of an environmentally caused illness? How would you study chronic diseases arising from a presumptive environmental agent?
3. Compare and contrast the utilities of animal experiments and human epidemiological studies in predicting adverse impacts on humans.
4. What is the difference between a public health specialist and an environmental health specialist?
5. How does "sprawl" lead to environmental health problems, and how do we combat them?

References

ACS. 2005a. American Chemical Society. CAS Registry. (Available at http://www.cas.org/EO/regsys.html; accessed March 4, 2005.)

ACS. 2005b. American Chemical Society. The latest CAS Registry Number® and substance count. (Available at http://www.cas.org/cgi-bin/regreport.pl; accessed March 4, 2005.)

ATSDR. 1997. Agency for Toxic Substances and Disease Registry. *Toxicological Profile for White Phosphorus*. Atlanta: U.S. Department of Health and Human Services. (Also available at http://www.atsdr.cdc.gov/toxprofiles/tp103-p.pdf; accessed March 15, 2005.)

ATSDR. 2001. Agency for Toxic Substances and Disease Registry. *Toxicological Profile for Asbestos.* Atlanta: U.S. Department of Health and Human Services.

Beck, E. C. 1979. The Love Canal tragedy. *EPA Journal* January. (Available at http://www.epa.gov/history/topics/lovecanal/01.htm; accessed March 15, 2005.)

Bown, P., R. Clapp. 2002. Looking back on Love Canal. *Public Health Rep* 117(2): 95–98.

Burns, R. S., C. C. Chiueh, S. P. Markey, et al. 1983. A primate model of parkinsonism: selective destruction of dopaminergic neurons in the pars compacta of the substantia nigra by N-methyl-4-phenyl-1,2,3,6-tetrahydropyridine. *Proc Natl Acad Sci U S A* 80(14):4546–45450.

Carta, P., G. Aru, P. Manca. 2001. Mortality from lung cancer among silicotic patients in Sardinia: an update study with 10 more years of follow up. *Occup Environ Med* 58(12):786–793.

CDC. 2005a. Centers for Disease Control and Prevention. Cancer clusters. (Available at http://www.cdc.gov/nceh/clusters/about_clusters.htm; accessed March 6, 2005.)

CDC. 2005b. Centers for Disease Control and Prevention. Cross-sectional exposure assessment of environmental contaminants in Churchill County, Nevada. (Available at http://www.cdc.gov/nceh/clusters/Fallon/factsheet.pdf; accessed March 15, 2005.)

CDC. 2005c. Centers for Disease Control and Prevention. Building blocks for primary prevention: Protecting children from lead-based paint hazards. Atlanta: U.S. Department of Health and Human Services.

Cherniack, M. G. 1992. Diseases of unusual occupations: an historical perspective. *Occup Med* Jul-Sep;7(3):369–384.

Duffin, J., C. R. Hayter. 2000. Baring the sole. The rise and fall of the shoe-fitting fluoroscope. *Isis* 91(2):260–282.

EPA. 2005a. Environmental Protection Agency. Major environmental laws. (Available at http://www.epa.gov/epahome/laws.htm; accessed March 15, 2005.)

EPA. 2005b. Environmental Protection Agency. CERCLA Overview. (Available at http://www.epa.gov/superfund/action/law/cercla.htm; accessed March 15, 2005.)

Frumkin, H., L. Frank, R. Jackson. 2004. *Urban Sprawl and Public Health: Designing, Planning, and Building for Healthy Communities.* Washington, DC: Island Press.

Guerrant, R. L., B. A. Carneiro-Filho, R. A. Dillingham. 2003. Cholera, diarrhea, and oral rehydration therapy: Triumph and indictment. *Clin Infect Dis* 37:398–405.

Hodge, A. T. 2002. *Aqueducts and Water Supply*. London: Duckworth & Co., pp 19–47.

Kilbourne, E. M. 1992. Eosinophilia-myalgia syndrome: Coming to grips with a new illness. *Epidemiol Rev* 14:16–36.

Langston, J. W., P. Ballard, J. W. Tetrud, I. Irwin. 1983. Chronic Parkinsonism in humans due to a product of meperidine-analog synthesis. *Science* 219:979–980.

Morgan, J. P., T. C. Tulloss. 1976. The Jake Walk Blues. A toxicologic tragedy mirrored in American popular music. *Ann Intern Med* 85(6):804–808.

Muirhead, C. R. 2003. Studies on the Hiroshima and Nagasaki survivors and their use in estimating radiation risks. *Radiat Prot Dosimetry* 104(4):331–335.

Mullan, F. 2000. Don Quixote, Machiavelli, and Robin Hood: Public health practice, past and present. *Am J Pub Health* 90(5):702–706.

Rachamin, G. Use of toxicological data in evaluating chemical safety. In: Bingham, E., B. Cohrssen, C. H. Powell, eds. 2001. *Patty's Toxicology*, 5th ed, Vol. 1. New York: John Wiley & Sons, Inc., pp. 381–414.

Rosenstock, L., K. M. Rest, J. A. Jr. Benson, et al. 1991. Occupational and environmental medicine. Meeting the growing need for clinical services. *N Engl J Med* 325(13):924–927.

Rost, A. 2004. Break through: The genesis for the Parkinson's Institute. *Cupertino Courier* August 11. (Also available at http://www.svcn.com/archives/cupertinocourier/20040811/cu-cover1.shtml; accessed March 9, 2005.)

Tinkle, S. S., P. W. Schwitters, L. S. Newman. 1996. Cytokine production by bronchoalveolar lavage cells in chronic beryllium disease. *Env Health Perspec* 104(S5):969–971.

Wax, P. M., M. D. Ford, G. R. Bond, et al. 2003. The core content of medical toxicology. *Ann Emerg Med* 43(2):1–6.

Wedeen, R. P. 1989. Were the hatters of New Jersey "mad"? *Am J Ind Med* 16(2):225–233.

Weis, C. P., A. J. Intrepido, A. K. Miller, et al. 2002. Secondary aerosolization of viable bacillus anthracis spores in a contaminated US Senate office. *JAMA* 288:2853–2858.

Woodard, H. Q., N. L. Higinbotham. 1962. Development of osteogenic sarcoma in a radium dial painter thirty-seven years after the end of exposure. *Am J Med* 32:96–102.

CHAPTER 11

Industrial Collaborative Solutions

Nelson L. Nemerow

Forward

After working in this (what is now known as the environmental engineering) field for a half century, I feel the need to present a suggested industrial collaborative solution to the never ending pollution problem. No solution is perfect, and certainly this one has its flaws. But not only is it the logical outcome of rational thinkers, but also it is ultimately economically sound and needs innovative minds and courageous actions. Its implementation will require suggested solutions to the never-ending pollution problem. No solution is perfect, and certainly manufacturing choices must still be made. If followed completely, it will require an entire restructuring of our industrial location and operation engineers—for they must consider other industry needs besides their own in making production choices. Society, however, will benefit from lower prices for goods and less pollution of its environment. What better goals to work for! The "waste utilization engineer" will replace the conventional "waste treatment engineer." Cleaner air and waters will replace dirtier environments. Lower costs with perhaps higher standards of living will replace more expensive products. All it takes is the courage of our convictions and an innovative mind. I've done my part by suggesting the primrose path to follow. Now it's up to you to read, assimilate, and act on the contents of this chapter.

Prologue

In the 1940s, 1950s, and even into the 1960s, we patterned industrial waste treatment after those used to treat municipal wastes. In the 1960s and 1970s we favored combining industrial wastes with municipal ones. During the 1970s and 1980s we began to treat industrial wastes separately with more exotic and efficient methods and to seek exterior markets for waste residuals in other industries. All of these methods and systems from the 1940s to the 1980s were only partially successful in attaining sufficient contaminant removal and certainly inappropriate and generally outdated by the 1990s.

During the 1950s–1980s period, industry discharged industrial wastewaters into municipal sewers and sewage treatment plants. Industry was aware of the accompanying problems as well as the inefficiencies that resulted from this practice. However, industry overlooked these and succumbed to this relatively easy method of disposal. Industrial costs were lower and legal responsibility for environmental degradation was avoided or at least jointly shared with the municipality.

Major requirements had to be met before industrial wastes could be treated along with municipal sewage. These requirements are valid even today, although more difficult to meet now than in earlier years (Nemerow and Agardy 1998a). They include technical, economic, legal, and managerial requirements.

Technical

Industrial wastes must be compatible (treatable) with the sewage. Industrial wastes must be equalized and proportioned to the flow and pollution load of the sewage. Industrial wastes should not contain any material toxic or detrimental to the operational objectives of the sewage treatment plant. Industrial wastes should not contain any substances hazardous to the operating personnel nor those of the environment near the effluent discharges. Industrial waste by-passes to the treatment plant must be provided if and when these wastes fail to meet the above four requirements. Proper treatment of the by-passed industrial waste must be used to ensure environmental protection.

Economic

The cost to industry for combining its waste with municipal sewage must be low enough to provide sufficient incentive for them to use this system. Such provisions have usually been attained by the imposition of a sewer service charge. The charge often is based on the industrial pollution load being treated and/or removed by the treatment system. However, the variation in economic benefit to industry has been tremendous from one city to another. It has varied from the free use of the sewer and sewage treatment plant or a nominal charge based on flow similar to householders to as prohibitive as a charge based on all extra pollutants. The free or nominal use of the municipal system has long been an unofficial "boon" to industry. Even the prohibitive charge to industry has often been economically acceptable to industry because it avoided legal and/or managerial responsibilities on the part of the industry itself.

Legal

Once an agreement between the city and its industry has been reached, industry avoids at least part, if not all, of the lawful responsibility of any external environmental damages. Many times this alone has provided sufficient incentive to industry for this mode of solution to its waste problem. Industry often used this system when its wastes were difficult to treat or even hazardous when discharged separately to a watercourse.

Managerial

To industries of the 1940s–1970s era treatment and disposal of its wastes represented a managerial burden. Because of this industrial attitude, combined treatment with municipal sewage was viewed as a welcome solution.

Conditions for Rejection

Despite its general use and acceptability there were and still are situations where combined treatment is not recommended. I recall the two most important ones.

1. Perhaps most important of these in the long run is the lack of cooperation and understanding between municipal and industrial officials. As a result of a lack of a good understanding, and cooperative relationship between them, technical problems will not be resolved satisfactorily.
2. Physical problems of industrial plant location, pumping, and hazardous waste nature may make this system unworkable.

All other adverse situations often can be overcome, if these two causes for rejection are absent.

Methods of Industrial Waste Reuse

Because of the problems and ineffectiveness of combined treatment as well as the situations which automatically call for its rejection, industry has sought other solutions. The most logical and most used solution is that of reusing its wastes. The method generally used is reusing wastes within its own industrial plant. When possible, in-plant reuse is most economic and effective. However, such situations are ideal and seldom encountered in actual practice. As one plant manager put it, "Our wastes are wastes because they are not useful to us in any way."

The next potential for waste reuse is to contract with a "scavenger" collector to transport such wastes to a large, central industrial waste disposal plant.

Such systems are costly to an industry for small volumes of wastes and also impractical for larger volumes. They may, however, relieve the industrial plant of managing liability of disposal discussed earlier.

The last waste reuse technique involves the direct marketing of waste as a resource for another industrial plant. To use this method industry must go through the familiar systems of selling its waste like it does its products. Selling is never easy, not guaranteed, and may involve revealing more waste product characteristic information than it would like. Finding a suitable buyer for its waste, in fact, may even be more difficult than for its primary products. Waste exchanges that publish industry waste needs and prices assist plants to find buyers. Disclosure of waste character for sale remains a stumbling block for sellers using the direct method.

The Ultimate for Industrial Waste Reuse—The EBIC

All of the previous descriptions of combined treatment lead to the main conclusion of this prologue. Because of the aforementioned inadequacies of combined treatment a new solution is evolving. I have designated its terminology as the *environmentally balanced industrial complex* (EBIC). This new system has evolved not only because the previous solutions were inefficient, but also because industry and society have grown exponentially to a degree that our environmental resources are imperiled beyond their capacity.

The EBIC can be described simply (Nemerow 1995) as a selective collection of compatible industrial plants located together in one area (complex) to minimize (or eliminate) both environmental impact and industrial production costs. These objectives are met by utilizing the waste materials of one plant as the raw materials for another with a minimum of transportation, storage, and raw material preparation. When a manufacturing plant neither treats its waste, nor stores or pretreats certain of its raw materials, its overall production costs must be reduced significantly.

Elimination of waste treatment costs alone may be sufficient to influence industrial managements to continue to produce their products in the highly competitive world market. It should be our dual obligation to minimize waste treatment costs and maximize protection of the environment.

Reuse costs within these complexes can be absorbed easier into production costs than end-of-the-line waste treatment costs. Despite the advantages of EBICs, many factors must be identified, clarified, and answered properly before such a system can be accepted. For example, reasonable matches of waste quantities and raw materials must be established. Also, we need to clarify the type of labor and worker numbers available in an area as well as the marketing of products from the area. The key to feasibility for any complex lies finally in production economics and environmental protection.

There are many potential EBICs which will yield several salable products and result in no wastes reaching the external environment (Nemerow

1995). Another important advantage which is readily apparent to the environmental engineer is that generally no waste treatment is necessary within the complex. When, in some cases, waste treatment is required to render it directly reusable, another product may also result.

There is an opportunity for industry and municipalities to collaborate in the future to produce industrial products directly from municipal wastes. In such cases the solid contaminants contained in municipal sewage would be converted within the treatment plant to industrial products for sale instead of for disposal on the land.

For example, settling tank sludge can be rotary dried, pulverized, amended, bagged, and sold as fertilizer to the agricultural industry. Floating matter from this same settling basin can be skimmed and rendered by steam heat treatment to produce, with certain additions, animal feed for this same agriculture industry. We practice both these treatments today to some degree and in a few instances. However, a concerted effort needs to be made to design municipal treatment plants to include industrial production as an integral part of its operation. This also requires a closer collaboration of industrial and municipal services rather than a haphazard afterthought following municipal sewage treatment. The reader is directed to discover solutions of this latter type by reviewing carefully the systems described in Chapter 4.

Another example is that of the fish food industry coupled with municipal sewage treatment. In this case the sewage treatment plant would be designed not only to purify its wastes, but also to produce fish (such as tilapia) for sale from this combined activity. Effluent from such a plant would be recirculated into algae ponds, which serve as food for fish pond production. Excess effluent would be sold also to the agricultural industry for irrigation water.

As you continue to read this chapter let me challenge you—municipal, federal, industrial, and environmental engineers—to "think" design and operation of truly combined treatment so that no effluents reach our fragile and disappearing environment, and at the same time producing valuable industrial products at lower cost.

Rationale of Environmentally Balanced Industrial Complexes (EBICs)

Industry's contribution to the era of waste utilization is highlighted by its acceptance to use the ultimate in waste utilization: the environmentally balanced industrial complex (EBIC) system. The system comprises two or more compatible industrial plants located in close proximity to one another in one complex. Each plant uses the waste of another plant as part or all of its raw material. No wastes are discarded into the environment outside the complex. No wastes have to be transported or sold outside the complex. The environment is protected from environmental degradation. And people (consumers) benefit from lower-priced products. Industry also gains a competitive price advantage over other plants which chose to operate in separate locations and encumbered with waste treatment costs.

Although the real measurable cost of industrial environmental pollution control remains relatively small when compared to total production or value-added costs, it can be a significant amount when considered by itself. In fact, the amount may be enough to influence industry management to consider whether to produce or discontinue the manufacture of specific goods. Though environmental engineers are usually not involved in that decision, the goal should be to reduce treatment costs to a minimum while protecting the environment to a maximum.

In conventional industrial solutions to waste problems, industry uses separate treatment plant units, employing physical, chemical, and biological systems. These separate treatment systems increase manufacturing costs. These costs are also easily identified and—even if relatively small when compared to other production costs—are opposed by industry. On the other hand, reuse costs, if any, in an EBIC will be difficult to identify and more easily absorbed into reasonable production costs.

Large, water-consuming and waste-producing industrial plants are ideally suited for location in such industrial complexes. Even though their wastes—if released to environment—might cause pollution, such wastes may be amenable to reuse by close association with satellite industrial plants using wastes and producing raw materials for others within the complex.

Examples of such major industries are steel mills, fertilizer plants, sugarcane refineries, pulp and paper mills, and tanneries. Cement plants may also produce the ideal product to allow a perfect match for the phosphate fertilizer plants in a balanced industrial complex.

One needs to choose the proper mix of industries of the appropriate size and locate them in a specific area isolated from other municipal, industrial, or commercial establishments. These choices will be highly influenced by marketing and socioeconomic factors.

Since 1977 I have proposed several typical complexes for tannery, pulp and paper, fertilizer, steel mill, sugarcane and textile industries (Nemerow et al. 1978; Nemerow 1980a,b; Tewari and Nemerow 1982; Nemerow 1984; Nemerow and Dasgupta 1984; Nemerow and Dasgupta 1985; Nemerow et al. 1987; Nemerow and Veziroglu 1988). Such complexes have the presumed advantages of minimizing production costs and adverse environmental impacts. Optimization of these advantages will meet the objectives of both industries and environmentalists.

Although the advantages of this type of complex are obvious, there are certain difficulties to overcome. One involves compatibility. There is no evidence, to date, that waste and product compatibility necessarily mean industrial working compatibility. Other plant operating requirements such as labor availability, marketing of products, and taxes may not mesh as easily.

Another involves optimal mass balances. Again there is no evidence to date to show that all plants within such a complex can operate at or near their optimum production required for economic purposes. However, lack of evidence is no reason to discard the principle, but rather reason for more complete investigation and trials.

In the middle to late 1990s the field of "industrial ecology" became recognized as one step in promoting "waste utilization" rather than "waste treatment." An ardent supporter of this concept is Suren Erkman[1] who has surveyed this field and written extensively on the modern interpretation of industrial ecology. In one of his latest contributions to this concept, he summarizes that

> industrial ecology aims at looking at the industrial system as a whole. Industrial ecology does not address just issues of pollution and environment, but considers as equally important, technologies, process economics, the inter-relationship of businesses, financing, overall government policy and the entire spectrum of issues that are involved in the management of commercial enterprises. As such, industrial ecology can provide a conceptual framework and an important tool for the process of planning economic development, particularly at the regional level. Also, industrial ecology may offer options, which are not only effective for protecting the environment but also for optimizing the use of scarce resources. Thus, industrial ecology is especially relevant in the context of developing countries, where growing populations with increasing economic aspirations should make the best use of limited resources. (Erkman 2001)

I encourage you to find solutions of this kind as put forth in Chapter 12 of this book.

The reader should recognize the similar philosophies and goals of the "industrial ecology" concept described by Erkman and others and the EBIC concept proposed here. Both "beg" for implementation on a practical scale to verify and fortify their theoretical premises.

Procedure for Industry in Attaining Zero Pollution

Planning for an EBIC

In planning for an EBIC several subjects and associated steps need to be considered and taken prior to actually starting an EBIC system. They include, at a minimum, the following 11 subject steps.

1. Suren Erkman. Institute for Communication and Analysis of Science and Technology, Geneva, Switzerland.

1. Selecting and educating the EBIC developer
2. Location
3. Compatibility
4. Optimizing production sizing of participating plants
5. Informational meeting with participating plants and regulatory officials
6. Designing the flow diagram for the complex
7. Developing a computer program of varied inputs and operating conditions
8. "Dry run" (on paper) of complex including all potential variations
9. Review architectural and engineering plans of EBIC
10. Participate and observe construction of EBIC
11. Observe and consult with plants during startup of EBIC

1. *Selecting and educating the EBIC developer*: At the onset the developer and/or purveyor of the land must be sought out and indoctrinated with the principles of an EBIC. Preferably the developer should be shown schematic diagrams of industrial plants operating at separate locations with associated environmental damage. Then he should be shown a schematic overlay of the same industrial plants located adjacent to one another in the EBIC—with no adverse environmental damage. At the very least he should also be shown examples of EBICs with proven economic advantages of lowered real production costs. The developer should be convinced to seek suitable industries to relate the economic and environmental advantages of his EBIC. The developer is then fortified as a seller of a more advantageous land utilization system.

2. *Location*: The location of the proposed complex is an extremely vital component of the plan. The site must be acceptable to each of the industrial participants. Each will have its own preferences depending upon many components such as source of raw materials, market for its products and availability of economical utilities. Compromises by all participants will be necessary to arrive at a site agreeable to all of them.

One thing is certain, however, no longer will the managers of industrial plants need to be concerned with the effect of their wastes on the environment. This factor has been eliminated by the use of the EBIC. In fact, now the plants may be able to locate at a site more favorable to other production and marketing decisions. For example, a site adjacent to a metropolitan or residential area may be selected, since pollution of the city or homes is stopped by using an EBIC.

A site may eventually be selected which will be less costly to purchase and/or operate on once concern for external pollution is relieved.

3. *Compatibility*: Compatibility between plants will be sought. The developer must select proper industrial plants that fill the needs for raw materials of each other. Their wastes must be reasonably suitable for raw materials for one another with little or no alteration. All plants should

produce wastes which need reusing by others in order to avoid environmental contamination. The incentive—both moral and financial—would then exist for locating in an EBIC. In fact, economic incentives are major driving forces for such industries with social environmental incentives secondary for most industries. It is important to remember that all industries exist to produce a useful product "at a profit."

4. *Optimizing production sizing of participating plants*: The developer must search and obtain plants of the proper size so as to optimize production quantities of the participants. It will not do to select an industry which produces more waste (as a normal operation) than can be utilized by another participant (also as a normal requirement). Naturally there will be times when one participant will produce an excess of waste due to market demands for its product. The other industry should recognize this and be prepared to either increase its production or prepare to store some waste for future use. It is vital, however, that all of one plants' waste be utilized eventually by the other plant(s) in the complex.

5. *Informational meeting with participating plants and regulatory officials*: An informational meeting between all complex parties and the proper state environmental regulatory authorities should be held prior to the decision to locate, build, and operate the EBIC. At this meeting it should be made clear to the state government that the concept of no wastes reaching the environment is new and binding upon all plants in the EBIC. Flow diagrams, mass balances, and even production data may have to be prepared ahead of time and explained to them. The concept of waste utilization rather than waste treatment must be explained and defended to all regulatory authorities. This latter will facilitate receiving a permit to operate within the state.

6. *Designing the flow diagram for the complex*: It falls upon the consulting engineer for the EBIC to design the flow diagram and to prepare and present each plant's flow diagram of its product, water, and wastes (air, water, and solids). He will have to work first with each industry separately and then with personnel of all plants jointly in this venture. It must be explained to all participants that no wastes can be left out and all wastes must be reused by another plant. The consultant must substantiate that all wastes are reusable by another plant. Proof of these assertions may come from the literature, other plants' experiences or even pilot plant studies, if necessary. The consulting engineer's new role is that of designing a mass balance flow diagram that guarantees full waste utilization within the complex. In some cases, it may be necessary for the engineer to design certain waste alteration or amelioration systems to provide a directly acceptable and reusable waste for one of the plants.

7. *Developing a computer program of varied inputs and operating conditions*: It is now necessary to develop a computer program that contains the various inputs and operating conditions of the EBIC. There are many possible computer programs which can optimize the efficiency of the

EBIC. But, all of them must contain the objective of eliminating any unused wastes. One potential simple example can be expressed by the following equation:

$$(P1 + P2)(W1 + W2) = (P1 + P2)W3$$

where P1 = production units of plant one
P2 = production units of plant two
W1 = wastes of plant one utilized by plant two
W2 = wastes of plant two utilized by plant one
and W3 = total wastes unused (not reused) from both plants plus the reused wastes (W1 + W2)

By increasing or decreasing P1 and P2 both W1 and W2 will increase or decrease accordingly. The goal of a perfect EBIC will be to produce no excess waste (W3) within the complex regardless of the levels of P1 and P2.

This can only be accomplished when W1 + W2 = W3.

For example, suppose plant 1, when operating normally produces 100 units of goods and 20 units of wastes; while plant 2 produces 50 units of goods and 10 units of wastes with no unused wastes.

$$\begin{aligned} \text{then } (100 + 50)(20 + 10) &= (100 + 50)(30) \\ \text{or } 150 \times 30 &= 150 \times 30 \\ \text{and } 4500 &= 4500 \end{aligned}$$

And now plant 1 increases its production to 150 units and 30 units of waste. If plant 2 does not also increase its production by 50%, then

$$\begin{aligned} (P1 + P2) \times (W1 + W2) &\,/\!= (P1 + P2) \times W3 \\ (150 + 50)(30 + 10) &\,/\!= (150 + 50)50 \\ 200 \times 40 &\,/\!= 200 \times 50 \\ \text{and } 8000 &\,/\!= 10{,}000 \\ (150 + 50)(30 + 10) &\,/\!= (150 + 50)50 \end{aligned}$$

Thus plant 2 must also increase its production in order to reuse all of plant 1's waste. The program will reveal how much production each plant must have in order to completely reuse all of both plant's wastes.

8. *"Dry run" (on paper) of complex including all potential variations*: Prior to an agreement of the participants in the EBIC they should engage in a "dry run" on paper including all potential variations of possibilities. Each plant participant would begin by revealing its planned starting production quantity and raw materials needed along with amounts of liquid, gaseous, and solid wastes expected to occur from that production. At this level of operation each plant would then agree on the amount of each others' waste which would be acceptable as a replacement for a portion of its raw mate-

rial. Ideally all waste quantities could be used to replace part or all of the raw material requirements of other plants in the EBIC. This would then result in zero discharge of any waste external to the EBIC.

If and when the proportions of wastes to raw material is not an exact match, participating plants must make adjustments in production quantities so that all wastes are reused.

If adjustments in productions cannot be made satisfactorily, participants would have to agree on how to store and/or handle the wastes rather than discharge them externally.

Regardless, discussions should ensue concerning all potential variations in production possibilities. These should include no production situations as well as greatly accelerated rate of production requirements. All potential situations should be discussed at this session so that no surprises or unusual events will occur under real conditions.

Emphasis should be placed on making certain that no situations arise where unusable waste cannot be handled without discharging outside the complex.

9. *Review architectural and engineering plans of EBIC*: Architectural and engineering plans for the EBIC are now in order. The engineering consultant for the EBIC should prepare preliminary plans for the location of the piping, plumbing, and mechanical and electrical equipment of the plants. These should be coordinated with those of the architect/engineer for each of the plant participants. These latter people have the responsibility to prepare the design and specifications and location for all production equipment of the plants. The engineering consultant, on the other hand, is responsible for all piping, pumping, and other treatment units for the water supplies and liquid, gaseous, and solid wastes arising from the production of manufactured goods in the complex. The two groups of consultants should work together closely—especially where production and wastes locations interface. In that manner decisions involving responsibilities for design and rationale for location of equipment can be discussed and agreed on. The final output from these two groups should provide a master plan containing the complexes' entire layout of production and wastes reuse, piping, pumping, and equipment.

It is recommended that all plant engineers and/or managers be present at meetings with the two groups of consultants. After the final master plan has been presented and approved by all plant administrators these plant managers would be familiar with reasons for and operations of all manufacturing units. This is extremely important if and when any unforeseen EBIC changes in operations are necessary.

Our overall aim is to produce a team of EBIC plant managers capable of making and following through on all production as well as environmental decisions for all plants in the complex

10. *Participate and observe construction of EBIC*: The participation and observation of the construction of the complex is of utmost importance.

Once the master plan for the complex has been prepared and approved construction permits must be obtained from local governments as well as state environmental agencies. The environmental consultant as well as engineers for all participating industries in the complex should supervise its overall construction.

Of prime importance during this phase of the project is that of making certain that the construction follows the plans accurately. The flow of all liquids, gases, and solids to and from all plants should exist as designed. Pipe and pump sizes should be ample to carry waters and wastes several times that of average flows. The consultant should pay special care that no liquids overflow or seep into the ground in the complex, and that no solid wastes be placed on bare ground while awaiting movement to be reused within the complex. Without any doubt construction should insure that neither liquid, gaseous, nor solid wastes arising from the participating plants can possibly escape the complex into any external air, land, or water environment.

11. *Observe and consult with plants during startup of EBIC*: The startup of operation should be observed by all plant key operating personnel. The environmental consultant should be onsite during the initial startup of the plants in the EBIC. Timing has to be such that all plants begin production simultaneously. Questions of waste reuse and utilization by the plants will be expected to occur. The consultant's presence provides assurance that these wastes—regardless of quality concerns—will be used entirely within the complex. Modifications may be necessary during the startup period to meet these requirements. Despite all previous planning, innovations may be deemed necessary. For example, some solid wastes may have to be altered or divided among production locations prior to acceptable reuse. Some liquid waste may require filtering, heating, disinfection, etc., so that production can proceed properly. Once again, the environmental consultant is on hand to answer questions, offer suggestions, but most of all to make sure that no external environments are impacted by accidental or purposeful discharge of wastes.

Realistic Implementation of the EBIC System

Perhaps the most important step and most difficult to instigate is that of implementing the EBIC system. I propose that it is most likely to commence in any of the following manners.

1. By state statute
2. By business acumen of a property developer
3. By either of the above, but with the aid of local governmental provision of tax-free land and/or services

There have been precedents for all of the above in environmental matters. However, some discussion of each of these procedures is in order.

1. *By state statute*: The states have administered water quality standards since the last half of the 20th century. These included not only establishing the standards, but also advising and reviewing the planning and procedures as well as enforcing the rules and regulations. In our case states would have to be empowered to require all waste-producing industries to locate their plants in complexes so that their wastes are reused as raw materials by ancillary plants. The states would then be assured that no wastes from these industries would reach the environment outside the complex.

2. *By business acumen of a property developer*: Ambitious entrepreneurs may acquire a large piece of property and seek and induce compatible industries to build and operate within the property complex. Such property owners/managers must understand and be champions of environmental protection. They must also be able to find and convince these industries that their manufacturing costs would be minimized within this complex, while at the same time their environmental pollution concerns would be eliminated. The support and encouragement of local and state governments would enhance land developers in obtaining industries for the complex. They might even join forces with environmental engineers to advise these industries on waste utilization and water quality. This would be a selling point for the developer and permit him to lease at a premium industrial property within the complex.

3. *By either of the above, but with the aid of local governmental provision of tax-free land and/or services*: The enactment of state statutes and/or the enterprise of entrepreneurial land developers would be greatly beneficial as local inducements to industry. In the 1940s–1960s it was the common practice of small governmental jurisdictions to set aside tax-free land to induce desirable industrial plants to build there. Often they also provided utilities such as water and sewage at reduced rates. They also built pipelines to accept their wastes and roads or railroad spurs to facilitate transporting of raw materials and finished products. Obviously the combination of 1 and 2 with 3 above would be the most desirable approach to implementing the EBIC concept.

From a practical standpoint, however, it will probably take an unusual, urgent event of some kind before the logical use of EBICs will commence. Industrial enlightenment usually occurs as a result of critical situations such as dangerous environmental pollution or disastrous economic conditions. When these occur, and coupled with the three procedures listed above, we will see the day of the natural EBIC begin as standard operating practice (SOP).

Economic Justification For Industrial Complexes

Changing from a "waste treatment" to a "waste utilization" culture involves overcoming many obstacles. The most important of these is proving to industry that by using an EBIC production dollars spent—as well as society's dollars used—will be less. Even then changes come slowly. Human nature seems to be comfortable with tried and proven practice, while, on the other hand, it abhors changing to an unknown one.

In the long run history has shown that, if we can "build a better mousetrap" at less cost, eventually industry will follow like sheep in the field. It is not within our ability to alter human nature, but we as engineers and scientists can show that economics of production and conservation of natural resources favor the EBIC approach.

Industry is—and always was—in business to make a profit, and the more profit the better. This is not to say that industry does not recognize, and is even sympathetic to the environmental cause. But, in its eyes the bottom line comes before any other consideration. Industry believes that it must concentrate on the bottom line to "stay in the game." And "staying in the game" is absolutely essential. It also has become increasingly aware that reusing water—a valuable diminishing and costly resource—is as important as reusing wastes themselves.

I present one recent (1999) decision made by industry to substantiate the above principle. The College Retirement Equities Fund (CREF) was established and continues today to invest its teachers' retirement money in stocks of industries to produce the greatest growth in profit (the bottom line) to them. A group of CREF participants presented a proposal in November of 1999 to the board that it divest itself of its holdings in a particular metals-producing industry stock. The participants proposal was based on the fact that the industry "created and continue to pose unreasonable or major environmental, health, or safety hazards with respect to the rivers that are being impacted by the tailings, the surrounding terrestrial ecosystem and the local inhabitants."

CREF's board rejected the proposal saying "were we to divest from a specific company because some participants object to that company's environmental or social record, there would be no reason why a multitude of such types of requests could not be made." That was CREF's reasoning; but here is what lies behind that reasoning. It continued to state, "It would be difficult to fully consider those requests *and run an effective investment program* for participants *who wish their investments to be based primarily upon financial analysis*." In other words money once again drives decisions and once again even at the expense of the environment—which belongs to everybody.

The concept of EBIC was originally proposed for the pulp and paper industry by this author in 1978 (Nemerow et al. 1978). In the next 27 years we have published many papers describing potential industrial complexes for a number of other industries. The majority of these are described in the author's 1995 book (Nemerow 1995).

Rationale for EBICs

The field of industrial waste treatment as practiced from the 1940s through the 1980s is now evolving from treatment to waste utilization as we begin the 21st century. Society is calling for lower manufacturing costs along with less environmental degradation. The use of EBICs is not only the logical answer, but the only rational response to society's demands. This system reuses one plant's waste as another's raw material, thus simultaneously reducing raw feed costs and eliminating waste treatment costs. However, the EBIC system depends on the inclusion of compatible industrial plants. Such a system completely changes our concept of industrial manufacturing. No longer should we locate industrial plants based solely upon the economic marketability of our product, but rather we must consider the usefulness of wastes as raw materials for the ancillary plant. To discharge wastes untreated or partially treated into the environment is no longer an alternative. And, to completely treat the same waste prior to discharge is too costly for both the industry and society. Simple logic dictates that this waste be utilized directly by another manufacturer to save operating capital for the plants and, simultaneously improve the quality of the receiving environment for society. When industry also reuses water as well as wastes, it satisfies another of its objectives as well as that of environmentalists.

In prior days industry was more concerned with production problems and costs at a particular site rather than importation of raw material costs. This is no longer the case. When production costs become competitive, and often prohibitive industry is concerned with importation costs of its raw materials. In an example of a recent decision involving these specific costs Tyson Foods stated, "Company officials say they will shift production of its bacon brands, which include Thorn Apple Valley and Colonial to more modern plants closer to its suppliers in the Midwest. Tyson says it isn't sure how much money the move will save, but it is clear that the cost to this town (Holly Ridge, North Carolina) is huge" (*Wall Street Journal*, 2002).

In a recent research project with these two industries these costs were identified as shown in Table 11.1 and delineated more fully in the next section (Case Study of Economic Proof of Industrial Complexes).

More specifically, the damage costs identified for these two industries were attributed to the following.

1. Unsightly collection and storage of the voluminous sludges on increasingly valuable lands
2. Leaching of contaminants from the sludge piles or stacks, which adversely affect drinking waters and fish life downstream
3. Grinding and burning dusts from the cement plants affect all three environments

TABLE 11.1
Comparison of Real Production Costs (1995 Dollars) of Free Standing Fertilizer and Cement Plants with EBIC Costs (1995 Dollars) per Ton of Fertilizer

Industry	*Real Cost includes the 3 damage Costs*	*EBIC Cost*	*Savings in Production Costs when Using EBIC*	*%Savings due to EBIC*
Fertilizer	$245.76	$183.95	$61.81	25
Cement	$67.09	$34.00	$33.09	49

Direct reuse of phosphogypsum and slime sludges in the cement plant within the complex and reuse of the dust by an adjacent fertilizer plant will lessen or eliminate all damage costs from these plants (as shown in column 3 of Table 11.1). A clean environment surrounding the plants will also be achieved.

Case Study of Economic Proof of Industrial Complexes

In this section I offer practical data proof, in one instance, that it is substantially less costly to produce two industrial products and reuse all wastes within one complex than to produce the same products at separate locations and discharge the untreated wastes to the surrounding environment. The original researchers[2] of this project have attempted to use real data to put a monetary value on the indiscriminate and wanton discharge of these plants' production wastes into the environment. When this value is added to the manufacturing costs of the two products, the real total production cost becomes 37% higher than that of the industrial complex. This finding should encourage you to consider the EBIC concept solution whenever and wherever possible.

Introduction

Industrial complexing is an innovative attempt to improve environmental quality, while at the same time lowering production costs. The ultimate goal of both the environmental and production engineers is to attain zero

2. S. V. Krishnan, N. L. Nemerow, T. N. Veziroglu, P. Khanna, and T. Chakrabarti, in 1996, under a National Science Foundation grant (University of Miami, Florida) and the National Environmental Engineering Research Institute (NEERI) (Nagpur, India).

pollution and a minimum, manufacturing cost. Before reaching this goal, industry **must include all environmental damage costs—direct, indirect and intangible—as part of the cost of manufacturing**.

The following is our attempt to measure all the environmental damage costs, add them to normal production costs, and obtain true and real costs of manufacturing a product. When armed with these true costs, industry can decide whether to cease deteriorating the environment by using EBICs.

Fertilizer and Cement Production

In conventional practice, phosphate fertilizer and cement are manufactured as shown in Figure 11.1.

The two major wastes impose environmental damage on the surroundings. Figures 11.2 and 11.3 present basic schematics of fertilizer and cement plants EBIC. In it, all four wastes are reused. Some raw materials are substituted by these wastes. Transportation of some raw materials is eliminated, resulting in cost saving. This complexing system thereby reduces production costs even without considering the benefits of abating all environmental damages.

Environmental Consequences of Complexing

Two significant damages caused by phosphate fertilizer plants' phosphogypsum wastes are:

1. Unsightly collection and storage of the voluminous sludges on increasingly valuable land
2. Leaching of contaminants from the sludge piles or stacks, which adversely affect drinking waters and fish life downstream

Direct reuse of phosphogypsum and slime sludges in the cement plant within the complex can eliminate the above stated two damage costs.

In addition, the dust reaching air around cement plants—both from the grinding and the burning operations—damage the environment. Direct reuse of these dusts by an adjacent fertilizer plant will lessen damage costs from them (see the schematic configuration shown in Figure 11.1).

Environmental Damage Costs

Three types of environmental damage costs result from industrial wastes.

The three categories of benefits accruing to society from using the environmental complex principle are presented here below (Nemerow and Agardy 1998b).

1. *Primary benefits*: those affecting the industrial plants themselves, resulting in direct costs

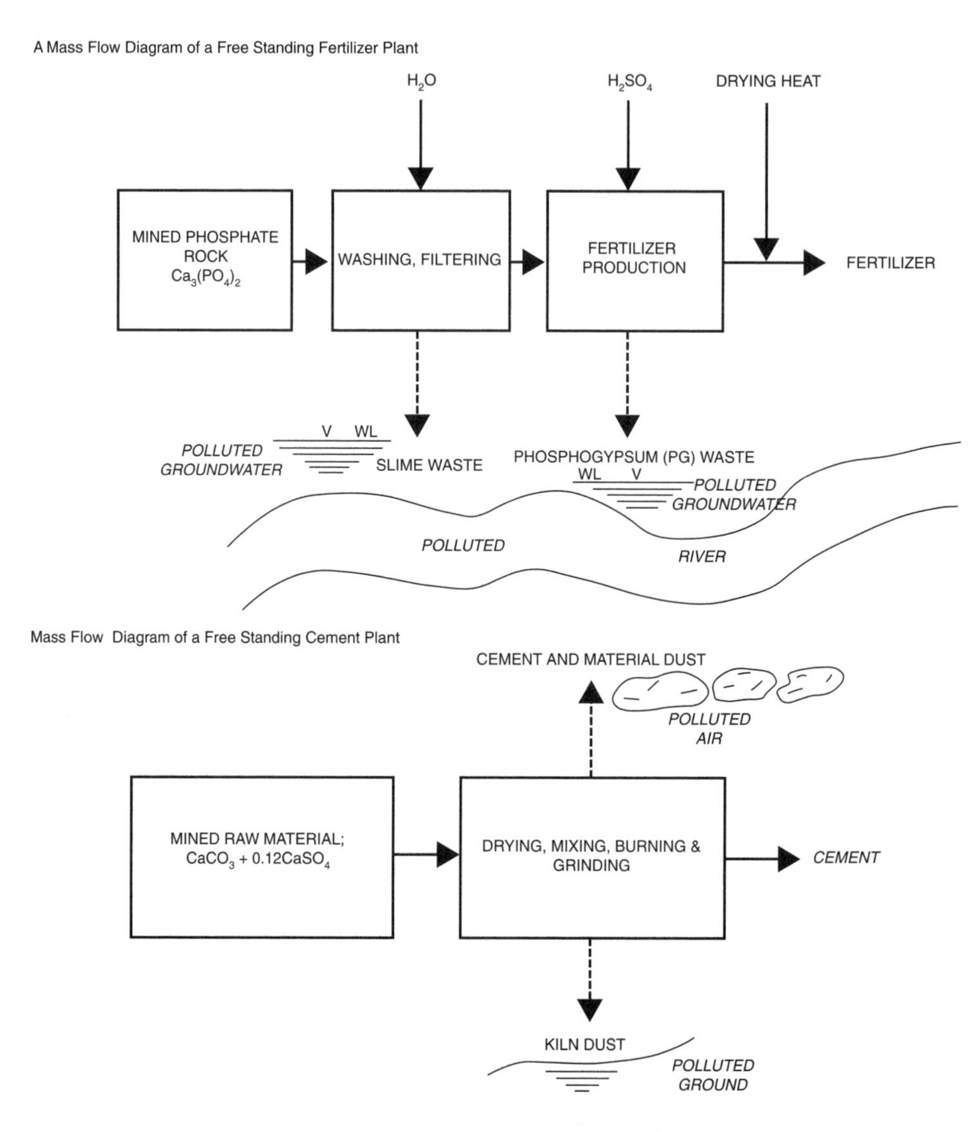

FIGURE 11.1. Mass flow diagram of a free-standing fertilizer plant.

2. *Secondary benefits*: those affecting the people surrounding the plants, resulting in indirect costs

3. *Intangible benefits*: those affecting society as a whole, and are difficult to quantify, resulting in intangible costs

When costs are tied to these benefits and added to normal plant production costs, one obtains the real cost of manufacturing a product.

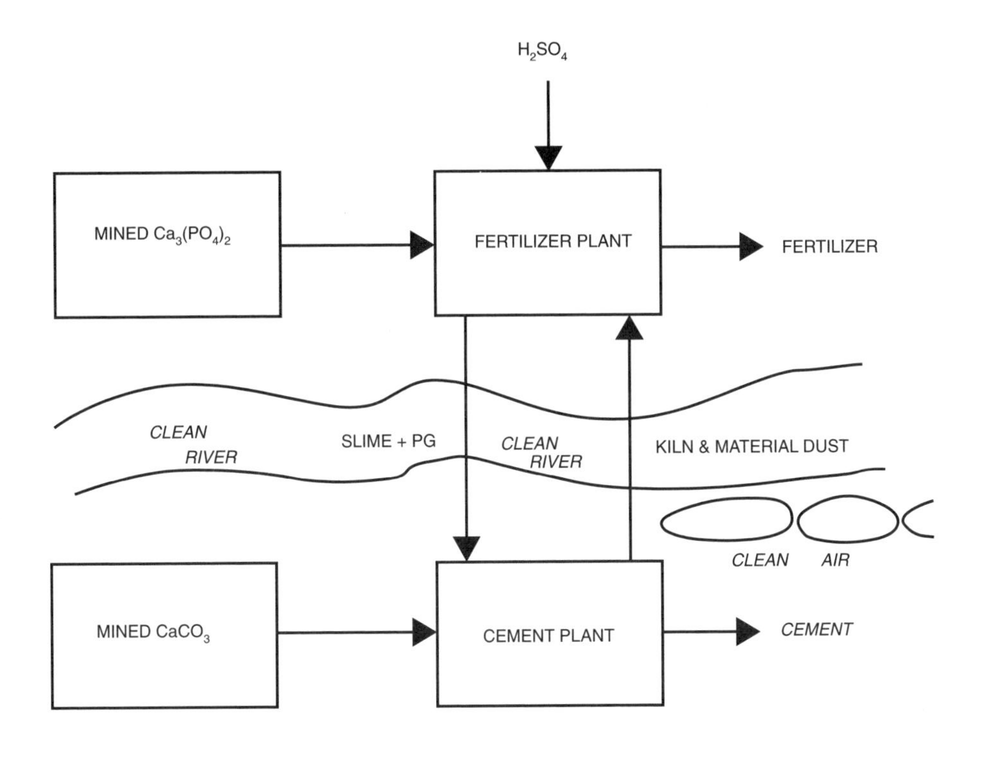

FIGURE 11.2. Mass flow diagram of a fertilizer—cement complex.

Real Production Costs

When a phosphate fertilizer plant is located alone, its production costs include the usual operation costs: maintenance, materials, and labor (Fc); the direct costs of required waste treatment (Fwt); the indirect cost of environmental damage to nearby owners (Fnd); and the intangible costs of environmental damage away from the plant and to the public at large (Fxd). In summation, the real production cost (Fr) becomes

$$\mathrm{Fr} = \mathrm{Fc} + \mathrm{Fwt} + \mathrm{Fnd} + \mathrm{Fxd} \tag{1}$$

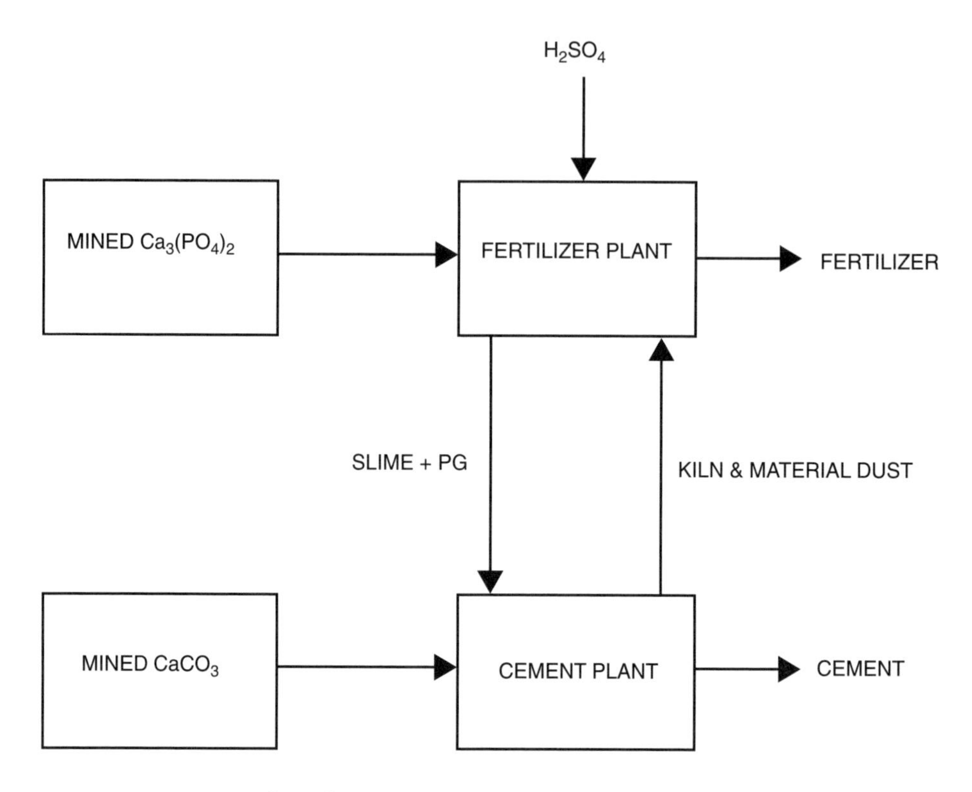

FIGURE 11.3. Mass flow diagram of a fertilizer—cement complex.

Each of the quantities in Equation (1) will now be identified for the fertilizer plant.

Fc is the Classical Production Cost

In 1995, 831,607 tons of phosphate fertilizer were manufactured by the average fertilizer plant in the United States.[3] In addition, the average annual production cost in a fertilizer plant is \$169.66 million. Therefore, the 1995 fertilizer production cost (Fc) was \$169,660,000/831,607 or \$204.02 per ton.

Fwt Is the Direct Cost of Waste Treatment

In this example, this cost includes that spent in preventing phosphogypsum piles from reaching the environment; in storing and preventing the escape of tailing wastes; in abating air contaminants from the production of fertilizer from sulfuric acid reactors; and other types of waste treatment. Some real examples of plant expenditures for these waste treatments include the following.

3. Private communication, 1995.

IMC-AGRICO spent $1 million in 1994 for water testing and new wells as a neighborly gesture, but denies that its mining caused water pollution and subsidence problems (Satchell 1995). The company has voluntarily spent $6.8 million to plug a sinkhole and control the spread of contaminants to the ground water. Assuming the life of a stack to be 10 years, this cost becomes $0.83/ton of fertilizer (Fwt2).

CORGILL fertilizer placed at least 18 inches of compacted clay over a layer of at least 15 feet of natural clay at its Hillsborough County, Florida mine (Newborn 1992). The company spent $22 million for lining the base of the new stack, plus another $5 million to close the existing gypsum stack. These expenditures for top coverings, when projected to 1995 costs, resulted in $3.73/ton of fertilizer (Fwt3).

IMC completed lining a new stack at its New Wales mine in Polk County, Florida with 20 million square feet of plastic for a cost of $70 million. This expenditure for bottom lining amounted to $9.67/ton (Fwt4). Therefore,

$$Fwt = Fwt1 + Fwt2 + Fwt3\ Fwt4 = \$15.46/ton$$

Fnd Is the Indirect Cost of Environmental Damage to Nearby Owners

In 1987 the State of Connecticut established values of fish killed by acid leakage; this situation amounted to $1,082 (Discover Magazine 1988) (Fnd1) or $0.01/ton of fertilizer.

On December 21, 1994, IMC-AGRICO agreed to pay $1.1 million to settle a lawsuit filed by the EPA, which had charged the company with violating water pollution limits at nine locations due to slime waste contaminating ground water. The cost of the legal proceedings came to $1.35/ton of fertilizer (Fnd2).

In 1994 the phosphate industry contributed $100,000 to some 25 "green ground societies." In 1988 the Audubon Society received $42,500 from IMC alone. These indirect costs amounted to $0.19/ton of fertilizer in 1995 dollars (Fnd3).

The industry also contributed to candidates for state and local offices. Total contributions were $160,000 for 1994. Donation indirect costs for election campaigns amounted to $0.20/ton of fertilizer (Fnd4). Therefore,

$$Fnd = Fnd1 + Fnd2 + Fnd3 + Fnd4 = \$1.75/ton$$

Fxd is the Intangible Cost of Environmental Damage

These costs have been elusive to pin down and may very well be the greatest of the three types of damage costs. In our research effort described in this chapter we established the following specific values.

Florida phosphate mining companies have paid about $1 billion in taxes to the state in the past 25 years, hence causing some influence on them. This amounts to a mean yearly value of \$51 per ton per year, assuming that 50% of them are small and operate on a small margin of profit thereby being adversely affected by state action to enforce pollution control measures.

Assuming, again, that 50% of these plants are then estimated to have closed down because of waste treatment pressure, the public lost 25% of the \$51 million of revenue taxes due to these plant closures. Hence,

$$\text{Fex1} = \$15.33/\text{ton of fertilizer}$$

Land value in the Tampa, Florida industrial area decreased from about \$1,250 per acre to \$600 per acre mainly due to fertilizer pollution over a period of 10 years, or \$65 per acre per year.[4] Therefore,

\$65 × 1,000 acres × 17 industries × 5 gypsum stacks/industry
divided by 831,607 tons per year
equals \$6.64 per ton of fertilizer
equals Fxd2

The State of Florida recently (1996) purchased 38,251 acres of land along five riverbanks for a price of \$21,270,000 or \$547.77 per acre (Browning 1996). This purchase is aimed at purifying these lands from the wastes mainly from fertilizer plants over a 10-year period. Therefore,

$$\text{Fxd3} = \$2.56 \text{ per ton of fertilizer}$$

The total intangible costs we were able to identify was calculated to be

$$\begin{aligned}\text{Fxd} &= \text{Fxd1} + \text{Fxd2} + \text{Fxd3} = \$15.33 + \$6.64 + \$2.56\\ &= \$24.53/\text{ton of fertilizer}\end{aligned}$$

Substituting our computed values in Equation (1), we obtained a real production cost of

\$204.02 + \$15.46 + \$1.75 + \$24.53 = \$245.76, when the major measurable environmental damages were considered.

Using the same procedure to compare cement plants, the cost relationship becomes

$$\text{Creal} = \text{Cc} + \text{Cwt} + \text{Cnd} + \text{Cfd}$$

4. G. M. Lloyd, personal communication, May 26, 1995.

Cc is the Conventional Cost of Cement Production
An average of 984,000 tons of cement were produced per cement plant in the United States in 1995 at an average production cost of $49.2 million.[5] Therefore,

$$Cc = \$49{,}200{,}000/984{,}000 \text{ tons} = \$50/\text{ton of cement}$$

Cwt Is the Cement Waste Treatment Direct Costs
Cement kiln dust is a major waste to the air environment and needs collection and disposal to protect the surrounding air. Kessler (1995) reports that typically each percent of dust wasted increases the specific heat consumption by about 0.7% and decreases clinker production by 0.5%. He gives the dust losses costs as Cwt1 = $4.08/ton of cement, due to the loss of raw material; Cwt2 = $4.59/ton by feed crushing, conveying, and drying and grinding costs; Cwt3 = $1.02/ton for transporting, conveying, handling, and dedusting; Cwt4 = $3.06/ton for landfill maintenance, monitoring, pile maintenance, and closing.

Total waste treatment costs for kiln dust only becomes

$$Cwt = Cwt1 + Cwt2 + Cwt3 + Cwt4 = \$4.08 + \$4.59 + 1.02 + \$3.06 = \$12.75$$

per ton of cement

CNN is the Indirect Cost of Cement Plant Waste Environmental Damages
In 1992 the EPA fined Lafarge's Michigan and Alabama cement plants $1.8 million for violating air emission operating rules (Ferguson1993). The cost (Cnd1) was $1.94/ton of cement.

Lafarge's switch to power plant fly ash from ground shale may have saved as much as $600,000 in fines levied by a U.S. District Court (*Engineering News Record* 1994). The additional cost (Cnd2) resulting from environmental violations was $0.62/ton of cement.

In 1992 a local customer sued Lafarge for $1 million over improper disposal for chromium tainted materials over a 5-year period. The cost of legal compensation (Cnd3) for affected victims was $0.22/ton of cement.

These three indirect costs amounted to

$$\begin{aligned} CNN &= Cnd1 + Cnd2 + Cnd3 = \$1.94 + \$0.62 + \$0.22 \\ &= \$2.78 \text{ per ton of cement} \end{aligned}$$

5. Private communication, 1996.

Cad Is the Intangible Cost of Cement Plant Environmental Damages

Since 1990 Hillary Clinton has been a director of Lafarge Cement Corporation, one of the largest operators of cement kilns fueled by burning hazardous wastes (Zweig 1992). In 1991 the First Lady earned $30,000 from Lafarge. It may have been inferred by some that this may indirectly influence the state and federal policies on Lafarge. The cost of potential influence (Cad) in this case was $0.03/ton of cement.

All cement damage costs equal

$$\text{Cwt} + \text{Cnd} + \text{Cxd} = \$12.75 + \$2.75 + \$0.03 = \$5.56/\text{ton of cement}$$

The real cement plant production costs equal

$$\text{Creal} = \text{Cc} + \text{Cwt} + \text{Cnd} + \text{Cxd} = \$50 + \$12.75 + \$2.78 + \$.03 = \$65.56/\text{ton}$$

EBIC of a Phosphate Fertilizer and Cement Plant

When these two plants are located in the same complex, all direct costs of waste treatment, indirect costs of environmental damages to nearby neighbors, and intangible costs to the general public are eliminated. In addition, transportation costs of replaced fertilizer raw material and cost of replaced fertilizer raw material decreases the real production costs.

The transportation cost of the replaced raw material (Fr) was computed to be $20 per ton of fertilizer and the cost of the replaced fertilizer (Fr) raw material was computed to be $0.07 per ton of fertilizer. Likewise, corresponding values for the cement plant were

$$\text{Car} = \$6.00 \text{ per ton of cement}$$

and

$$\text{Cram} = \$10 \text{ per ton of cement}$$

Calculations were made for the production of one ton (total) of EBIC product as shown schematically in Figure 11.3. Since 3.5 tons of cement can be produced for each ton of fertilizer product, each ton of EBIC product would consist of 0.22 tons of fertilizer and 0.78 tons of cement.

To illustrate the truth of the economic value of transportation of wastes away from an industrial site, take the case of the American Waste Transport Corporation. This company is one of the major contract transporters of solid waste in southern California and southwest Arizona. In fiscal 1999, it had revenues in excess of $12 million according to a company statement (Clark 2000). It supplies materials to processors that convert waste into fertilizer and compost or burn it to produce energy. In our complex system we would eliminate this massive transportation cost and perform the same or similar reuses within the complex. To continue, the CEO of U Biomes said "they currently transport 1,400 tons per day of green wastes" and that "green waste recycling is expected to *grow dramatically as California begins to comply with its state* recycling law, which requires diversion of 50% of waste from landfills." This rush to transporting and reusing solid wastes is also occurring in many other states to comply with similar laws. Our proposed system, EBIC, eliminates these unnecessary and burdensome transportation costs.

Conclusions[6]

It is evident that real societal costs of products are more than the classical costs. In the case of fertilizer, it is 20% greater, and in the case of cement it is 34% greater. By industrial complexing, so that waste products of one industry become raw materials for the other, environmental damage costs can be eliminated and also the raw material and transportation costs can be reduced. These result in a cleaner environment and lower product cost. In the case of fertilizer and cement plants—in a weighted sense—there is a 20% saving as compared with real societal cost and 37% saving as compared with real cost. The author wants these industries to become aware of the advantages of industrial complexing, both financially and environmentally, so that whenever feasible such two or more industry complexes can be achieved to the benefit of the manufacturer, the consumer and the environment we live in.

Marketing Unused Waste Resources as a Solution

There may be instances where the EBIC will not be able to reuse all the waste materials completely. These cases could occur despite the concerted efforts of the industrial plants to reuse all wastes as raw materials. In these rare situations the EBIC participants should market these wastes in an efficient economic manner.

This author recommends marketing these excess wastes using a system I have been advocating officially since 1969 (Nemerow 1985). I maintained that the assimilative capacity of the environmental resource should play a major role in determining the price industry should pay for polluting it. The more assimilative capacity available the lower the unit sale price of the wastes. As the assimilative capacity of the resource (air, water, or land) becomes limited, the higher the unit cost to discharge the

6. The author acknowledges the financial assistance from the National Science Foundation, in the form of partial funding for this project. Thanks as well to Mr. G. Michael Lloyd of the Florida Institute of Phosphate Research and to Mr. Victor Turiel of Pennsuco Cement Corporation for assistance in providing significant data.

waste. Instead of discharge or treatment, of course, I recommend that another external buyer purchase these excess units of waste at the price predetermined by its detrimental effect on the environment. In that way the EBIC participants will not be forced to pay for polluting and the external environment will remain clean. If a buyer cannot be found, then the EBIC members have a choice either to pay a fair value for its untreated discharge into the environment or pay for its treatment before discharge.

In a recent book by Lomberg's, the reviewer[7] points out that "Clearly regulation has worked to improve these common areas (air and streams): our air and streams are cleaner than they were. But there is good evidence that assigning property rights and market mechanisms to such resources would have resulted in a faster and cheaper cleanup."

Further discussion of using this marketing system was also presented in our recent book (Nemerow and Agardy 1998c). As an example, you may determine that the beneficial damage of the industry's BOD pollutant was $10 per pound; in which case the industry would have the option to buy a certain number of BOD pounds "rights" to discharge at that price or treat its waste to remove that number of pounds. As described in the previous references (Nemerow 1985; Nemerow and Agardy 1998c), as the available limit of BOD diminishes the market price of a unit of BOD increases. The reason for this is to protect the water quality level of the receiving water. Industry is discouraged from using the last available BOD units to preserve that water quality level. Since the cost of buying BOD rights increases as the available resources decrease, industry must treat its waste—usually at a lower unit cost than buying rights. Of course, the possibility of buying lower cost BOD unit rights from another industry exists in the free market system.

The free market system could also be used for buying land unit rights for solid wastes and air capacity rights for air pollutants. The reader should be aware that this system of buying and selling "pollution rights" should be based on the benefit lost of the resource by adding an incremental pollutant load. The environmental benefit lost becomes greater, and hence does the pollutant right cost, as the available environmental resource gets used up. This method of pricing, although more difficult to compute unit costs, is preferable to an arbitrary price placed on the right by some overseeing agency. It is even preferable, in my opinion, than a truly free market pricing system since that is based on more tangible and measurable environmental damage costs.

Although Rinda E. Vas, the editor of *Environmental Technology*, found that "there are several kinds of market-like mechanisms that might be employed in environmental regulation," she doesn't quite include *a market charge based on damage costs* as one of them (Vas 2000).

7. The Skeptical Environmentalist by Bjorn Lomborg; and reviewed by Ronald Bailey (Wall Street Journal, October 2, 2001, p. A-17).

In evaluating current emission trading systems Bryner gets closer to my proposal of a system by concluding that

> Emission trading programs should lead to other, more powerful regulatory innovations that will more effectively encourage ecologically sustainable activities. Emission trading programs should be designed as a transition to a system of emission fees or taxes and other efforts to reflect true costs in prices and to create more powerful incentives to reduce and prevent pollution." (Vas 2000; emphasis added)

He concluded that "the ultimate test of an emission trading program is its contribution to a more fundamental shift in practices aimed at reducing pollution, improving efficiency, and conserving resources." All of these above practices are incumbent in my market pricing system proposed in 1985, 1998, and again in this book. Solomon and Lee (2000) wrote that "Despite the success of these trading systems in affordably reducing emissions, they have been criticized by several environmental organizations for allegedly creating toxic hot spots (local areas with excessively high emissions or concentrations of a hazardous air pollutant)."

Utility business is similar, in certain instances, to the environmental resource business. Units of power can be sold at a price based upon its real market value. The real market value can be reached by adding the existing kilowatt hours (kWh) charge to a unit local societal monetary loss to arrive at a total real value.

Available power supplies are decreasing fast, especially during peak power demand periods—the San Diego, California area is an example. Rose (2000) wrote "some suggest that power companies are holding back (construction of power plants), perhaps waiting for a crises that would provide them with the financial incentive to build." Rose also quoted Edwin Guiles (President of San Diego Gas & Electric): "We are in favor of all solutions being considered—new generation, demand-side alternatives, distributed generation—but we have to make sure there is a solution that can deliver in the time period we have."

I suggest that the purchase of kWh be based on benefit costs of not having power units above the basic level that exists. Such benefit losses include the following.

- Lower standard of living from lack of adequate air conditioning and heating
- Loss of industry production increase due to unavailability of power
- Lack of municipal growth due to inadequate power, etc.

One can then put added values on each excess of these and sell kWh to all consumers. The added dollar kWh charge can be used by power companies as an incentive to build and produce more kWh capacity. When extra kWh capacity is met, the dollar extra kWh charge can be dropped until demand exceeds supply again.

As recently as July 2001 the Op-Ed editor of the *Wall Street Journal* (2001) questioned whether pricing emissions is possible and advocated its use. The editor wrote that "by providing flexibility and financial incentives, a *cap-and trade program* [his term for selling resources] will result in more abatement from those firms who can do it at relatively lower cost. The net will be the same amount of overall pollution reduction, but achieved at lower cost than would obtain under traditional regulation."

The editor referred to the Energy Information Administration as stating that the cost of power plant CO_2 reductions according to the requirements of the Kyoto treaty agreement could be as much as 4% of the GDP. Whereas "in a scenario offered back in 1998 by the Clinton administration's Council of Economic Advisors, if the U.S. buys permits for its excess emissions-so that it doesn't have to reduce by very much its own emissions— the cost would be only 0.1% of GDP."

With these facts in mind the editor recommends that the Bush administration propose a domestic *cap-and-trade program* for CO_2 that could, of course, be expanded to Canada and Mexico, and later to Latin America and the rest of the world.

Realistic Industrial Complexes

In some cases I have gathered more industrial production and waste data than others. I classify these as *realistic industrial complexes* and are presented for illustration purposes in this chapter of the book. Later I propose other possible industrial complexes about which little operating data has been amassed. These EBICs are classified as *potential industrial complexes*.

Five EBICs are depicted here as realistic (Figs. 11.4–11.8). Some of these I have reported in earlier publications, and some have been fortified with additional data.

Potential Industrial Complexes

Introduction

There are many possible examples of potential industrial complexes which may accomplish the objectives set forth in Chapter 4. I list some here and provide brief descriptions of their configuration mostly in schematic ways. The reader may also be aware of others which may also meet our requirements for workable EBICs. Those in this chapter do not contain mass balances nor any other detailed operating data editions.

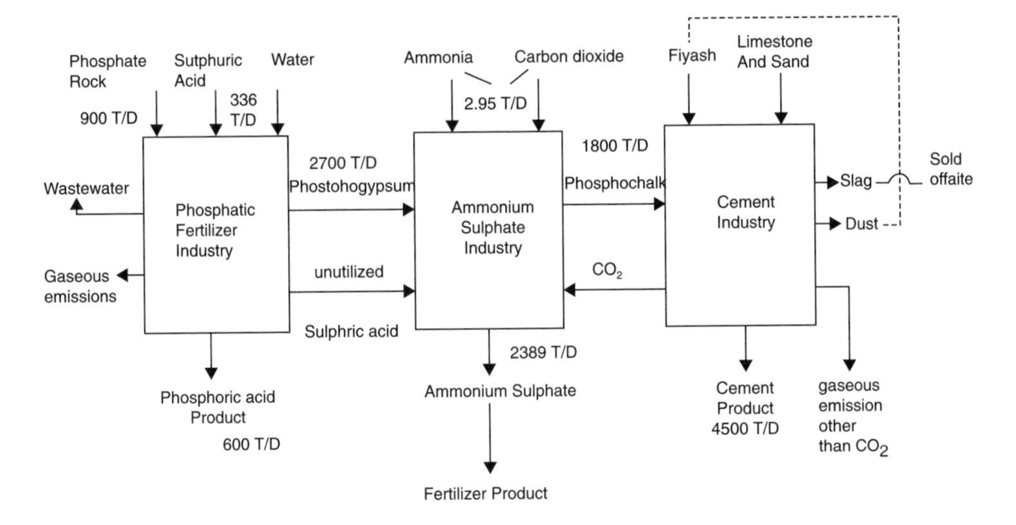

FIGURE 11.4. Schematic diagram of environmentally balanced phosphatic fertilizer cement industrial complex.

I am not so presumptuous as to aver positively that the complexes described here are (or ever will be) feasible. I present them here, only to stimulate the reader's thinking about these possibilities as solutions. If, in fact, it turns out that these potential complexes actually become reality, then I can look back and take some comfort in the fact that practice sometimes results from theory. If, on the other hand, they never become reality, I hope they at least provided the reader with the incentive to innovate. The principle of industrial complexing is an art of (a) finding a troubled waste-producing industry and (b) matching it with another industry that can alleviate the trouble by consuming its contaminating waste. Once you develop the technique of mastering this principle you will have utilized successfully the concept of *environmentally balanced industrial complex*.

The following are some of the potential EBICs known to the author and proposed to the reader at this time. Look for the wastes of one plant to serve as raw materials for the adjacent plant in the complex. In my next book I intend to present detailed schematic diagrams of such complexes.

Wood—papermill complex
Steel Mill—coke and gas and fertilizer plants
Finished metals—plastic plant complex
Organic chemical—wood processing plant complex
Waste waters—power plant complex
Steel mill—fertilizer—cement plants complex
Coal power plant—cement—concrete block plant complex

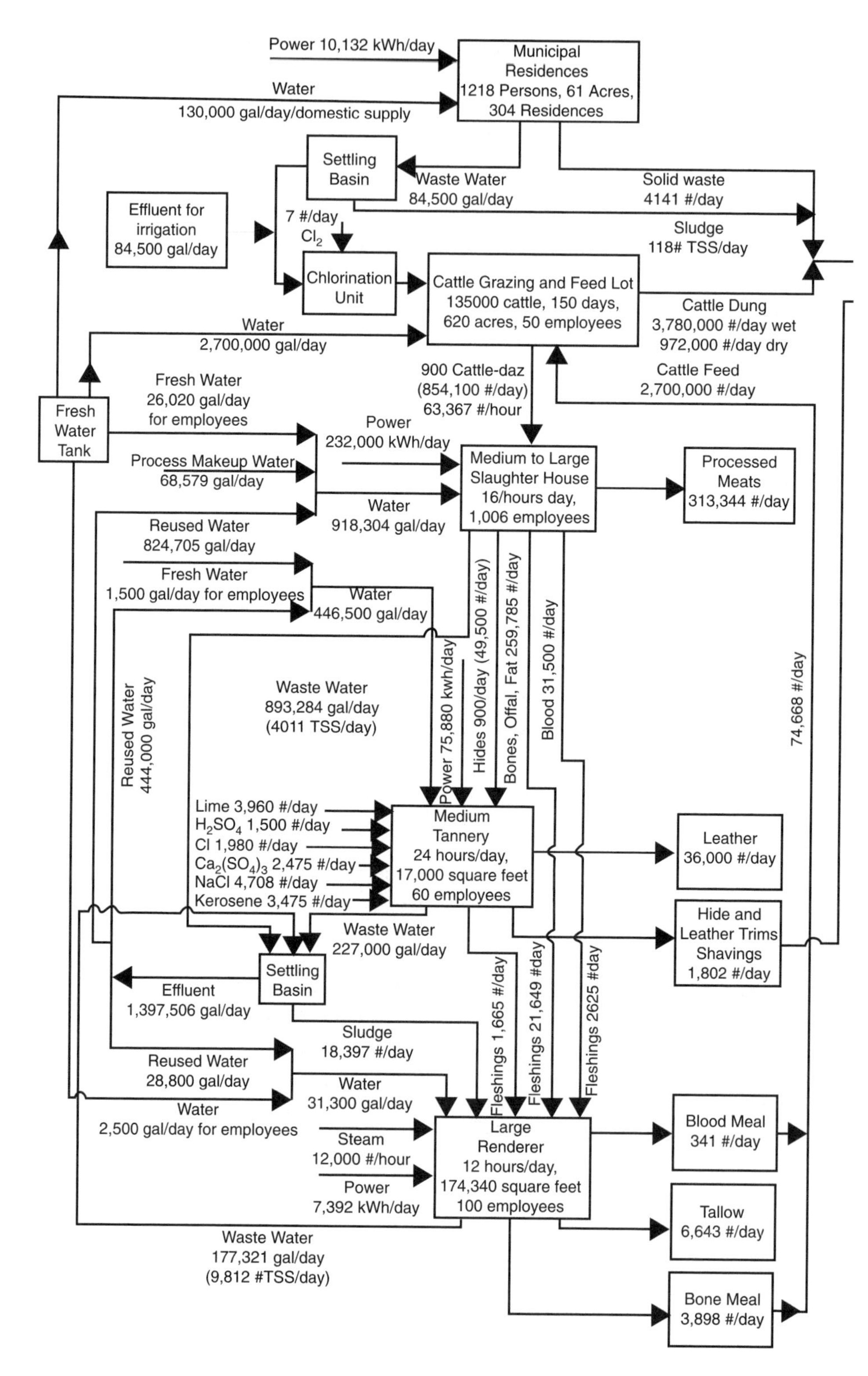

FIGURE 11.5. Three industry complex (tannery—slaughterhouse—rendering).

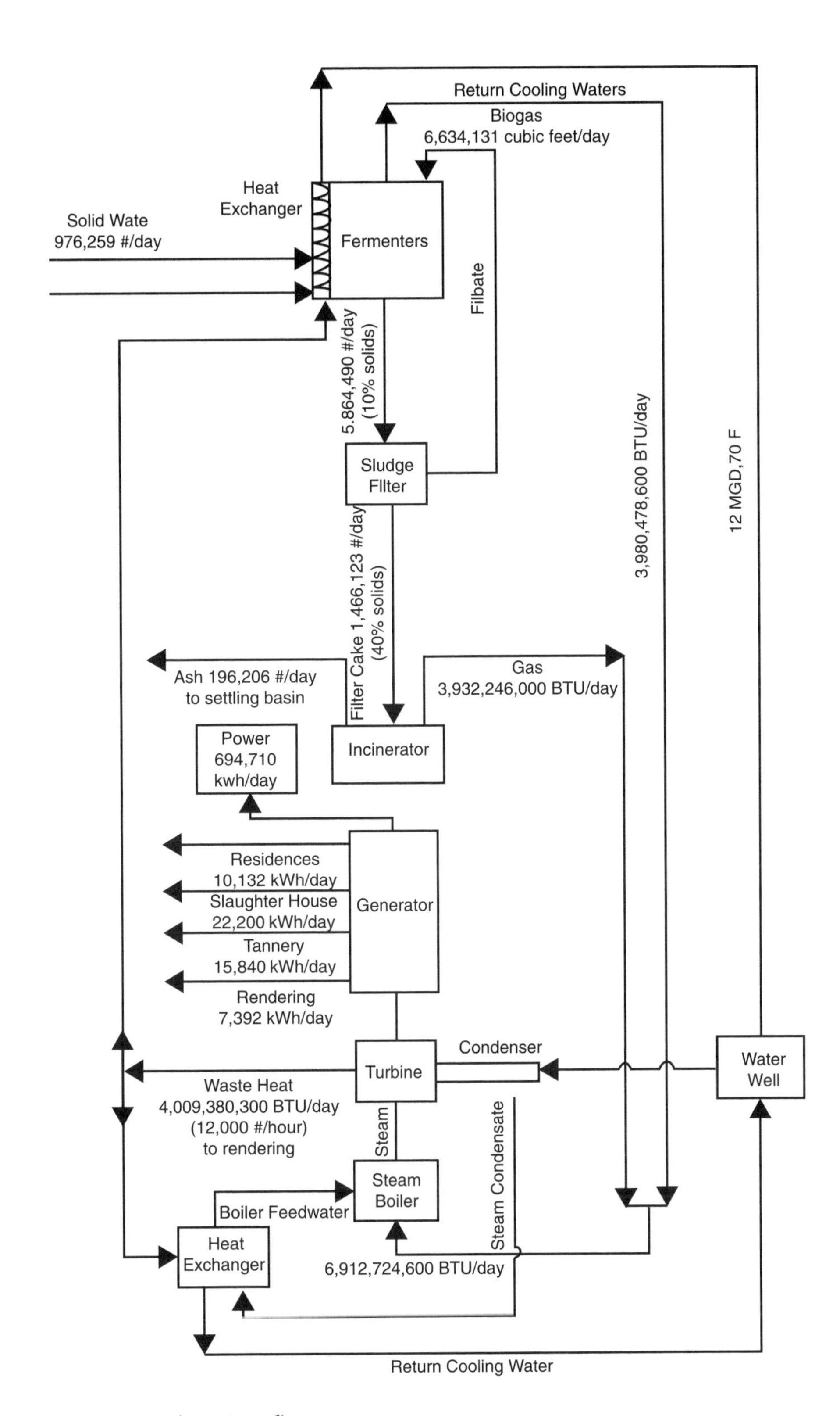

FIGURE 11.5. (*continued*)

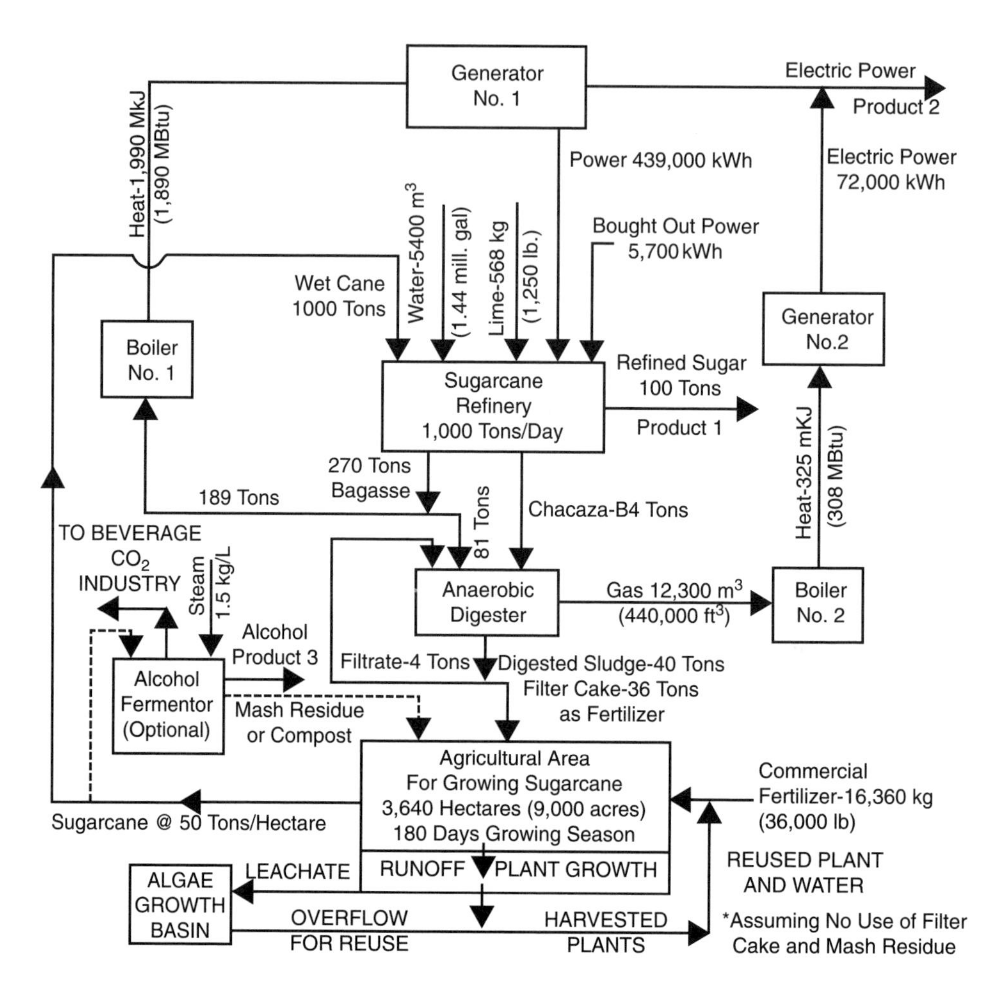

FIGURE 11.6. Sugarcane refinery-based EBIC (sugarcane–power–alcohol complex).

Plastic plant complex
Cement—lime—power plant complex
Wood—lumber mill complex
Power plant—agriculture complex
Cannery—agriculture complex
Nuclear power—glass block complex
Animal feedlot—plant food complex
Coke (steel mill)—tar—benzol plant complex
Wood—ethanol plant complex
Water—electricity—chlorine—lye plant complex

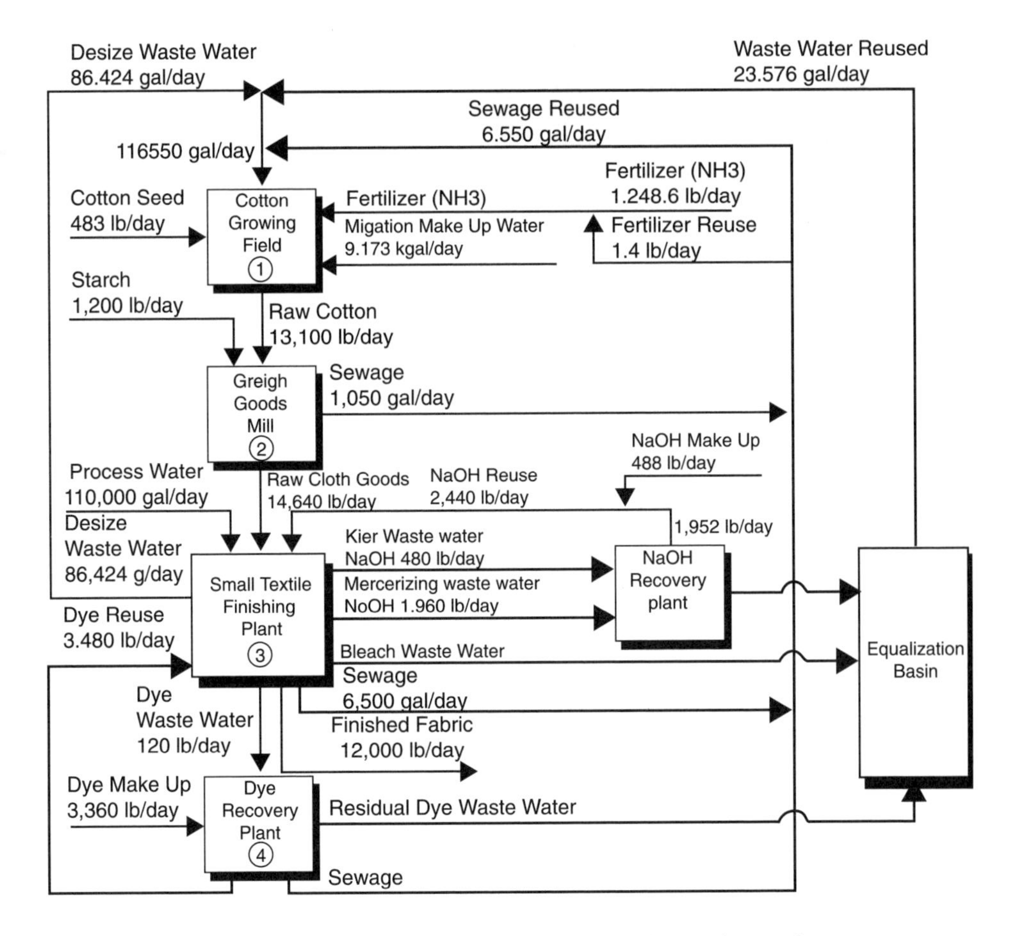

FIGURE 11.7. Diagram of the integrated five plant industrial complex.

Aluminum—electricity—red brick plant complex
Corn growing—alcohol producing plants complex
Restaurant—paint manufacturing complex
Oil drilling offshore—seashore recreation
Metal plants—dry cleaning plants complex
Electrical storing and/or converting voltage wax
Manufacturing complex
Nuclear power plant—waste reprocessing—cannery complex
Electric power—drinking water plant complex
Vegetable pickling cannery—inorganic chemical—chlorine plant complex
Sugar—ethanol—gasoline plants complex
Reclaimed cell phones—cement plant—concrete products complex

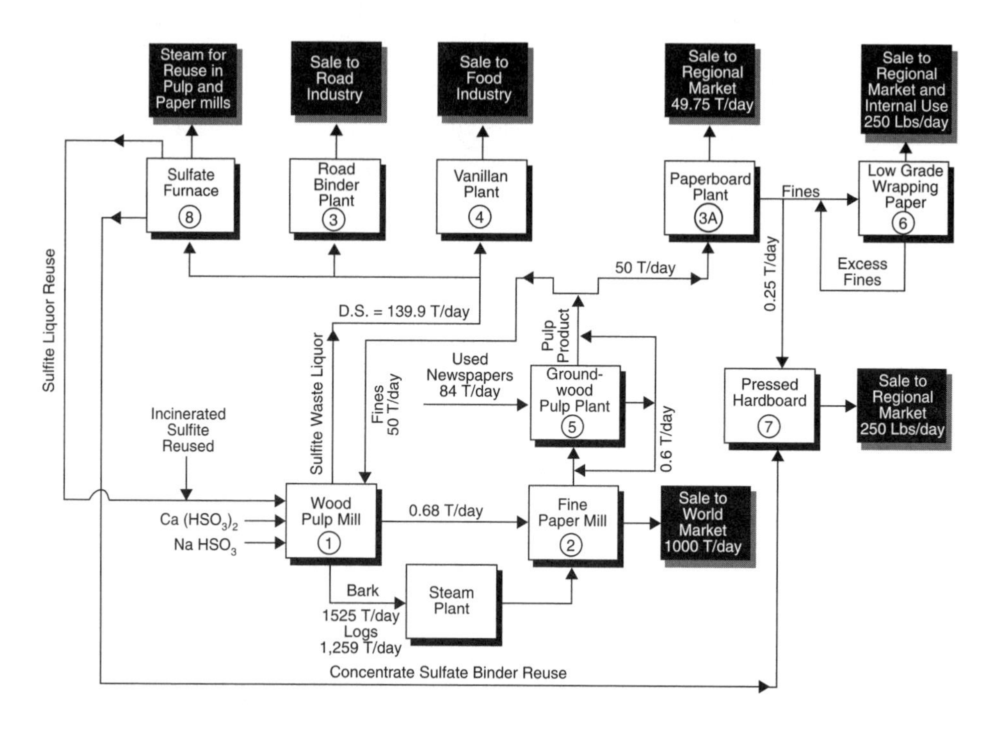

FIGURE 11.8. Pulp and paper mill complex.

Potential Municipal Industrial Complexes

Introduction

For many years municipalities have been cooperating with industry by permitting them to utilize its public plants to dispose of local factory wastes. I first wrote about this "joint treatment" in 1963. These systems allowed participating industry to contract with its municipal sewer service agency to accept certain amounts and types of wastes into their systems with and without payment provision. Thus, precedent has been established, provisions made, and experiences gained from these prior associations of city and industry. Some of the past arrangements of combined treatment turned out to be a boon to both parties while others resulted in an equal bane to both. The reader is urged to read my works and other discussions of joint treatment to aid in comprehending the problems of the past and the rational for optimism for what we now recommend for the future.

For the present and the future I recommend the municipal—industrial complex concept similar to what we review above in Potential Industrial Complexes. In order to accomplish this type of "pollution solution" the industrial plant must be located at the site of the municipal wastewater treatment plant. This should really not be a great burden to the industry since it would find it advantageous to operate at the lower elevation as well as secluded location usually selected by cities.

We will consider now only one of many potential municipal—industrial complexes in which municipal solid wastes are recovered and reused within the complex by two industrial plants

Municipal Solid Wastes—Industrial Complexes

Municipal wastewaters normally contain approximately 5% to 10% settleable suspended solids. In addition some of these solids are grease-like in nature and will separate from the denser, settleable solids. Usually cities use large settling tanks to remove both types of solids-one by the process of sedimentation and the other by floatation. Both types of solids are eventually usually wasted into the environment or treated extensively before some type of ultimate disposal is used. Both the treatment and/or the discharge into the environment are costly and damaging.

In Figure 11.9, I present a schematic concept of one municipal—industrial complex in which both types of solids are recovered and used by industries in the complex to make additional products.

In this complex the municipal sewage's settled solids are rotary-dried to produce a 5% to 10% cake which is then conveyed to the agricultural growing area to enhance the growth of selected fruits and/or vegetables.

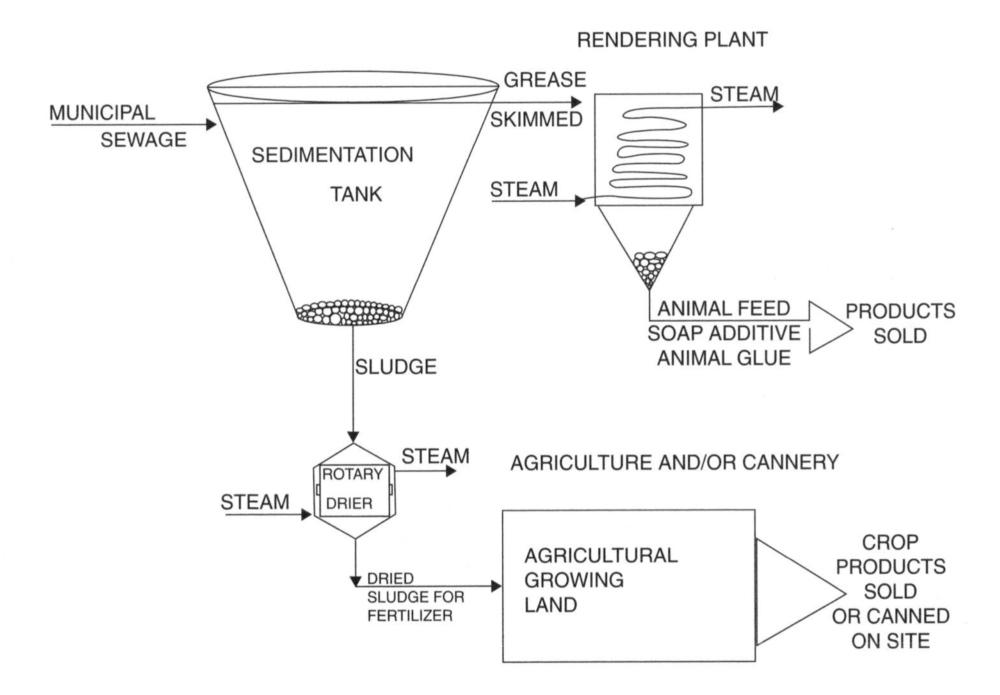

FIGURE 11.9. Municipal Sewage—agriculture growing—rendering plants.

These food crops are then harvested and sold to outside canners or canned, if possible, onsite of the complex.

The lighter-than-water solids (grease) are skimmed from the settling tank surface and conveyed directly to the onsite renderer which then concentrates and converts these solids by enclosed heating to an edible animal food additive, natural animal glue, or a soap base for sale as products.

Transportation of the solids to distant industrial factories is avoided as is the importation of raw materials for industrial production. No municipal solids are released to the air, water, or land environments before or after costly treatments.

In addition, other advantages may exist for such a complex system. For example, the wastewater effluent from the municipal system may be reused to irrigate the agricultural growing area rather than discharging it to a nearby watercourse.

Municipal Wastewater—Industrial Complex

Municipal wastewaters also contain about 1000 parts per million of dissolved and colloidal solids most of which are organic in nature. These solids are costly to remove by effective treatment (usually biological). Hence, some, if not all, of these solids are discharged into the environment causing degradation of our watercourses.

An ideal solution to this problem would be to incorporate the municipal treatment plant into a complex with other industries which can utilize these solids to evolve a product for commercial use. One such complex is shown in Figure 11.10. In this complex illustration two industries and the municipality are involved: fishery food, agriculture crop food, and municipal sewage.

The sewage is first treated by sedimentation. The supernatant wastewater is directed to an algal production pond. Here algae grow aided by natural sunlight and mineral and other nutrients remaining in the sewage. After sufficient detention time the algae-laden overflow is discharged to the fish culture pond. Starter fish fry and air are introduced into this pond to enhance the growth of fish such as tilapia for sale as animal and/or human food. The fish growth is enhanced by the amount and nature of algae fed to the culture pond. Some fish pond culture is re-circulated to the algae production pond to stimulate the growth of more algae.

Excess culture pond effluent is distributed on a field crop growing area. Crops such as corn, beans, and tomatoes can be grown in this field. The crop growth is also aided by applying (spraying) settled sewage sludge from the sedimentation basin. Following harvesting and crop irradiation the food products can be sold again for animal and/or human consumption.

By using a complex system such as this one no municipal sludge solids nor contaminated liquid effluent reaches the land, air, or water environment outside the complex.

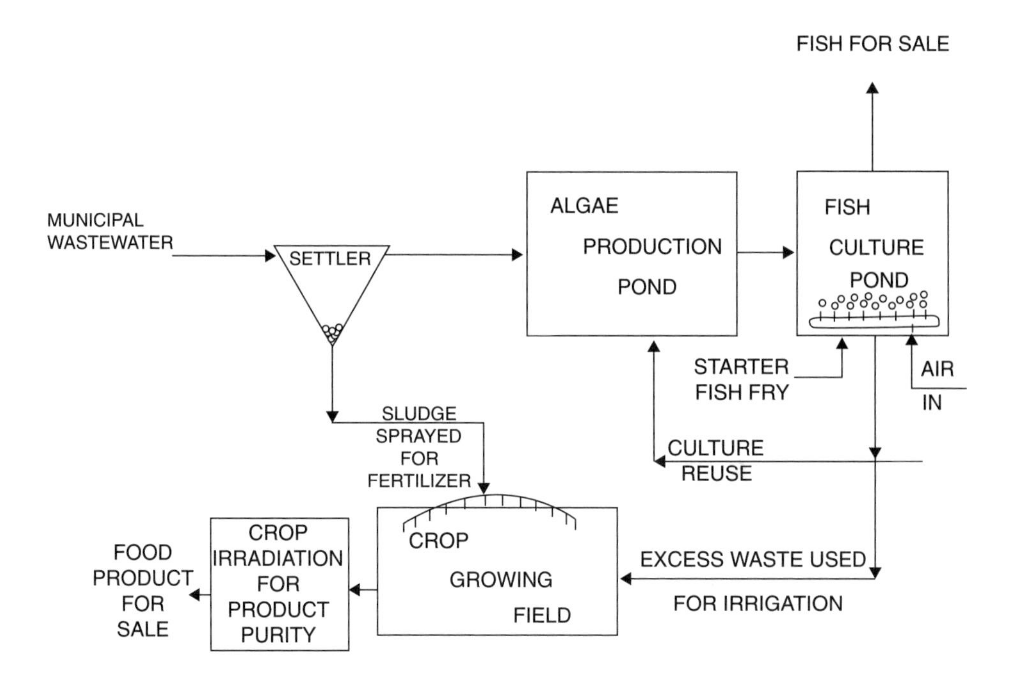

FIGURE 11.10. Municipal wastewater—agriculture (food) complex.

This complex represents just one of several which can acccomplish this same objective while evolving other industrial products.

Lake Industry—Villagers Complex

Lakes can serve as valuable industries for its users. For example, some lakes can be used by boaters, fishermen, and even by bathers and swimmers. If, at the same time, these lakes are used by downstream owners for generating electricity and for irrigation water by farmers, conflicts over the use and operation of lakes may exist.

We are just beginning to amass information about ways and means of using and operating such multipurpose lakes. This author has proposed economic methods for allocating water usage in lakes as far back as 1970 and repeated in summarized form (Nemerow, Farooq and Sengupta, 1980). However, until now, administrative decisions largely determine which water use gets the priority and how much water it receives. Administrative decisions are usually made by political pressures rather than by rational economic values.

For example, downstream water users may sometimes exert enough influence on upstream lake property owners to release (or lower) water levels in the lake, thus interfering with lake-industry uses. I suggest one technical (or non-administrative, non-economic) method whereby such situations can be ameliorated.

Usually lakes are surrounded by nearby villages or cities whose citizens use the lake for recreation. When lake levels are lowered due to release of water for downstream users, village lake users lose some or all of the recreational benefits available to them. One way of overcoming this dilemma, to some extent, is to treat and discharge its domestic wastewater (sewage) at the headwaters of the lake.

This will add about 100 gallons of water per person per day to the lake instead of sending the sewage water immediately downstream as an eventual waste. A village of only 1000 people, for example, could increase the water volume of the lake by about 6.35 million gallons each year ($1000 \times 100 \times 365$). This may, by itself, be enough to preserve boating and fishing activities in the lake. In a manner of speaking, this is a form of municipal wastewater reuse which may not be as objectionable to the public as other uses, such as irrigating edible crops and street cleaning. Such a complex is depicted in Figure 11.11.

In this complex the lake bordering village treats and pumps its sewage to the upper part of the lake instead of discharging it below the lake dam

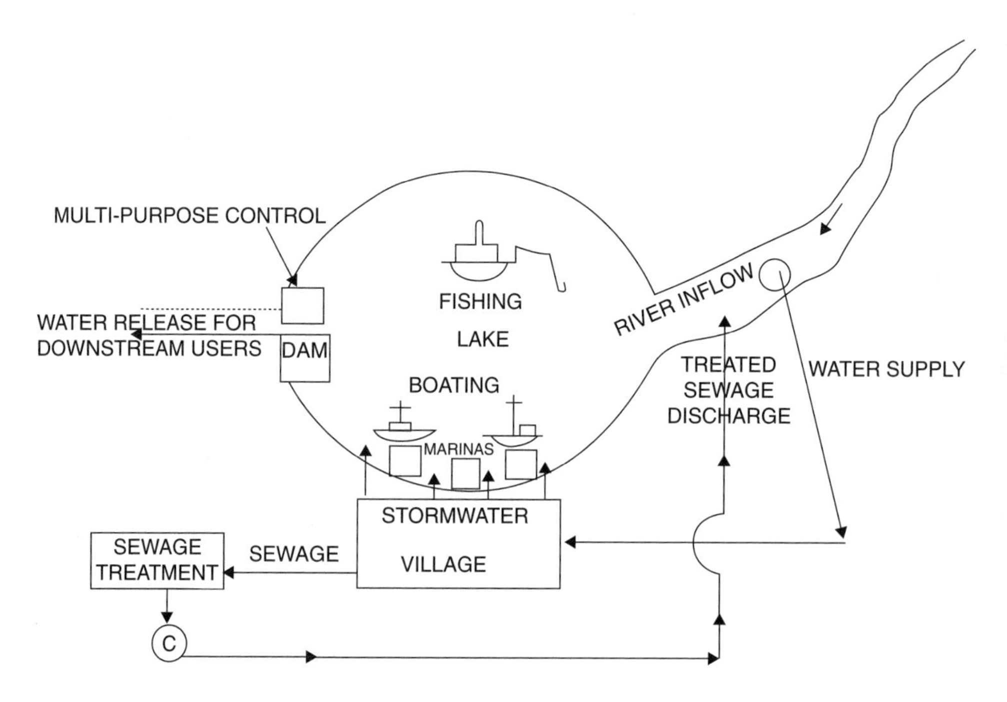

FIGURE 11.11. Lake industry—village complex.

to downstream river users. This in no way affects the villages' upstream drinking water quality or quantity. The lake's fishing, boating, and recreation industry will be enhanced, especially when downstream water users are lowering dam overflows to serve their interests.

Thus, we have a type of complex where the lake industry is aided by the wastes of the villagers. An illustration of a situation in which this complex solution might be used is in Detroit, Oregon. The Detroit Lake boat business evaporated because competition for water prompted by the near record drought in Northeast (Gavin 2001). This caused the U.S. Army Corps of Engineers to decide against bringing the water level high enough to support recreational lake activities.

The Ultimate Natural Resource Conservation and Resource Preservation Plant

This section is for those readers whose imagination has been stimulated sufficiently to envision the ultimate in this concept.

Why not utilize all liquid and solid wastes of a municipality to produce some of the products that municipality needs to survive and grow good electricity and water. It can be done—and again with no wastage of material to harm the surrounding environment.

If you doubt that it is possible, I refer you to Figure 11.12 as proof of the concept. In this depiction of an ultimate EBIC all municipal wastewater and combustible refuse represents the EBIC plant input. The wastewater is settled to produce grease and sludge which are fed to the fermenter to produce methane gas for burning and steam formation in the power plant for electrical energy.

The burnable refuse is also fed directly to the power plant for burning, steam formation, and power plant electricity. The settling basin effluent, after chlorination is fed to an algae growth pond fortified with tricalcium phosphate fertilizer and carbon dioxide from the power plant stack gas. The algae are introduced into the fish growth pond to produce an edible fish product for sale after irradiation.

The entire fish pond effluent is distributed on a crop-growing field to irrigate the agricultural growth product. The field is also supplemented with waste digested sludge from the fermenter. The agricultural product is periodically harvested, irradiated, and sold back to the city folk.

Meanwhile, all the excess wastewater overflow which was fed to the growth field for irrigation is filtered through the underground soil and becomes a source of reusable groundwater for reuse by the municipality.

In summary, all municipal wastewater and burnable refuse has been treated and reused as a source of water, electricity, and food for the same municipality.

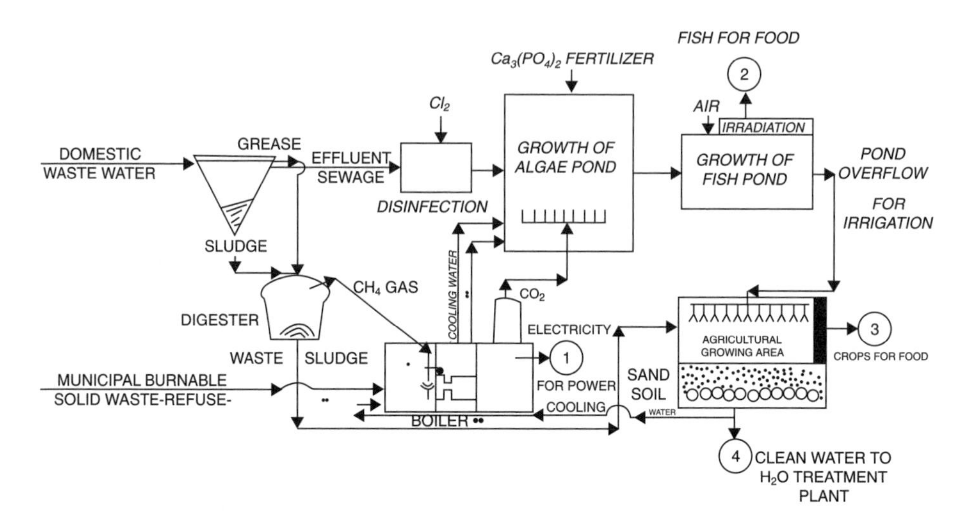

FIGURE 11.12. Food—electricity—water production: The ultimate natural resource, conservation, and preservation plant.

Benefit-Related Expenditures for Industrial Waste Treatment

I cannot complete this chapter without some discussion to clarify the position of how much industry without industrial complexing could afford to pay to treat its industrial waste. I must make it clear to you that waste treatment represents a significant cost of doing business. That cost must not only be predetermined by industry, but also it must decide on the amount it is willing and able to spend. Understanding this expenditure will aid the reader in reaching the conclusion that regardless of the accepted cost of waste treatment, the EBIC is less costly, more efficient, and leaves no residual environmental pollution.

Thirty-three years ago I made the following public statements which are still valid today (Nemerow 1972).

Industrial waste treatment is a necessity to preserve our water resources. Economical stability of our society is also equally vital to our well-being. Waste treatment may cost more than an industrial plant is willing or even able to spend. This is true especially in situations where the stream resources are limited and intense competition exists between water users and consumers. Unfortunately for us these latter situations are becoming more and more prevalent. What then is industry to do in these cases? Move? Cease production? Enter into a legal maneuver in order to delay or prevent excessive costs for waste treatment? None of these is really desirable for industry or society. What should governmental regulatory agencies do in these critical situations? Force industry into one of the above alternates or ignore the need to protect the stream resources and allow the plant to continue to pollute? Neither of these positions is satisfactory. How then do we solve the problem of apparently conflicting interests of two factions of our society? In this chapter some answers should become apparent to you.

Lest any reader question whether treatment costs are justifiable I refer you to the following list of primary, secondary, and intangible benefits of industrial waste treatment.

Primary Benefits

- Savings in dollars to the industrial firm by the reuse of treated effluents instead of fresh water
- Savings in dollars resulting from compliance with regulatory agencies, i.e., avoidance of legal and expert fees and time of management involved in court cases
- Savings in dollars from increased production efficiency, made possible by improved knowledge of the waste-producing processes and practices

Secondary Benefits

- Saving in dollars to downstream consumers from improved water quality and hence lowered operating and damage costs
- Increase in employment, higher local payroll, and greater economic purchasing power of labor force used in construction and operation of waste treatment facilities
- Increased economic growth of the area due to the commitment of industry to waste treatment and potential for expansion at the existing plant
- Increased economic growth of area with more clean water available for additional industrial operations, which in turn yield more employment and money for the area
- Increased value of adjacent properties as a result of a cleaner, more desirable receiving stream
- Increased population potential for the area since cleaner water will be available at a lower cost: the limiting factor of water cost and quantity have been pushed back further into the future
- Increased recreational uses, such as fishing, boating, swimming, as a result of increased purity of water; recreational opportunities previously eliminated are available again

Intangible Benefits

- Good public relations and an improved industrial image after installation of pollution abatement devices

- Improved mental health of citizens in the area confident of having adequate waste treatment and clean waters
- Improved conservation practices, which will eventually yield payoffs in the form of more clean water for more people for more years
- Renewal and preservation of scenic beauty and historical sites
- Residential development potential for land areas nearby because of the presence of clean recreational waters
- Elimination of relocation costs (for persons, groups, and establishments) because of impure waters
- Removal of potential physical health hazards of using polluted water for recreation
- Industrial capital investment assures permanence of the plant in the area, thus lending confidence to other firms and citizens depending on the output produced by the industry
- Technological progress, resulting from the conception, design, construction, and operation of industrial waste treatment facilities

The most obvious and prominent observation from the listing of benefits is that one must quantify these in some manner in order to arrive at a specific level of justifiable expenditure. I have made an attempt to do just that above in the Case Study of Economic Proof of Industrial Complexes. However, at this point I would like to express my opposition to the view expressed by some that all industrial waste treatment costs are justifiable to protect the stream resources. Advocates of this position make light of any attempt to quantify benefits because of their foregone conclusions. These advocates further believe that wastewater resources engineers are "poaching" on other fields in applying economic measures to treatment decisions. What these over exuberant conservationists fail to consider is that our economic ability to ameliorate society's ills is limited by not being able to afford to do everything to improve the environment instantly. Therefore, someone has to establish priorities setting forth the proper amount of waste treatment required. We are obliged to provide government with formulas or at least methods for making more objective decisions in pollution abatement situations.

Quantification of Benefits

We can begin the process of quantification by defining benefits as a willingness to pay or the value of avoiding payment of a given number of dollars at the given water quality by actual and potential water users (Nemerow 1970).

The dollar benefit of a water resource at a given quality may be determined by listing all the uses affected by water quality, by valuing each use individually, and by summing the resultant values. The major uses which are affected by water quality may be grouped in the following categories:

- Recreation uses
- Withdrawal water uses
- Wastewater disposal uses
- Bordering land uses
- In-stream water uses

The value of these uses may be estimated by taking surveys of the users to determine the extent of demand for each use and the amount each user is willing to pay for a unit of use or the unit benefit. Annual dollar benefits for a given use are the product of the total demand times unit benefit. Total annual dollar benefit at a given water quality is the sum of these benefits for each use.

Total annual dollar benefit at a higher water quality may be estimated by determining the probable demand for beneficial water uses at the new quality. This demand may be estimated by surveying the present need for comparable uses at a nearby lake or stream with this new water quality: or it may be estimated by questioning potential water users to determine latent demand likely to be present at this new quality for possible beneficial uses which are presently being foregone.

An expanded description of these five receiving water uses is given below.

Recreation uses include sightseeing, walking and hiking, swimming, fishing, picnicking, boating, hunting, camping, water skiing, canoeing, sailing, and skin and scuba diving. These recreation uses may be valued by including all the expenditures of the average recreations as a measure of his willingness to pay. These include the costs of equipment, food, travel, and recreation area user fees.

Withdrawal water uses include municipal water, industrial water, and agricultural and farmstead water supplies. The water quality benefits reflected in municipal water supply may be estimated to be at least equal to the cost of water treatment by chemical coagulation, sedimentation, and rapid sandfiltration. Water quality benefits for industrial water supply may be estimated by using water treatment costs, not to exceed those for municipal treatment. Industrial costs to produce ultra-pure water are not assigned as water quality benefits, since these costs are more related to overhead costs of as particular manufacturing process, in contrast to the cost of a normally supplied public utility. Agricultural and farmstead water use benefits may be estimated as negative values, if damages have occurred to irrigation, poultry and livestock watering, farmstead family or dairy uses.

Wastewater disposal benefits may be estimated to be the total annual costs for waste treatment required to meet existing minimum stream or effluent standards. The difference in annual costs between the existing level of treatment and the level required to meet the minimum standards may be considered a present benefit to the waste discharger. These costs include those for the common waste treatment plants and also the costs of industrial wastewater reduction practices, interceptor sewers, water quality surveillance, stream low-flow augmentation and possibly in-stream aeration.

Bordering land value benefits at a given water quality may be estimated for a given land use by comparing the per acre market value of shoreline property with the value of nearby non shoreline property. These market values may be estimated by using local tax records and the tax equalization rates. The difference in these per acre values will then reflect the unit benefits or damages of the shoreline location. Values at a higher water quality may be estimated by applying this technique to a nearby lake and projecting the ratio of shoreline to non-shoreline per acre values back to the original lake.

In-stream water uses include commercial fishing, barge and ship navigation, flood control, and hydroelectric power generation. The value of fish caught commercially may be taken as a benefit while the other uses involve damages or negative benefits.

Further economic analyses can be found in this book in Chapter 4.

Summary

I have attempted to build the case for a change in industrial manufacturing from one consisting of *waste treatment* to one encompassing *waste utilization.* This significant change is made possible by using the concept of the environmentally balanced industrial complex (EBIC) instead of free-standing industrial plants. EBICs are designed so that each industrial plant utilizes some wastes from other adjacent plants with the result that no wastes escape into the exterior environment. The ultimate results are elimination of industrial environmental pollution and lower industrial production costs. Many examples, both planned realistic and potential, of such EBICs are given throughout the book. In Chapter 12 Erkman and Ramaswamy present some complexes that evolved over time rather than being planned initially.

I hope the case studies as well as background rationale for EBIC will provide more clarity and proof that the new method of industrial production is not only practical and economical, but also is necessary and vital to our future.

Review Questions

1. How would you go about setting up an EBIC?
2. Why is the EBIC the best and ultimate solution to the industrial pollution dilemma?
3. List the benefits of an EBIC for our Pulp and Paper industry and are there any disadvantages which must be overcome?
4. Why isn't simple recovery and reuse sufficient to abate industrial waste pollution?
5. How can the real cost savings to industries in an EBIC be quantified and proven?

References

The published references of the author's industrial complexes are included here for referral for readers of this book. For further reading, I recommend Chapter 19 in my book Zero Pollution for Industry: Waste Minimization Through Industrial Complexes, *New York: Wiley, 1995.*

Browning, M. 1996. Glorious riverfronts, Crystal Waters, Florida. *Miami Herald,* March 3, p. 6B.

Clark, B. E. 2000. USA biomes to buy American waste transport. *San Diego Union Tribune,* February 17, p. C-3.

Discover Magazine. 1988. That sucker's going to cost you. May, p. 6.

Erkman, S. 2001. Industrial ecology: a new perspective on the future of the industrial system. *Swiss Medical Weekly* 131:531–538.

Ferguson, J. 1993. Cement companies go toxic. *The Nation,* March 8, p. 307.

Engineering News Record. 1994. Cement maker finds cleaner product mix. December 12, p. 27.

Gavin, R. 2001. Missing: Little lake in Oregon. *Wall Street Journal,* October 3, p. B1.

Kessler, G. R. 1995. Cement kiln dust (CKD): Methods for reduction and control. *IEEE Transactions on Industry Applications,* March/April, pp. 407–408.

Nemerow, N. L. 1963. *Theories and Practices of Industrial Waste Treatment.* Reading, MA: Addison Wesley Publishing, Chapter 15.

Nemerow, N. L. 1972. Costs of water pollution control. National Symposium, North Carolina State University, Raleigh, April 6–7, pp. 230–245.

Nemerow, N. L. 1980a. Environmentally optimized industrial complexes [lecture]. Bound Proceedings of the National Environmental Engineering Research Institute, Nagpur, India.

Nemerow, N. L. 1980b. Preliminary balanced industrial complex: A three stage evolution. A Report to United States Environmental Protection Agency, Contract No. 68-02-3170 RTP, North Carolina, June.

Nemerow, N. L. 1984. *Environmentally Balanced Industrial Complexes: The Biosphere: Problems and Solutions*. Amsterdam: Elsevier Science Publisher B.V., pp. 461–470.

Nemerow, N. L. 1985. *Stream, Lake, Estuary and Ocean Pollution*. New York, Van Nostrand Reinhhold, pp. 303–309.

Nemerow, N. L. 1995. *Zero Pollution for Industry*. New York: Wiley-Interscience, p. 109.

Nemerow, N. L., F. J. Agardy. 1998a. *Strategies of Industrial and Hazardous Waste Treatment*. New York: Wiley, Chapter 16, pp. 231–242.

Nemerow, N. L., F. J. Agardy. 1998b. *Strategies of Industrial and Hazardous Waste Management*. New York: Wiley, p. 81.

Nemerow, N. L., F. J. Agardy. 1998c. *Strategic Management of Industrial Wastes*. New York: Wiley, pp. 91–93.

Nemerow, N. L., A. Dasgupta. 1981. Environmentally balanced industrial complexes. 36th Annual Purdue University Industrial Waste Conference Proceedings, p. 416.

Nemerow, N. L., A. Dasgupta. 1984. Zero pollution: A sugarcane refinery-based environmentally-industrial complex. 57th Annual Conference of Water Pollution Control Federation, New Orleans, LA, October.

Nemerow, N. L., A. Dasgupta. 1985. Zero pollution for textile wastes. 7th Alternative Energy Conference, Miami, FL, December.

Nemerow, N. L., R. C. Faro. 1970. Total dollar benefit of water pollution control. *J. San. Eng. Div.*, Proceedings of the American Society of Civil Engineers, June, SA3:7323.

Nemerow, N. L., T. N. Veziroglu. 1988. U.S.-India joint research on industrial complexing: A solution to phosphogypsum fertilizer waste problem. Washington, DC: National Science Foundation, March.

Nemerow, N. L., S. Farooq, S. Sengupta. 1978. Industrial Complexes and their relevances for pulp and papermills [presentation]. Seminar on Industrial Wastes, Calcutta, India.

Nemerow, N. L., S. Farooq, S. Sengupta. 1980. Industrial complexes and their relevance for pulp and paper mills. *Journal of Environmental International* 3(1):133.

Nemerow, N. L., T. D. Waite, T. Tekindur. 1987. Industrial complexing and ferrate treatment for reuse of wastewater of small textile mills. Proceedings of 8th International Conference on Alternative Energy Sources Session on Environmental Problems, Miami Beach, FL, December 15.

Newborn, S. 1992. Phosphate regulations proposed. *The Tampa, Florida Tribune*, September 30.

Rose, C. D. 2000. Power in peril. *San Diego Union Tribune*, February 13, p. I-1.

Satchell, M. 1995. Sinkholes and stacks. *U.S. News and World Report*, June 12, pp. 53–56.

Solomon, B., L. Russell. 2000. Why are some population segments more exposed to pollutants than others. *Environment*, October: 34.

Tewari, R. N., N. L. Nemerow. 1982. Environmentally balanced and resource optimized kraft pulp and papermill complex. 37th Annual Purdue University Industrial Waste Conference Proceedings, May 12, p. 353.

Vas, R. E., compiler. 2000. An assessment of market-based regulatory tools. *Environmental Technology*, Jan-Feb, pp. 15–17; from an October 1999 report authored by Gary C. Bryner.

Wall Street Journal. 2002. Plant closing shows intensity of cost cuts, June 24, p. A2.

Wall Street Journal. 2001. Emissions impossible? July 23, Op-Ed, p. A-14.

Zweig, J. 1992. Cement shoes. *Forbes Magazine*, May 11, p. 20.

CHAPTER 12

Industrial Ecological Solutions

Suren Erkman and Ramesh Ramaswamy

A New Environmental Agenda for Developing Countries

View from India

Background

India became conscious of environmental issues around the 1960s and the first legislation relating to the environment, the Water (Prevention and Control of Pollution) Act, was enacted in the 1974. Since then more legislation covering pollution of other media and more comprehensive legislation for protection of the environment have come into force. However all these laws are directed largely at "controlling pollution" coming from industrial activity. The inspiration for most of these enactments and the related rules has been the legal system in the developed countries such as the United States and Germany. A huge establishment to monitor such emissions has been set up.

A drive through any part of India would show that these strategies have not worked. The quality of air and water are unbelievably poor. The ground water is often contaminated and is depleting. This is a serious concern as a majority of the population uses untreated groundwater for all their needs. As per the estimates of 1996 [WHO 2003] the number of deaths per year in children under 5 years old from diarrheal diseases was 840,000, a direct result of poor water quality and from acute respiratory infections 600,000, often due to poor unsanitary living conditions and polluted air.

It becomes very necessary to raise many important questions. Could there be something wrong with the environmental or development policies that are being followed? Are the policies directed at the most important polluters? Are the laws adequate to deal with the situation? Is there a problem with implementation of the laws? Do the citizens feel involved in policy implementation process? Is there a better way?

The answers to these questions could point to new and better strategies to manage the environment in India. The answers to these questions are not easy. Industrial ecology can possibly point a way to create a basis to answer some of these questions.

Industrial Ecology and the Kalundborg "Trap"

Industrial ecology explores the assumption that the industrial system can be seen as a certain kind of ecosystem. After all, the industrial system (at least its material substratum), just as natural ecosystems, can be described as a particular distribution of materials, energy, and information flows. Further, the entire industrial system relies on resources and services provided by the biosphere, from which it cannot be dissociated.

The key elements of an industrial ecology perspective are:

(a) It is a systemic, comprehensive, integrated view of all the components of the industrial economy and their relations with the biosphere.
(b) It emphasizes the biophysical substratum of human activities, i.e., the complex patterns of material flows within and outside the industrial system, in contrast with current approaches that mostly consider the economy in terms of abstract monetary units, or alternatively on energy flows.
(c) It considers technological dynamics, i.e., the long-term evolution (technological trajectories) of clusters of key technologies as a crucial (but not exclusive) element for the transition from the actual unsustainable industrial system to a viable industrial ecosystem.
(d) By contrast to usual environmental strategies, which are targeted toward individual companies and organizations, industrial ecology implies the collaboration, at a large scale, of many different economic and social partners, who traditionally just ignore each other.

One of the inspirations for industrial ecology has been the "symbiosis" in Kalundborg, Denmark, where a few industries started sharing their resources and their wastes to mutual benefit.

The history of Kalundborg virtually began in 1961 with a project to use surface water from Lake Tissø for a new oil refinery to save the limited supplies of ground water. The city of Kalundborg took the responsibility for building the pipeline while the refinery financed it. Starting from this initial collaboration, a number of other collaborative projects were subsequently introduced and the number of partners gradually increased. By the end of the 1980s, the partners realized that they had effectively "self-organized" into what is probably the best-known example of a working industrial ecosystem, or to use their term—an *industrial symbiosis* (Ehrenfeld and Chertow 2002; UNEP-DTIE 2001).

Although Kalundborg has been a wonderful example of industrial ecology at work, the excessive accent on the symbiosis there has led to the danger of industrial ecology being narrowly identified as a synonym of industrial symbiosis or more narrowly of "waste exchange."

The scope of industrial ecology goes well beyond waste exchange. The larger message from Kalundborg is that one of the key elements of policy that could be both economically viable and environmentally sustainable is the optimization of resources flowing through the economic system.

This is immediately relevant in developing countries where the availability of resources to the people is very poor. A policy platform that is based on the optimization of resources would also appeal to every citizen in these countries and this would ensure their involvement—so critical to the implementation process.

Industrial Ecology and the Development of an Environmental Agenda

The question of how environmental agendas should be developed in the developing world needs a serious review. Following from this, important issues that need immediate attention are:

- Should these be developed internationally or locally? At what level should the agenda be developed—by each country/state/community?
- Should these goals be different for developing and developed countries?
- What are the processes that would be needed to develop such goals and objectives?

In our opinion, lack of clarity about these issues is something that contributes seriously to the ills of the environment programs in developing countries.

Although there are some broad global priorities, the accent that each country or indeed each community, chooses to place on the different items on the environmental agenda, will decide whether locally relevant issues get addressed at all or not. At the moment, in our view, since much of the development funding comes from the rich countries or the institutions based there, the developing countries are thrust priorities that do not truly reflect the actual needs of these poor societies. The result could be that enormous effort and budgets are being spent on programs that are relatively trivial in the local context and issues critical to the local community are often ignored.

A balance has to be struck between globally important issues and the needs of local communities while framing an environmental agenda. Industrial ecology could help. A material flow analysis (MFA) or more aptly a resource flow analysis (RFA), a basic methodology elaborated within the framework of industrial ecology, could be useful for "understanding the system," so essential to finding systemic solutions that are locally relevant.

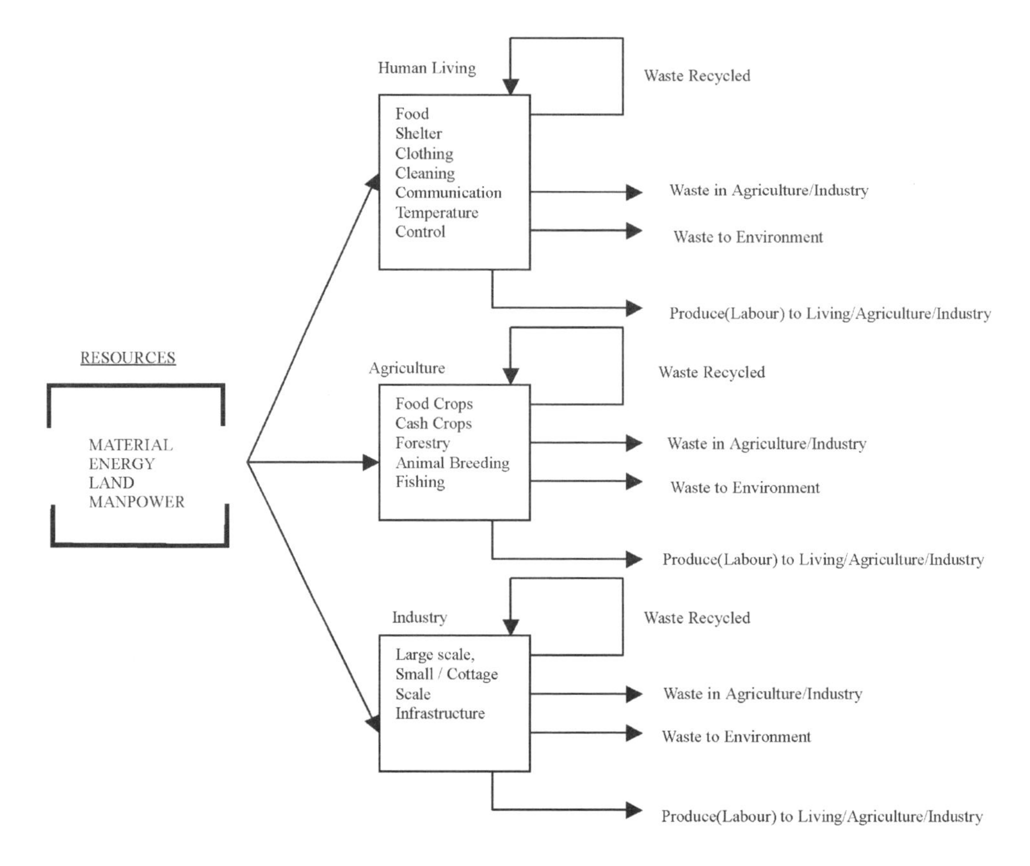

FIGURE 12.1. Flow of resources through an economic system.

The socio-economic system could be depicted as in Figure 12.1 where materials or resources flow through different segments of the system. An understanding of the quantities of such resources flowing through different segments of the system, even if accurate figures are not available, could help in setting priorities.

Ground Realities of Developing Countries

Among the many specific aspects of developing countries that have to be born in mind, is the fact that the pattern of resource flows in developing countries and hence the resultant environmental threat could be very different than what it is in the industrialized West.

Typically, the flows of materials through the large, organized manufacturing facilities in the developing countries could be very small in relation to the overall material flow. Table 12.1, showing the comparative

TABLE 12.1
Fresh Water Drawals in Different Countries

Country	*% age Agriculture*	*% age Industry*	*% age Domestic*
U S A	27	65	8
India	92	3	5
Sri Lanka	94	1	5
Bangla Desh	86	2	12

Source: World Development Indicators, World Bank 2002.

withdrawal of water in different countries by different segments of the socio-economy, is very revealing. If any action has to be taken to preserve water in India, for example, or stop the deterioration in its quality, the action may have to go far beyond the large, formal manufacturing facilities. Similarly an understanding of the other resource flows through the system would immediately point to directions for action.

Another point that needs elaboration is the definition of "industry." In the developing world, the small, informal "industry" plays a key role and forms a very significant portion of the economic activity. Table 12.2 gives the relative importance of the small-scale sector in different countries. Typically, such units:

- Do not use high technology and cannot afford sophisticated pollution abatement systems
- Are too numerous to be effectively policed by the state
- Employ large numbers of people (and no democratic government would risk any potentially disturbing action)

In fact, even reliable data about the number and activities of such small industries is often lacking. As a consequence, the environmental laws are not strictly enforced on the small industries, although the collective consumption of material by the small industries and the resultant threat to the environment could be much higher than the large industrial units.

In India, there were an estimated 3.37 million modern SSI units as of the end of March 2001, providing direct employment to around 1.86 million persons. This does not include the multitude of unregistered units (SIDBI 2001).

Figure 12.2 attempts to list a number of possible characteristics of developing countries that may need to be taken into account while preparing an environmental agenda and action strategy.

It is important to understand the context in developing countries while developing strategies. For example, what good would it be, to make elaborate laws if they are not implemented or cannot be implemented? Possibly,

TABLE 12.2
Indicators of SMEs in Selected Economies in the Mid 1990s

Country	*As % of total enterprises*	*% share in Employment*	*% Contribution to GDP*	*% share in total exports*
Japan (1994)	99.0 (excl. primary industry)	78.0	56.0 (of total value-added in mfg. Sector)	14
Korea, Rep. Of (1996)	99.1	78.5		42
Taiwan Prov. Of China (1994)	97.8	81.1		56
People's Republic of China	98	70 (mfg. only)		40–60
Indonesia	97	42		10.6
Thailand	98	n.a.		10
Malaysia	96	40		15
Philippines	99	45	28% of mfg. Value added	20
Singapore	89	42		16
Vietnam	83	67		20

Source: Mikio Kuwayama, 2000. "E-Commerce As A Tool Of Export Promotion For SMEs: Comparison Between East Asia And Latin America." Kuala Lumpur: Asian and Pacific Development Center.

after getting an understanding of the system, strategies that are suitable for the local context need to be developed.

The Appeal of an Environmental Agenda

If the state of the environment and indeed the quality of life is to be improved in the developing countries, it is critical to involve the people and find an appeal that touches their lives. For example, a person who is starving is not likely to be worried about saving the world food supply for the future. He wants a meal right now.

The following could be some of the possible appeals of any environmental agenda:

- To save the planet
- Community health
- To improve profitability of companies
- Part of corporate social responsibility of companies
- To conserve community's resources

Characteristics of Developing Countries

Obviously, the Industrial Ecology conceptual framework, which was originally formulated in the USA, does not directly apply to the vastly different context of the developing world. In contrast to industrialized nations, characteristics of developing countries may include:

Land-related issues

High population density, which makes land a very vital resource
Low per capita availability of arable land
Low agricultural yields

Water-related issues

Low per capita availability of freshwater (either surface or groundwater)
Lack of treated drinking water
Lack of an adequate sewerage system
High cost of central water treatment and disposal systems
Widespread dependence on untreated groundwater.

Infrastructure

Poor Transport network
Poor telecom links
Non-availability of reliable data due to the existence of a huge informal sector
Accelerated obsolescence of infrastructure due to climatic conditions and population pressure

Labor

High levels of unemployment
Low labor cost
Low labor productivity
Poor work ethic
High level of daily wage earners with no job guarantees
Low level of skills
Poor working conditions and inadequate social security

Economy

Restricted availability of raw materials caused by limited financial resources
Low levels of technology
Smaller scales of manufacturing
Existence of millions of informal businesses
High inflation
High cost of capital
Restrictions on imports
Volatile foreign exchange rages
High need to export and earn hard currency
Low brand equity with many small units doing job work for large domestic or foreign companies
Perverse subsidies, which often encourage wastage of resources

Social

Low levels of education and, consequently, poor awareness of health hazards from pollution or Indus trial accidents
Sometimes less concern for social issues (as jobs are often more important than whether a long-term environmental problem is caused)
Low concern for global issues that do not have an immediate bearing on the society
Lower 'social cost' of law breaking
Often-high levels of corruption among law-keepers

Legal

Laxity in laws governing environment and worker safety, and hence low costs of disposal of wastes
Lax enforcement of laws
Ineffective or slow legal system

FIGURE 12.2. Some characteristics of developing countries.

In poor countries, the basic requirements of life are hard to come by. Even some of the major cities in India such as Chennai (formerly known as Madras) have such acute water shortages that the lives of the common people revolves around getting their daily requirements of water through tankers, as it is very long since they have seen water in their taps!

We believe that the appeal that would involve the people in the environmental agenda is the conservation of the community's resources—not necessarily for future generations but for use during their own lifetimes.

The perceptions of environmental issues in the industrialized countries could be far different from the needs in the poor countries. While the developed countries see the object of an environmental agenda as preservation of the planet for future generations, the additional accent in the poor countries is to preserve resources for the comfortable survival of the present generation of people.

Concrete Examples: Three Case Studies

We would like to briefly present three case studies that highlight some of these issues (Erkman and Ramaswamy 2003). These case studies were carried out during the period 1996–1998. Although the data has not been updated, the core issues remain unchanged. The case studies of the Tirupur textile industries and the leather industry, illustrate how redefining the problem from a perspective of resource conservation and on the basis of resource flow data could point to totally new directions for strategy planning. The case study of the Damodar Valley region amplifies the importance of looking beyond formal industry to solve an environmental problem. It shows that even for globally critical programs such "climate change program," in developing countries, it is just not enough to estimate the emissions from the formal industrial sectors.

Case Study of Tirupur Town

A resource flow analysis (RFA) was undertaken for the town of Tirupur, a major textile cluster in the south of India, which could serve as an example of how a Regional Resource Flow Analysis could be effectively used. The RFA for Tirupur is shown in Figure 12.3.

Tirupur is a major center for the production of knitted cotton hosiery. The town is located in the State of Tamil Nadu and has a population of about 300,000. The 4000 small units in the town specialize in different aspects of the manufacturing process. The aggregate annual value of production in the town is around US$700 million. Much of the produce is exported, bringing in very valuable foreign exchange.

Water is scarce in the area and the wet processing of textiles has rendered the ground water unusable. A large quantity of salt is used in the

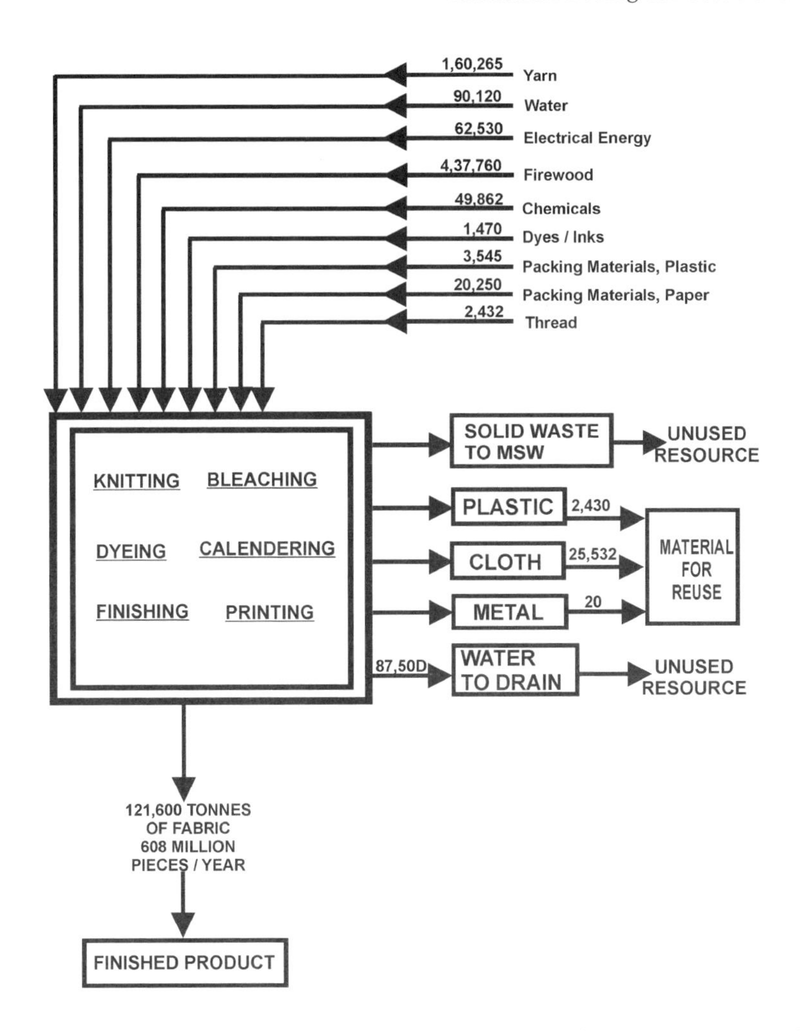

FIGURE 12.3. Resource flow analysis for Tirupur town. (Units: water, 1000 liters *per day*; electrical energy, 1000 kWh per year; others, tons per year.)

dyeing process and the process wastewater (90 million liters per day) is highly saline and is contaminated with a variety of chemicals. As there is hardly any source of fresh water nearby, trucks bring in water from ground water sources (which are yet to be polluted) as far as 50 kilometers away at an enormous cost.

The response of the regulatory agencies to the public outcry about the destruction of water resources was to ask the industries to treat the water before discharge through traditional treatment systems. Since the individual industries were too small to afford such systems, a massive US$30 million project was set up to treat the wastewater at Central

Effluent Treatment Facilities. After such expensive treatment, the water will still be unusable, as the facility does not include any system for desalination of the wastewater.

A detailed RFA was carried out for the town. Only when the figures were aggregated did the industrialists realize that they were collectively spending over US$7 million annually on buying water and in addition, the annual maintenance cost of the effluent treatment plant would be an enormous burden. The aggregate figures immediately showed that water could be recycled profitably. On the basis of the study, a private entrepreneur developed a water recycling system, which could be installed in each dyeing unit. The system used the waste heat from the boilers already working in the dyeing units for the recycling process. This is a relatively low cost system, which is gaining popularity in the town.

The second outcome of the study was that the study highlighted the fact that the calorific value of the solid waste (garbage) was high as it contained large quantities of textile and paper wastes. This could be used effectively to partially replace the 500,000 tons of scarce firewood being used in the town (there is grave concern over rapid deforestation in India). Since the use of the firewood is distributed over nearly 1200 points, it was not obvious that such large quantities of firewood were being used. The possibility of setting up a central steam source (needed by some of the industries) is also under serious consideration in order to reduce the consumption of firewood.

Case Study of the Leather Industry in Tamil Nadu

This case study is intended to highlight the option of strategic relocation of an industry segment to ensure its long-term survival.

Tamil Nadu, a state in the south of India, is the premier center in India for the processing of leather. Water is extremely scarce in Tamil Nadu. The leather industry has been flourishing in the region for decades. Its growth has been possibly due to the fact that Madras was a major trading center during the British rule in India. The industry is a major foreign exchange earner and important to the economy of the state. The strict enforcement of environmental regulations in the developed countries have also helped the leather industry to grow, as the buyers in the developed countries prefer to source their tanned products from India. Compliance with strict environment regulations has rendered the processing very expensive in the developed world.

The leather industry (which is made up of thousands of small industries) is a major user of water, as each ton of hide/skin needs 30,000 to 50,000 liters of water for processing. This is a large volume, as the average per capita water availability for human settlement in India is estimated at around 30 liters per day. The industry has been under pressure from the pollution control authorities and many have subscribed to Central Effluent

Treatment Plants. The water after treatment continues to be unusable, as it is vary saline. The sludge from water treatment, estimated at 250 kg per 1000 kg of hide processed, continues to be a problem. The sludge is carelessly dumped and the pollutants leach into the groundwater. The industry often buys water in trucks at a high cost.

A detailed study in the context of industrial ecology helped in redefining the problem. The problem until now, was only viewed as a pollution control problem, where the effluents did not meet the specifications laid down by the pollution control authorities. Many academic studies have been undertaken to ensure that the effluent quality "comes as close as possible" to the standards.

However, the problem is much more serious. The industry is using a resource, water, which is extremely scarce in the region. It is also contaminating ground water in the region, which is causing great hardship to the population, as it is depriving them of desperately needed water. The industry has been using the slow judicial process in India to survive. However, it will not be long before the social pressure brings the industry to a halt.

In the long term, an alternate solution will need to be found. One of the options that could be considered in the context of industrial ecology, would be to relocate the entire industry along the coast, where the industry draws sea water, desalinates it for use, treats the waste water and discharges the saline waste water into the sea. The process of desalination is expensive. In order to reduce the cost of desalination, it may be possible to set up a thermal power plant and use the waste heat for desalination. The sludge from the process could also incinerated and the energy used in the desalination process (Figure 12.4).

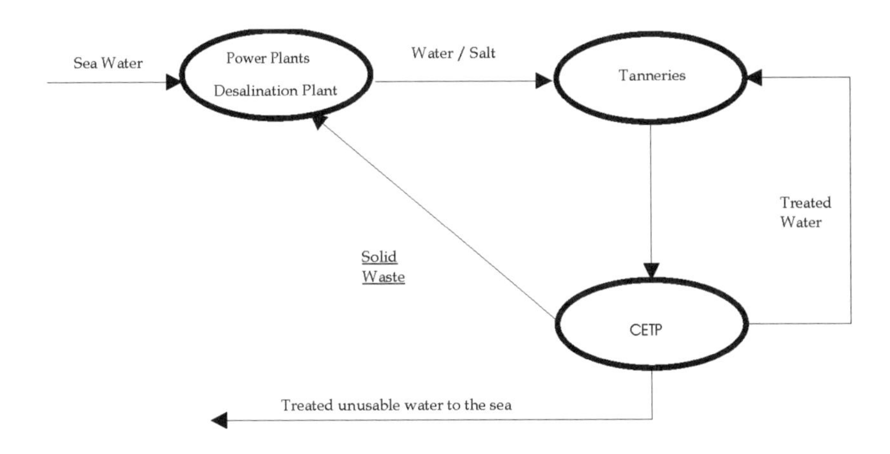

FIGURE 12.4. A conceptual picture of an ideal system built around the leather industry.

A systemic view gave rise to new directions that could not have been possible within the traditional perspective of end-of-pipe pollution control.

Case Study of the Damodar Valley Region

The basin of the River Damodar, in the eastern part of India, covers a vast area. This mineral-rich region near Kolkata (formally known as Calcutta) is the source of much of the coal produced in India. Coal is a major energy source in the country. Many large power utilities and steel plants are located here, in addition to industries associated with coal, such as coal washeries and coke ovens. The region is considered very highly polluted.

An industrial metabolism study was undertaken in the region. The quantities of the flow of two of the major local resources, the waters of the River Damodar and coal, were studied. The results of the study gave a good overview of how the waters of the river and coal are used in the system (SIDBI 2001).

Since agriculture consumes nearly 85% of the waters of the river, it is critical to estimate the impact on the agricultural produce, of the thousands of tons of potentially toxic wastes dumped into the river, resulting from the high levels of industrial activity upstream. All along, to reduce the high levels of air pollution, the policy of the regulatory authorities had been to focus on the "major" polluters, which in their opinion were the steel and power plants. These plants have access to some of the best available technologies for controlling their pollution.

However, a study of the flow of coal gave surprising results. Huge quantities of coal are consumed in millions of homes and in the informal sector. In this sector, coal is used in very inefficient combustion systems, obviously without any pollution control systems, which makes the whole area extremely polluted. It was obvious that if the air had to be clean, a new fuel policy would have to be evolved. Some new systems of transportation of coal also need to be evolved to minimize the spillages during transportation, a major contributor to the dust levels of the region.

Even a broad understanding of the flow of the resources could serve as a guide to the policy maker and gives a new perspective and a direction for policy making.

The Road Ahead

An organized attempt is urgently needed to re-think environmental priorities and think of new strategies in the developing countries to safeguard the comfortable survival of millions of people in these countries. The legal, social, and the technological issues need to be harmonized to evolve such strategies. The systems approach advocated by industrial ecology is critically necessary to optimize the use of resources in these countries. For a discussion of the use of industrial ecological systems in developed countries the reader is directed to Chapter 4.

Traditionally, because of the high consumption patterns in the industrialized countries, they have been the focus of most environmental programs. With rapid globalization, it is time that the material flows in the developing countries are taken seriously as well. More than 80% of the world's population lives in the developing world and the planet will have to contend with the higher consumption associated with the growing material aspirations of this population.

The approach in the developing countries has to go far beyond organized industry and has to include many other facets of the socio-economy such as the public services. The exercise cannot stop at the manufacturing step and has to include consumption in industry, agriculture and households, services, etc. Industrial ecology, in particular thanks to the methodology of MFA, can definitely contribute to find new solutions to environmental problems, by helping to look beyond industrial parks and the formally organized sector. The effort has to focus on resource optimization at a systemic level, at the scale of municipalities and regions, which will ensure the involvement of the local populations.

Review Questions

1. What are the main characteristics of a developing country like India, regarding economic activities and resource consumption?
2. What is the problem with the traditional approach to environmental and pollution problems in a developing country like India?
3. What are the main characteristics of the industrial ecology approach?
4. How does the industrial ecology strategy differ from the traditional pollution control approach?
5. How is the industrial ecology strategy relevant for developing countries? Discuss the narrow industrial ecology approach (Kalundborg, eco-industrial parks) vs. the broader industrial ecology vision.

References

Ehrenfeld, J., M. Chertow. 2002. Industrial symbiosis: The legacy of Kalundborg. In: R. U. Ayres, L. W. Ayres, eds, *A Handbook of Industrial Ecology*. Cheltenham, UK: Edward Elgar.

Erkman, S., Ramaswamy, R. 2003. Applied Industrial Ecology—A New Strategy for Planning Sustainable Societies. Bangalore, India: AICRA Publishers.

SIDBI. 2001. Report on small scale industries sector. Small Industries Development Bank of India (SIDBI), Economic Planning and Research Cell, New Delhi, India.

UNEP-DTIE. 2001. United Nations Environment Programme, Division of Technology, Industry and Economic, Environmental Management for Industrial Estates. Information and Training Resources. Prepared by ICAST (Colin Francis and Suren Erkman) for UNEP-DTIE, Paris. (United Nations Publication, ISBN: 92-807-2078-3, also available at: http://www.uneptie.org/pc/ind-estates/home.htm.)

WHO. 2003. World Health Organization, Regional Office for South-East Asia. (Available at http://w3.whosea.org/index.htm.)

Part III

International Aspects

CHAPTER 13

Rural and Developing Country Solutions

Salah M. El Haggar

1. Introduction

One of the major problems facing developing countries is the environmental protection cost and return. The current practice of agricultural waste, municipal solid waste, industrial waste, municipal waste water, etc. can be considered disasterous all over rural communities and developing countries. Any solution should suit the rural communities and developing countries and should include the economical benefits, technological availability, environmental and social perspectives, otherwise it will never be sustainable. The objective of this chapter is to approach 100% full utilization of all types of wastes by recycling concept.

Waste Management

The selection of a combination of techniques, technologies, and management programs to achieve waste management objectives is called integrated solid waste management (ISWM). The hierarchy of actions to implement ISWM is reduction, reuse, recycle, treatment, and final disposal (Tchobanoglous et al. 1993).

Finding new sources of raw material are becoming costly and difficult. Concurrently, the cost of safe disposal of waste is escalating exponentially and even locating waste disposal sites are becoming more difficult. As a result, a new hierarchy for waste management to approach full utilization of waste is a must, which starts from reduction at the source, reuse, recycle and partial treatment for possible material recovery using cleaner production technologies (El Haggar 2002).

Reuse of materials involves extended use of a product (retrading auto tires) or use of a product for other purposes (tin cans for holding nails, soft drink bottles for holding water in refrigerators, etc.). Reusing the product does not return the material to the industry for manufacturing. While, recycling of material involves manufacturing of other products with less quality. Quality can be adjusted by additives. Recovery differs from recycling in that waste is collected as mixed refuse, and then various processing steps remove the materials. Separating oil from waste water effluent of oil and soap industry by gravity oil separator (GOS) is material recovery from waste. This material is then sold back to less quality soap industry or returned back to the industrial process. The difference between recycling and recovery, the two primary methods of returning waste materials to industry for manufacturing and subsequent use, is that the latter require a process to remove the material from the waste while the former does not require any processes for separation, sorting can be done manually.

In April 1988, the U.S. EPA published the waste minimization techniques, separating waste minimization into two categories: source reduction and recycling, both within the plant (onsite) and outside the plant (offsite). Source reduction can be achieved by product modification or sources control. The latter can be done by changing the raw material, changing the technologies or through good operating practices. It is obvious that waste minimization techniques are good business for industrial sectors.

Cleaner Production

The term cleaner production (CP) was launched in 1989 by the United Nations Environment Programme (UNEP) as a response to the question of how to produce in a sustainable manner. Its core element is prevention vs. clean-up or end-of-pipe solutions to environmental problems. Resources should be used efficiently thus reducing environmental pollution and improving health and safety. Economic profitability together with environmental improvement is the aim. Cleaner production typically includes measures such as good housekeeping, process modifications, eco-design of products, and cleaner technologies, etc. The UNEP defines Cleaner Production as, "the continuous application of an integrated, preventative environmental strategy to processes, products and services to increase eco-efficiency and reduce risks to humans and the environment" (UNEP 1997).

Cleaner production focuses on conservation of natural resources such as water, energy and raw materials and avoiding the end of pipe treatment. It involves rethinking for products, processes and services to move towards sustainable development.

By considering production processes, cleaner production includes conserving raw materials and energy, eliminating toxic raw materials, and reducing the quantity and toxicity of all emissions and wastes before they leave a process. For products, the strategy focuses on reducing impacts along the entire life-cycle of the product, from raw material extraction to the ultimate disposal of the product. Cleaner production is achieved by applying know-how, by improving technology, and by changing attitudes. Changing attitudes is the most challenging and the most important step in applying cleaner production concept.

The conceptual and procedural approach to production that demands that all phases of the life-cycle of products must be addressed with the objective of the prevention or minimization of short- and long-term risks to humans and to the environment.

One factor in defining cleaner production is therefore the reduction in production costs that results from improved process efficiencies. In terms of investment the key difference is that investment in end-of-pipe technologies is nearly always additional investment, whereas investment in cleaner production is usually, at least partly, in replacing existing systems or equipment. This has obvious implications for employment.

A useful definition of cleaner production needs to take account of the distinction between technologies and processes. For example, a process may be made "cleaner" without necessarily replacing process equipment with "cleaner components"—by changing the way a process is operated, by implementing improved housekeeping or by replacing a feedstock with a "cleaner" one. Cleaner production may or may not, therefore, entail the use of cleaner technologies. Investment in cleaner production via the implementation of clean technologies is clearly easier to identify than investment in cleaner production by any other means. Whatever the method employed to make production cleaner, the result is to reduce the amount of pollutants and waste generated and reduces the amounts of nonrenewable or harmful inputs used.

Most of the countries (developed and developing countries) are working toward zero pollution not only in industrial sectors but also in vehicle emissions to reduce the gaseous emissions to the allowable limits as well as other sectors such as construction sector and agricultural sector. To approach zero pollution, industry should prevent all pollutants from its effluent. The cleaner production hierarchy to eliminate all pollutants and approach zero waste/pollution should start from raw material selection to recycling, all the way to product modifications in order to avoid end of pipe treatment (El Haggar 2001c) as will be explained throughout this chapter.

The Benefits of Cleaner Production

Cleaner production can reduce operating costs, improve profitability, worker safety and reduce the environmental impact of the business. Companies are frequently surprised at the cost reductions achievable through the adoption of cleaner production techniques. Frequently, minimal or no capital expenditure is required to achieve worthwhile gains, with fast payback periods. Waste handling and charges, raw material usage and insurance premiums can often be cut, along with potential risks. It is obvious that cleaner production techniques are good business for industry because it will

- Reduce waste disposal cost
- Reduce raw material cost
- Reduce health-safety-environment (HSE) damage cost
- Improve public relations/image
- Improve company's performance
- Improve the local and international market competitiveness
- Help comply with environmental protection regulations

On a broader scale, cleaner production can help alleviate the serious and increasing problems of air and water pollution, ozone depletion, global warning, landscape degradation, solid and liquid wastes, resource depletion, acidification of the natural and built environment, visual pollution and reduced biodiversity.

Obstacles and Solutions of Cleaner Production

An industrial program in education must precede a successful reuse/recycling program by acquainting plant personnel with the potential value contained in the waste. Detailed qualitative analysis of wastes should be made available over a relatively long period of time (1 year).

Establishing another industry correlating as much industrial waste as possible with some other additives to produce a good quality product. This is the challenging point because industrial waste varies in quality and quantity from time to time according to changes of products types and amounts.

To overcome these obstacles, an assistance of the government/ or industrial development agency would be helpful in obtaining agreement for locating plants of these types with all the necessary licenses and support. In fact, the newly established industries based on industrial waste offers one of the most promising long-term solutions to today's environmental pollution problems as well as to many industrial economic problems of the future as a result of any damage might occur.

Waste treatment may cost more than an establishing industrial plant based on waste as a raw material. This will lead to economic stability. Resources are limited and there is a competition between users and consumers of these resources. What should an industry should follow? Cease production? Move to another sites?

Treatment

It is necessary for any establishment to treat its waste to be able to comply with environmental protection regulations. Some industry resisted compliance in order to avoid costs that would jeopardize. Now, new industries are accepting waste treatment as an integrated part of production cost. The added costs must then be passed on to consumers or deducted from the profits of the firm depends on market competition.

Treatment means converting harmful waste into less harmful waste. In other words, treatment means converting waste from one form to waste in another form. The direct cost of waste treatment is more than just the expense of capital equipment and running cost (maintenance, operation, and labor). This direct cost represents only a portion of the total cost. The other indirect cost may not be an easily identified and quantified. This includes the disposal cost and the cost related to adverse impact of the waste on the environment contaminating air, water and land.

If the disposal is not done properly, or the disposal facility is not well designed and constructed, the adverse impacts will be significant. Who will pay for the disposal facility? Who will manage the disposal facility? Who will run the disposal facility? Government, society, or industry?

Unfortunately, indirect costs are external costs to an industry, in that they must be borne by society often at some distance from the manufacturing plant. Some can be quantified, but most are costs of environmental damages which can not evaluated accurately.

In the contrary, the less the waste treatment provided by industry, the greater the cost of environmental damages. If industry will not provide waste treatment, environmental damage cost will be maximum. This will bring us to a very complicated formula, which is if no waste treatment, the damage will be high, and if no proper disposal facilities, the damage will be high too. Then, what is the solution, the solution is zero pollution, which is the subject of the next section.

Some industries claim that we cannot have both jobs and capital spending for growth and at the same time, clean air and water. This statement is not true for industry, because waste and emission was originally raw material and should be treated as a by-product not as a waste through reusing, recycling or recovery techniques.

Treatment should be modified in the hierarchy of cleaner production for zero pollution to recovery (or possible treatment for material recovery). In other words, to what degree of treatment is required to arrive at the optimum outcome for material recovery without damaging the environment. Thus, waste treatment can be partially or completely eliminated for cleaner production hierarchy (El Haggar 2001c).

The optimum approach that industries can use to eliminate completely the environmental damages is to weigh the pros and cons of each technique of the hierarchy. Economic indicators should be used through cost-benefit analysis, as a primary criterion in making the decision but the health-safety-environment (HSE) intangible benefits including the environmental monetary benefits of abating pollution should be considered.

The challenge of industry is to determine which techniques of the hierarchy including treatment to some degree (if applicable) should be followed. Although technical parameters such as quantity and quality of waste are the primary factors, economical, political, social and psychological factors are also extremely important.

Innovative solutions are required to solve the problem of industrial pollution through each of the cleaner production hierarchy techniques such as:

- Reduction at the source by
- Changing the raw material with better quality
- Product modification
- Reuse directly within a plant or indirectly by other industrial plants and/or recycle (onsite) the waste stream resource.
- Marketing of stream resources (offsite reuse or recycling) and mixed with another industrial waste to produce a valuable product
- Recovery of materials by partial treatment as the case of gravity oil separator (GOS) and dissolved air flotation (DAF) in oil and soap industry to recover fat and grease and comply the effluent with environment protection regulations

Zero Pollution

In the past, people's dream was to turn sand into gold. Today though, the gold's secret is not any more in sand but in our waste and pollution. This was a dream until the new hierarchy to approach zero pollution was developed (El Haggar 2001c) and the 6-Rs golden rule was initiated (El Haggar 2003c). That is, the rule aims at reducing, reusing and recycling of waste. Whereas, the fourth R of the 6-Rs golden rule emphasizes the recovering of raw materials from waste, leading to a partial treatment. The last 2-Rs are rethinking and renovation, where people should rethink about their waste before taking action for treatment and develop renovative techniques to solve the problem. Another R should be added at the top to the previous 6-Rs, which is regulation; without regulation nothing will be implemented. It is very important to add regulation through the system and enforce 6-Rs golden rule into the management system. Thus the 7-Rs golden rule encompass regulation, reducing, reusing, recycling, recovering, rethinking, and renovation is the basic tool for industrial ecology (El Haggar 2004a). The rule provides a theory to manipulate current activities to approach zero pollution and avoid landfill, incineration and/or treatment. Fortunately, with full success, the theory was practically applied in a pilot scale to most of the industrial sectors (El Haggar 2000, 2002, 2003a), numerous projects (El Haggar 2003b), and rural communities (El Haggar 2001b), as will be explained throughout this chapter.

Capital investment, running cost as well as adverse environmental impacts of landfills, incineration and treatment heartens the implementation of the 7-Rs. It is very simple, natural and not a newborn theory. Fundamentally, the theory depends on all kinds of recycling (onsite recycling, offsite recycling, partial treatment for possible recycling, etc). This is mainly because recycling is considered a pivotal income generated activity that conserves natural resources, protects the environment and provides job opportunities.

Zero pollution can be defined as the pollution generated within the allowable limits. A new hierarchy for waste management to approach zero pollution starts from reduction at the source to re-use and recycle and partial treatment for possible material recovery, etc. is the basic principal of 7-Rs. This approach is based on the concept of adapting the best practicable environmental option for individual waste streams and dealing with waste as a by-product. This 7-Rs for zero pollution (regulation, reduce, reuse, recycle, recovery, rethinking, and renovation) can be considered the cleaner production hierarchy for zero pollution (CPHZP). There is no need for treatment or landfill to approach zero pollution through the 7-Rs. We might need partial treatment for possible material recovery such as a gravity oil separator (GOS) as well as innovative techniques as will be discussed later in reject recycling of municipal solid waste.

The concept of zero pollution is not new; Professor Nemerow (1995) developed a methodology for the future industrial complexes or parks to approach zero pollution. Most of the countries (developed and developing countries) are working toward zero pollution not only in industrial sectors but also in all other sectors.

1.5 Cradle-to-Cradle Concept

Life cycle analysis (LCA) has been defined by the EPA as a way to "evaluate the environmental effects associated with any given industrial activity from the initial gathering of raw materials from the earth until the point at which all residuals are returned to the earth" or "cradle-to-grave." Several organizations have developed methods for LCA, each using a different analytic approach to this complex activity. Regardless of the approach, several generic difficulties challenge LCA, including poor quality data, weak reasons or procedures for establishing analytic boundaries, and diverse values inherent in comparing environmental factors with no common objective, quantitative basis. The current selection of products undergoing LCA has been chaotic; some products have been strongly scrutinized while others have been totally neglected.

Comparing existing methods for LCA gives insight into the conceptual framework used by researchers. The Code of Practice for LCA stands out currently as the most widely recognized procedural model. The code divides LCA into four distinct components: (1) scoping; (2) compiling quantitative data on direct and indirect materials/energy inputs and waste emissions; (3) impact assessment; and (4) improvement assessment. While variations exist, the theme of taking an inventory and performing an assessment based on collected data is common to all LCA approaches dating back to the early 1970s (Wernick and Ausubel 1997).

There have been a lot of different methods developed by researchers to obtain LCA. Though some methods for LCA receive approval for thoroughness and analytic consistency, these same methods have been criticized as requiring too much data, time, and money when each are in short supply. As an alternative method for assessing the environmental impact of products, researchers at AT&T have devised the Abridged Life Cycle Assessment Matrix, a method that couples quantitative environmental data with qualitative expert opinion into an analysis that conveys the uncertainty and multidimensionality of LCA and also yields a quantitative result (Wernick and Ausubel 1997).

The International Organization for Standardization has developed a series of international standards to cover LCA in a more global sense, such as ISO 14040 (LCA-Principals and guidelines), ISO 14041 (LCA-Life Inventory Analysis), ISO 14042 (LCA-Impact Assessment), and ISO 14043 (LCA-Interpretation).

The environmental and health impacts of landfill and incineration are becoming more dangerous and disaster for developing countries and rural communities as well as developed countries. Establishing industrial ecology within the industrial activity with LCA will avoid landfill, incineration and treatment and change the concept of "cradle-to-grave" into "cradle-to-cradle" by a full utilization of raw material and considering the waste as a by-product using the 7-Rs golden rule (El Haggar 2004b) as will be explained later in this chapter.

Municipal Solid Waste

Municipal solid waste is considered one of the major environmental problem all over the world, especially in developing countries. Solid waste in general can be classified into industrial solid waste, agricultural solid waste and municipal solid waste. Municipal solid waste (MSW) consists of hazardous and nonhazardous waste. It is a fact that solid waste composition differs from one community to another according to their culture and socioeconomic level. However, solving the problem in rural and developing countries is challenging because of two factors: the low socioeconomic level of the majority of the population and their lack of awareness of the problem size. In addition to the lack of the suitable technology platform needed to face the problem.

These problems can be classified into social, economic, technical and environmental problems. Social problems like sight pollution caused by the piles of garbage in the streets which in its turn causes psychological problems, which was proved to affect the individual work efficiency. In addition to this, there are economic problems concerning the cost of MSW management including cost of collection, sorting, incineration and landfill. Another economic problem arises from leaving MSW without recycling which is loss of resources and energy. Moving to the technical problem, the lack of technological platform in developing countries make it a must to import the technology needed which might be infeasible because of the cost and the need for adaptation of the foreign technologies to suit developing countries environment. Concerning the environmental problem, it evolves from the bad odor of the waste that attracts all kinds of flies and mosquitoes which carry diseases to humans causing health problems in addition to pollution.

The main objective of this section is to apply the 7-Rs and develop a suitable municipal solid waste management system in order to approach 100% recycling of nonhazardous MSW using the new hierarchy for zero pollution developed by El Haggar (2003b) and discussed above. This system will utilize the nonrenewable resources, eliminate the problems caused by the MSW, and decrease the cost of MSW management, and approach cradle-to-cradle concept.

Transfer Stations

With the low volume of solid waste generated from rural areas and large distances to transfer solid waste generated in rural areas, it is always not recommended to transport waste directly from the collection point to the recycling center, which is usually situated some distance away from the generation point. Transfer stations, which can be strategically located to accept waste from collection trucks can represent a suitable and more economic solution (El Haggar 2004c).

The transfer station can be divided into several workstations. The first workstation would include a conveyor belt as shown in Figure 13.1 where trucks are allowed to enter leaving all the waste to be sorted manually on the conveyor. A second workstation would include a glass crusher as shown in Figure 13.2, where glass is crushed into small particles ready for recycling. The third workstation would include a hydraulic press as shown in Figure 13.3 to compact the sorted solid waste such as plastics, metals, paper, textile, etc. The hydraulic press is used to decrease their volume for easy storage, handling and transport to other companies for recycling. An important consideration while designing the transfer station is taking into account the method of energy used and capital investment such as automatic sorting, manual sorting, magnetic separator, etc.

Recycling

Recycling is the first process for solid waste management to be considered. The process of recycling is used to recover and reuse materials from spent products. That is why recycling is made to recover as much as possible from the waste and other treating processes handles the remaining of waste. Recycling will reduce waste and the limited resources will be conserved for

FIGURE 13.1. Conveyer belt.

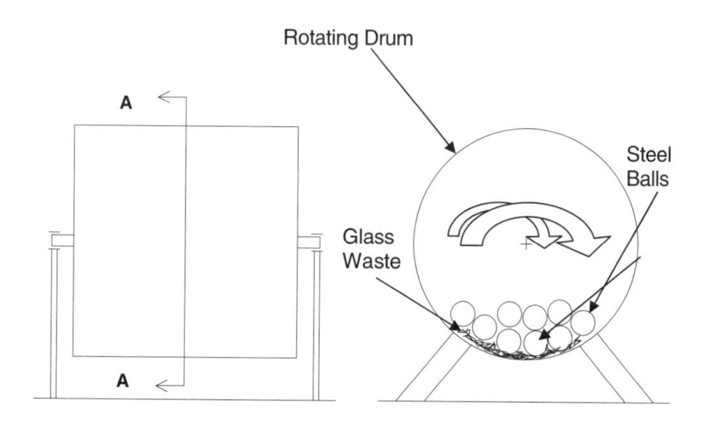

FIGURE 13.2. Glass crusher (Ball Mill).

future use since the scale of consumption of raw material is very large. Recyclable materials are plastics, papers, metals, bones, glass, and food wastes. Figure 13.4 illustrates the system to reach 100% recycling of non-hazardous MSW as will be explained in detail through this section.

FIGURE 13.3. Hydraulic press.

Recycling (material recovery) can provide a cost-effective waste management approach. This technique can help reduce costs for raw materials and waste disposal and provide income from a salable waste as well as protecting the environment. The type of wastes that is separated and can be recycled easily with high benefits including paper and cardboard; aluminum cans and tin cans; plastics; textiles, bones and glass. Organic waste or food waste recycling should be treated with special attention because it contains some rejects such as contaminated plastic bags and small pieces of glass, etc. There are many ways to recycle organic waste and convert it into soil conditioner (fertilizer) such as aerobic fermentation (composting), anaerobic fermentation (biogas), vermin composting and co-composting processes. Composting is the most commonly used method to recycle organic wastes from technical, economical, and environmental point of views.

Paper Recycling

The earth has limited natural resources, which must be conserved for the future of the coming generations. The scale of consumption of raw materials through forest and crop planting is very large in paper manufacturing. Hence one of the solutions to this problem is to recycle existing product's

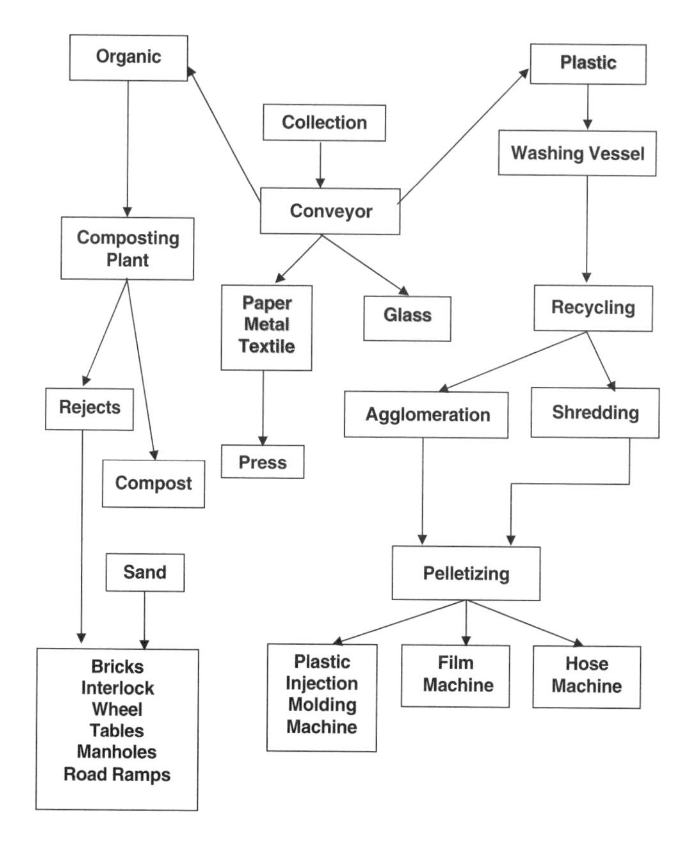

FIGURE 13.4. Proposed system for reaching 100% recycling.

waste. Recycling may enable the recovery of primary material for reuse in the paper manufacturing process.

The paper industry introduced the concept of paper recycling 280 years ago because recycling is considered to be more cost-effective than incineration or landfilling. This demonstrates the importance of paper recycling in paper manufacturing process. The worldwide paper recycling has been improving since 1986 as shown in Tables 13.1 and 13.2.

In Egypt, organic waste forms about 50% of solid waste, while inorganic matter forms 50% (Table 13.3). In high-income areas the percentage of inorganic matter such as paper, metals, plastics, and other recyclables is substantial, while in low-income areas the proportion of organic food waste to inorganic matter is higher. In the U.S. organic waste represents only 7% of solid waste and in Britain 20%. On the other hand, in urban areas of Egypt organic waste represents 46% of solid waste. Table 13.3 shows the percentage of municipal solid waste components in the U.S. (Cichonski and Hill 1993), the U.K. (Newel 1990), and Egypt (EEAA 1994).

TABLE 13.1
Worldwide Usage of Waste Paper [Mckinney, 1991]

Year	*Pulp & paper production (m ton)*	*Wastepaper consumption (m ton)*	*Apparent utilization Rate (%)*
1986	202	63	31
1990	237	85	36
1991	239	91	38
1992	246	96	39
2000	307	138	45

TABLE 13.2
World Recovered Paper Utilization [Kilby, 2001&CEPI, 1999]

Country	*Recovery, 000 tones*	*Utilization rate, %*
EU	34,988	44
USA	32,943	38
Canada	4,810	26
Japan	16,378	55
Brazil	2,295	35
Mexico	3,395	93
Australia	1,463	58
Others	9,759	

TABLE 13.3
Municipal Solid Waste in USA, UK and Egypt

Constituent	*% of total waste in USA*	*% of total waste in UK*	*% of total waste in Egypt*
Food	7	20	46
Paper	40	33	21
Plastics	8	7	4
Metals	8	10	2
Glass	7	10	2
Yard waste	18	**	**
Textile	**	4	2
Miscellaneous	12*	8	23***

* Mainly Rubber and textile.
** Not recorded.
*** Mainly dust.

In Egypt, recycling paper is very important as it is estimated that Cairo produces 8000 tons/day of municipal solid waste, 21% to 23% of which is paper and paperboard. Figure 13.5 shows the classification of the solid waste in Cairo, Egypt (El Haggar 2004c).

El Haggar (1996, 1998) and El Haggar and Baher (1999) stated that the main objective of the recycling process is to reduce the amount of solid waste present as well as to recover some of the primary materials. They found that papers used for newspaper or for educational textbooks are being imported in all developing countries, which explain the marked increase in the price of press and publication in general. This crisis occurred during 1995 when the imported newspaper prices had increased in Egypt and other developing countries by 40%. This will not only affect the press and publication but also the educational system where most of the government textbooks depend on this type of imported paper. This clearly shows the importance of implanting recycling in Egypt and other developing countries. They also argued that by applying recycling a huge reduction in both the volume of waste and the green house effect would result, in addition to saving water and energy, which will help in having a better environment.

The American University in Cairo (AUC) started a paper-recycling sub-program in 1994 within a program called "Industrial/Municipal Waste Management Program, IMWMP." A model of a paper recycling machine was designed and manufactured at AUC to test different factors affecting the recycling process and quality of produced paper (Fig. 13.6). A de-inking system was also incorporated into the design to remove the ink mechanically from the recycled paper pulp (Fig. 13.7). Overall, the system proved to

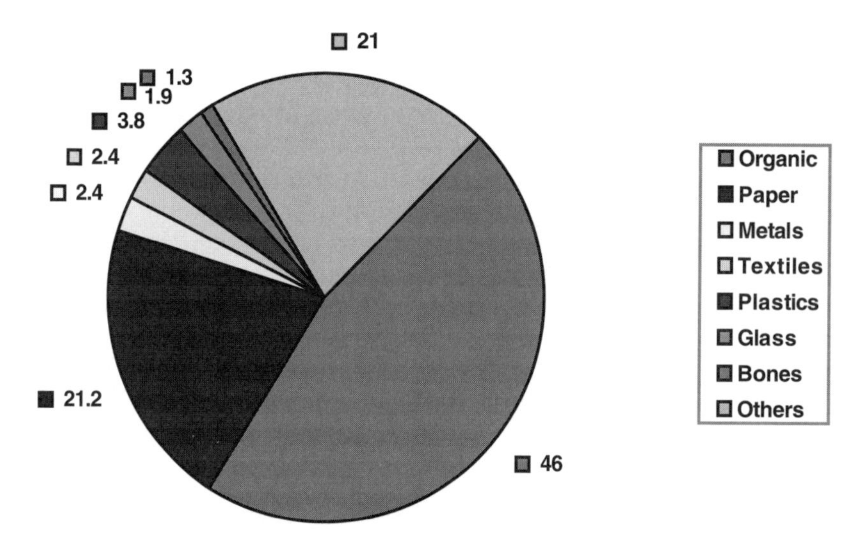

FIGURE 13.5. Municipal solid waste (MSW%) in Cairo, Egypt.

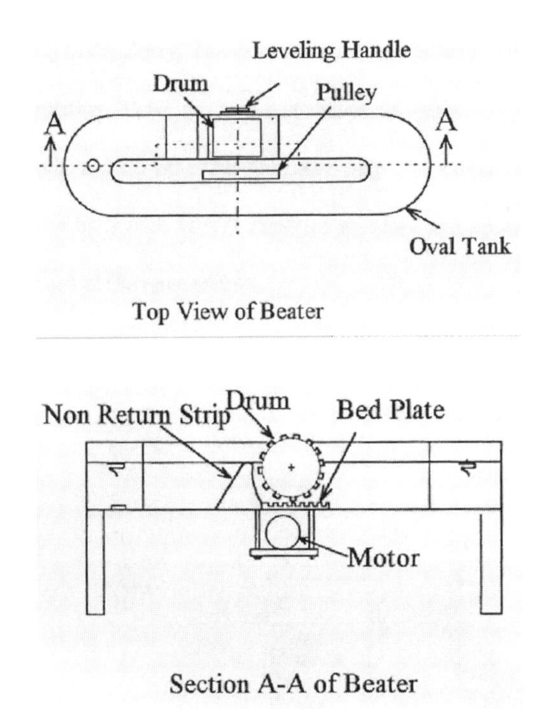

FIGURE 13.6. Schematic drawing of paper pulping (beater) machine.

be highly effective in producing quality paper and the focus of the research has turned to optimize the paper recycling process. Different raw materials were tested to optimize the mixing ratio for better product quality.

Categories of Paper

There are four main categories of paper: high-grade printing and writing paper, newsprint, corrugated/paperboard (including packaging), and tissue/towel products. The overall recycled content in each category varies, with tissue/towel containing the highest percentage of recycled material and printing and writing paper the lowest. So far, all pulp used for tissue/towel products are imported from outside as well as newspaper in most developing countries. These products could be produced in developing countries very easily through simple recycling technologies with high return because it does not require high technologies or sophisticated quality (El Haggar et al. 2001; El Haggar 2001a).

Market Value

Approximately 3.3 million tons of waste paper is dumped in Egypt every year. Paper constitutes approximately 21%–23% of the municipal solid waste generated in Cairo and offering a promising opportunity for the recycling industry. Waste paper sells for $50–$60 per ton.

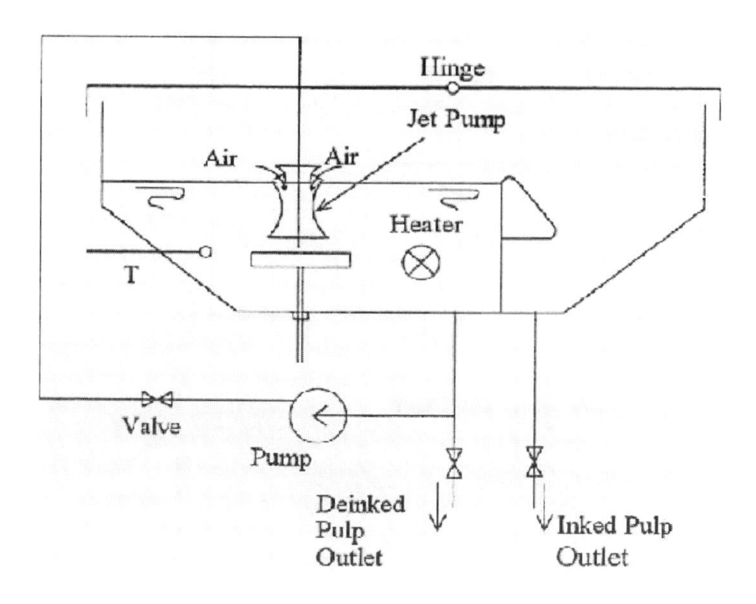

FIGURE 13.7. Schematic drawing of air injection flotation cell.

TABLE 13.4
Paper Imports in Egypt [CAPMAS, 1999]

Year	*1991*	*1992*	*1993*	*1994*	*1995*	*1996*	*1997*	*1998*
Value Million $	324	405	411	367	376	622	493	611

Egypt and most developing countries have a shortage of forests and so the country imports all of its pulp requirements. Eighty percent of the imported pulp is used in the paper making industry and the remaining 20% in other industries. Around $600 million worth of paper, paperboard and paper making materials were imported in 1998 and increasing every year as shown in Table 13.4.

At present, paper recycling in Egypt produces only low-grade quality paper. Reclaimed paper and cardboard are used extensively in the manufacture of local craft and cardboard. Board egg trays are produced mainly from paper waste, particularly newsprints. Table 13.5 demonstrate the consumption capacity during last 5 years. It is worth noting that the amount of production during the same period ranging from 40% to 45%. All newspaper is imported with an estimated cost of $70 million during 1997–1998.

TABLE 13.5
Consumption Capacities (1000 tons) of Paper in Egypt [CAPMASS, 1999]

Year	Consumption
1996/1997	319.5
1997/1998	333.9
1998/1999	348.3
1999/2000	362.7
2000/2001	377.1

Plastic Recycling

The problem of plastics wastes has increased tremendously since the use of plastics increased in most industrial, commercial and residential applications. Households and industry produce huge amount of plastic waste. Plastic waste causes severe environmental problem when incinerated or open burn on roadsides or illegal dumpsites. Also plastic bags are a major source of littering the residential areas, parks and even protected areas.

About 50% of the total volume of plastic wastes consists of household plastics refuse, which are mainly in the form of packaging wastes. Once rejected, plastics packages gets contaminated and while reusing them, a more serious problem appear which is the so-called commingled plastics, affecting in return the properties of the new recycled products (Wogrolly et al. 1995).

The recycling of thermoplastics, or plastics, can be accomplished easily with high revenue. Each type of plastic must go through a different process before being recycled. Hundreds of different types of plastics exist, but 80%–90% of the plastics used in consumer products are:

- PET (polyethylene terephthalate)
- HDPE (high-density polyethylene),
- V (vinyl)
- LDPE (low-density polyethylene)
- PP (polypropylene)
- PS (polystyrene)
- PVC (poly-vinyl chloride)

The most common items produced from post consumer HDPE are milk and detergent bottles and motor oil containers. Soda bottles, mineral water cooking oil bottles are made of PET.

Mechanical recycling: which involve cleaning, sorting, cutting, shredding, pelletizing and finally reprocessing by injection molding or extrusion. A simplified scheme for the recycling process was illustrated in Figure 13.4.

To recycle plastics three main stages are needed. The initial stage where the wastes are collected, sorted, separated and cleaned, this stage is mainly labor intensive requiring little capital investment and relatively no technical skills. The second stage is where the collected wastes are being prepared for reprocessing. In this stage, the wastes are reduced in size by undergoing cutting, shredding and agglomeration. The final stage is the reprocessing stage, where mixing, extrusion and product manufacturing takes place (Lardinos and Van de Klundert 1995).

Collection, Sorting, Separation, and Cleaning Stage

The first step in the initial stage is the collection, which is usually carried in developing countries by the informal sector and small individual enterprises. This stage is labor intensive and requires no technical experience. Collection usually takes place from the final dumping site or the transfer station, since it is the most practical method in developing countries.

The next step in the initial stage is sorting of plastics, which depends on the manufacturing demands. It could be sorted based on color or based on type. Sorting is usually carried out by women and children in developing countries because of low wages and no technical skills required. The working conditions at this stage are not hygienic because generally sorting is the first stage after collection and before cleaning which leaves the workers exposed to contaminated plastics and more subject to a fermented environment as well as continuous contact to dirt and chemicals (Lardinos and Van de Klundert 1995).

In sorting there are basic guidelines followed in rural areas and developing countries. They separate plastics films from rigid plastics in which each undergoes a further separation based on color and type. The waste plastics are first sorted and classified into bottles, transparent plastics, rigid plastics and flexible plastics; within these a further separation is done based on the type and color.

While sorting plastic we need to separate different kinds of plastics from each other because while reprocessing the plastics waste we need to know which type of plastics we are dealing with. This is because plastics types have different properties and reprocessing them together will produce a product with poor properties, unexpected durability and poor appearance. Some plastics may even look similar and need testing to find out its type. To have an identification of the plastic type a chemical test such as infrared analysis could be used. However, experience can help in this field but in case of doubt we have to revert to testing (Lardinos and Van de Klundert 1995).

Some basic tests include:

1. To distinguish between thermoplastics and thermosets, press a piece of wire just below red heat and try to press it into the material. If it penetrates it is a thermoplastics, if not, it is a thermoset (Vogler 1984).

2. To distinguish between the types of plastics, there is a primitive way which is to scratch the surface with the fingernail to see the material flexibility. However, this test needs an experienced person and is not always reliable since the material could have been exposed to many weathering conditions and could become brittle and rigid and cannot be scratched. A thin material may seem flexible and thick (Vogler 1984).

3. To separate plastics from non-plastics, or to separate between plastics types, we can use flotation test. This test is useful to distinguish between PP and HDPE, and between HDPE and LDPE. The test basically consists of differentiating between the different densities of the two plastics by mixing them with water and alcohol in certain proportions. It was found that in a mixture with an exact density of 0.925, HDPE will sink and PP will float and in a mixture of 0.93 of density, HDPE will sink and LDPE will float. However, for overlapping densities of plastics like PP and LDPE, the fingernail test and the visual inspection can be more reliable (Vogler 1984).

In order to distinguish between PS and PVC, another flotation test is carried using pure water and salt. The existence of salt in water forces the PVC and dirt to sink and the PS to float. However the amount of salts need not be measured, it is achieved by experience based on trials (Vogler 1984).

4. Burning test is a test used to differentiate between the different types of plastics based on the color and smell of the flame produced. The test is carried out as follows: cut a 5 cm long of a strip of plastic material, tapered at a point at one end and 1 cm wide at the other. Hold the sample over a stone and light the tapered end. However the person carrying this test should be careful not to be too close to the sample under tests and not to inhale the smoke coming out of it as it might contains hazardous substances (Vogler 1984).

The technique used in plastic testing differs between the developed and developing countries. In the developing countries, the technology and the expertise are not available in the informal sector. Whereas in the industrialized countries mechanical separation techniques are used. Instrumental analytical methods are becoming more available like infrared spectroscopy and thermal analysis (Lardinos and Van de Klundert 1995).

Some of the techniques developed by industrialized countries concerning the separation of plastics include a method of separating the plastic packaging materials where velocity plays an important role to identify about six packaging plastics. Velocity plays a minor role distinguishing between about 30 different plastics, mainly those made from engineering plastics (Wogrolly et al. 1995). Other methods include separating the plastics based on density, surface structure, ferromagnetism, conductivity, color, etc. And since the purity of the finished product depends on the sorting accuracy, we need to make a distinction between: sorting due to density, flotation, electrostatic sorting, thermal separation; and sorting of plastics by the hand picking method discussed earlier (Wogrolly et al. 1995).

Other methods for separations include the bottle sorting mechanisms, which can be either manually as described earlier, semi-automatic or automatic bottle sorting methods (Wogrolly et al. 1995). However, the problem of separation become sometimes more difficult when mixed plastics are involved. Therefore, the society of plastics industry in the United States has developed a coding system shown in Figure 13.8 using symbols and numbers for the types of plastics used. This system has also been introduced in Europe (Lardinos and Van de Klundert 1995).

The next step in the initial stage is the cleaning, which usually consists of washing and drying the plastics. It is important to wash the plastic before shredding it since it improves its quality. Usually washing takes place after cutting (cutting refers to splitting the plastics into two parts to ease the washing process) and sometimes even after shredding to obtain better results. Foreign materials like paper and covers are removed before cleaning. Washing can be carried out either manually or mechanically. Manually, the plastics are replaced in a drum cut in half and the water is stirred with a paddle. The water is heated if the waste is greasy; soap and caustic soda are added to help removing the more difficult grease (Lardinos and Van de Klundert 1995).

Like with washing, the plastics can be dried either manually or mechanically. Manually, the plastics are left to dry in the sun. This method works best for plastic films, which are hung on lines to dry. Shredded plastics, are centrifuged and left to further drying in the sun. If left to dry without further being centrifuged, they will need longer time and more often stirring and a space of around 15–20 m^2 for a 300 kg of shredded plastics (Lardinos and Van de Klundert 1995).

In developing countries, plastics are cut in two halves and washed in a hot water basin using burners. For one ton of plastics, 25 kg of caustic soda should be added to 2 m^3 of warm water. The plastics are then rinsed in cold water in two different basins to remove the caustic soda; it is then centrifuged and left to dry for 2–4 hours in the sun.

The Second Stage, Plastics Are Prepared for Further Reprocessing

This stage involves size reduction in which waste plastics undergoes cutting, shredding, agglomeration and pelletizing. This process increases

FIGURE 13.8. Plastic coding system.

the cost of plastics waste since it eases their use in manufacturing process and decreases transportation cost (Lardinos and Van de Klundert 1995). The smaller the size of the shredded plastics, the more regular their shape (as in the case of pelletizing) leading to wider market demand and higher price.

The first material transformation step is **cutting** the plastics into smaller pieces, as is the case with large bottles, cans and buckets. These waste plastics are usually cut by a circular saw or with a bawsaw (Lardinos and Van de Klundert 1995). In most developing countries, the sorted and washed plastics are cut into two pieces by large scissors fixed on a wooden base. It is estimated that an average of three workers can cut up to one tone of sorted plastics per day.

Waste plastics cut in two or more pieces are fed into a shredder for further cutting. The plastics is cleaned before **shredding** and sometimes gets further cleaning after shredding according to the manufacturer demand. The shredder machine used for thin film plastics is different from that used for rigid plastics. For rigid plastics a horizontal cutting machine is used where the blades are rotating on a horizontal axis and the shredded plastics passes through a grid into a collecting tray (Lardinos and Van de Klundert 1995).

The sizes of shredded plastics is from 5 to 10 mm. The motor that drives the shredder is approximately 30 HP and the shredded plastics are collected in bags to be further reprocessed or sold. Figure 13.9 shows a

FIGURE 13.9. Horizontal axis shredder.

horizontal axis shredder used for rigid plastics waste. The end products of shredding are irregularly shaped pieces of plastics depending on the required final product and the type of industry that will use it. The shredded plastics could undergo further washing to ensure cleanliness, especially that the shredded pieces are easily cleaned when they are in small sizes than in larger ones (Lardinos and Van de Klundert 1995).

In order to avoid feeding plastics films, bags and sheets directly in a shredder we use an **agglomerator** that cuts, preheats and dries the plastics. The agglomerator will increase the material's density and quality which will end up with a continuous flow in the extruder and hence better efficiency. It is better to clean the plastics before agglomeration since the foreign substances will be processed together with the plastics. In the process of agglomeration, heat is added indirectly through friction between plastic film and the rotating blade located at the bottom of the agglomerator. It is therefore important to rapidly cool the plastic film to obtain the crumb shape desired. This is achieved through adding a small cup of water. An example of an agglomerator is given in Figure 13.10.

In the second stages, pelletizing is considered to be the last one as shown in Figure 13.11. It is the process in which the shredded or agglomerated plastics are uniformly sized to produce a better product quality and increase the efficiency of the product manufacturing process, due to the higher bulk density of the pellets compared to the shredded and agglomerated one. To reach the final product of plastics pellets, the shredded rigid plastics or the agglomerated films pass through extrusion and pelletizing processes. In the extrusion phase, the plastics undergo mixing, homogenization, compression, plasticization and melt filtration. Some pigments can be added to the process to reach other properties (e.g., coloring pigments) (Lardinos and Van de Klundert 1995).

The material is introduced in the hopper to the rotating horizontal screw in which it is heated by the use of the heating elements which plas-

FIGURE 13.10. Aggolmerator machine.

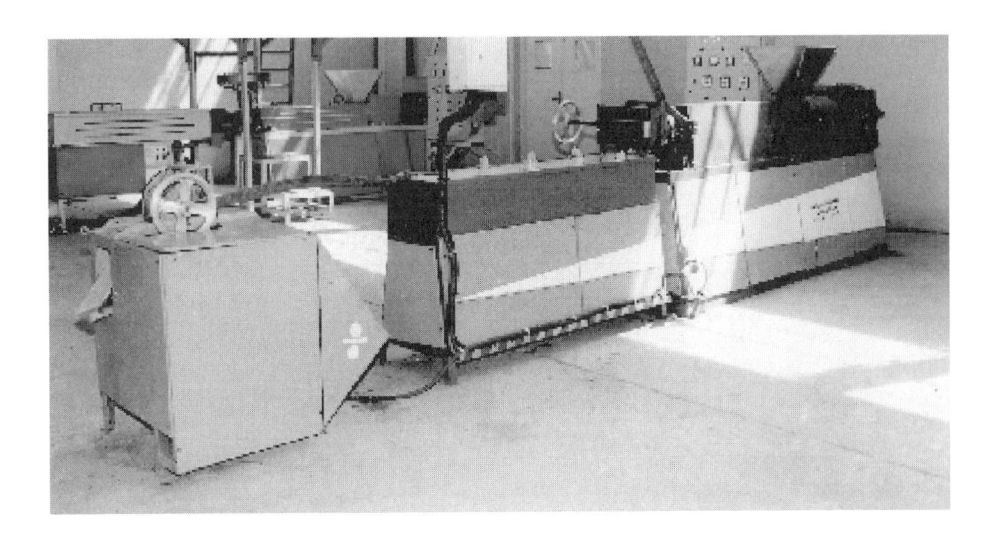

FIGURE 13.11. Pelletizing machine.

ticize the waste and the screw compresses it. The extruder screw is surrounded with water jacket to cool it. The heated mix then pass to a filter screen to remove solid particles and then to the extrusion head. This filter screen needs to be replaced every 2 hours to ensure better quality and avoid blocking it. The mix comes out of the die head as a hot spaghetti like shape that gets cooled by passing through a water basin. The strings are then drawn into the pelletizer by passing first through the supporting rollers. The pelletizer shops the strings into short, uniform cylindrical pellets that are packed in bags to be sold to the manufacturer (Lardinos and Van de Klundert 1995).

The following steps can improve the quality of pellets:

- Adding virgin plastic pellets depending on the quality needed
- Preheat the shredded plastics in a drying installation
- Reduce moisture content of reprocessed pellets by further extrusion through a finer filter screen
- Increase capacity of the process by introducing a rotating spiral gear wheel in the hopper that presses the material down to the screw and thus speed the process and increase production

Mixing, Extrusion and Product Manufacturing Stage

In the third and last stage of plastic reprocessing, manufacturing processes takes place, which include extrusion, injection molding, blow molding and film blowing. The **extrusion** processes is similar to the one used in the pelletizing unit except that the die takes the form of the output product. In case of tube production the die is made of a steel plate pierced with a hole. The extruded material has to be cooled to solidify by water bath. Similar to the extrusion, the **injection molding** process as shown in Figure 13.12 have a spiral screw which is fed through the hopper and presses the mix to a strong, split steel mold (Lardinos and Van de Klundert 1995).

The rotating screw pushes the plastic while the heating element plasticize it. The melt is pushed to a closed steel mold which is kept cool by a water jacket that accelerates the solidification process. Usually the process is manually stopped, this is achieved by allowing the mold to have one small hole allowing the exit of extra material that indicates a full mold. The worker then stops the process, opens the mold, gets the piece out and leaves it to cool. He then chops off all extra parts from the piece (Lardinos a Van de Klundert 1995).

The process of **blow molding** is used to produce bottles. The principle of the operation mainly takes place in two stages. First the plastic is extruded in the form of a tube then it passes through a mold which closes around the tube. Compressed air is then blown into a hollow mandrel which exists in the closed mold to force the tube to take the shape of the mold. The product is left to cool to solidify and later removed from the mold to repeat the process (Lardinos and Van de Klundert, 1995). Plastic bags can be used using the blow-molding machine as shown in Figure 13.13. The capacities of the blow molding machines (film) vary between 100 to 200 kg per day which depends on the power of the motor ranging from 10 to 15 HP, respectively.

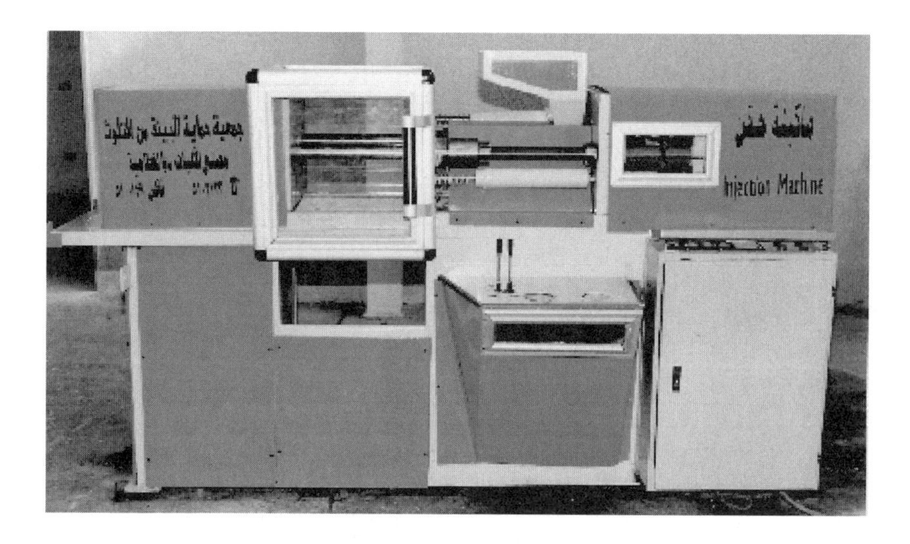

FIGURE 13.12. Injection molding machine.

FIGURE 13.13. Blow molding machine (film).

In film blowing, the plastics is first extruded from a tubular die and moves upward to a film tower with a collapsing frame, guide rolls and pull rolls driven by a motor. Air is compressed through the center of the die and inflates the tube. Air ring are mounted above the die to cool the outside surface. The tube is sealed and cut once it passes through the pull rolls.

Bone Recycling

Bone recycling is a simple process where useful products can be extracted. Minerals such as calcium powder for animal; feed are extracted from the bone itself. The base material for cosmetics and some detergent manufacturing needs are extracted from the bone marrow.

The bone recycling process passes through seven stages starting from crushing and ending with packing. Figure 13.14 gives a schematic diagram showing the bone recycling process which goes through the following steps:

1. **Crushing:** Bones are conveyed through an auger from the receiving area to the crusher where bones are broken into pieces of about 10 cm in length.

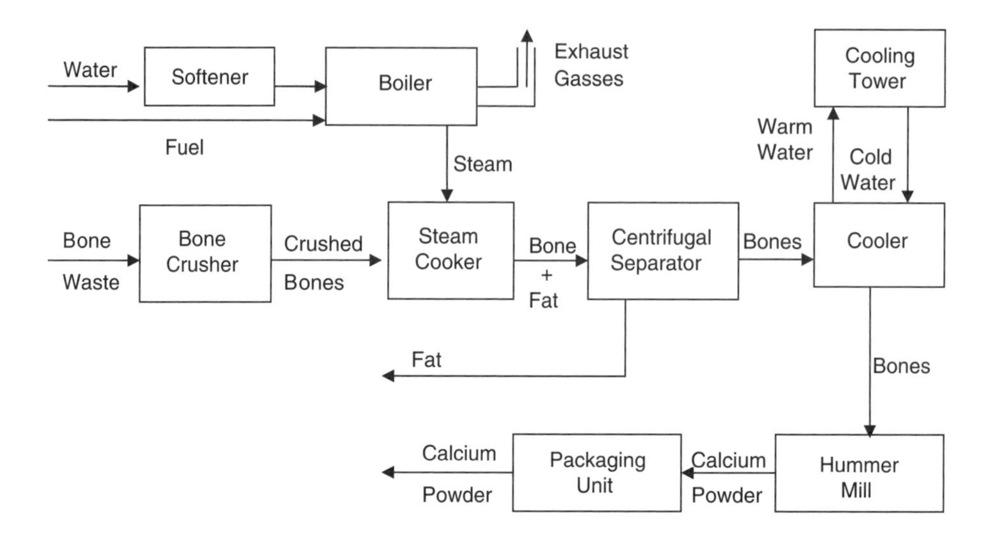

FIGURE 13.14. Bone recycling process.

2. **Cooking:** In the cooker, crushed bones are subject to saturated steam supplied from a fire tube boiler via the steam line to cook bones with fat and protein and kill any bacteria or pathogens.
3. **Centrifugal separator:** In the centrifugal separator unit, bone marrow and fats are expelled out of a perforated tank leaving the crushed bones in the bottom.
4. **Cooling:** Crushed bones are cooled by circulating water in a cooler hopper. The circulating water in cooled in a forced draft-cooling tower.
5. **Fine grinding:** The cooled bones are transferred from the cooler to the hummer mill using an auger to obtain finer grains of calcium powder.
6. **Cyclone separator:** A two-stage cyclone separator is used to separate the calcium powder before backing.
7. **Packing:** At this stage the fine grains are weighed, packed and are ready for marketing.

Glass Recycling

The consideration in glass recycling is color separation. Permanent dyes are used to make different colored glass containers. The most common colors are green, brown, and clear (or colorless). In the industry, green glass is called emerald, brown glass is amber, and clear glass is flint. For bottles and jars to meet strict manufacturing specifications, only emerald or amber cullet (crushed glass) can be used for green and brown bottles, respectively. The glass is color sorted and sent to a glass crusher or ball mill.

The cullet is then mixed with the raw material used in the production of glass. After the batch is mixed, it is melted in a furnace at temperature ranging from 1200°C to 1400°C. The mix can burn at low temperature if more culets are used. The melted glass is dropped into a forming machine where it is blown or passed into shape. The newly formed glass containers are slowly cooled in an annealing furnace.

Glass used for new bottles and containers must be sorted by color and must not contain contaminants such as dirt, rocks, ceramics, etc. These materials, known as refractory materials, have higher melting temperatures than container glass and form a solid inclusion in the finished product.

Aluminum and Tin Cans Recycling

Aluminum wastes are one of the most common items can be recovered through municipal solid waste because it provides higher revenues than other recyclable materials. The recycling of Aluminum cans uses 70%–90% less energy than producing them from virgin materials.

Steel food cans make up of 80%–90% of all food containers, are often called tin cans because of the thin tin coating used to protect the containers from corrosion. Some steel cans, such as tuna fish cans, are made with tin-free steel, while others have an aluminum lid and a steel body and are commonly called bimetal cans. All these empty cans are completely recyclable by the steel industry and should be included in any recycling program. The collection vehicle discharges the solid wastes into a hopper bin, which discharges to a conveyor belt shown in Figure 13.1 in the transfer station. The conveyor transports the cans past an overhead magnetic separator where the tin cans are removed. The belt continues past a pulley magnetic separator, where any tin cans not removed with the overhead magnet are taken out. The aluminum and tin cans, collected separately, are baled for shipment through a hydraulic press located in the transfer station as shown in Figure 13.3. At the aluminum foundry, aluminum scrap is melted in smelting. Molten metal is formed into ingots that are transferred to manufacturing plants and cut into disks, from which cans and other products are formed.

Textile Recycling

Textile recycling has a long history, not for making new textile or return textile back to its original fibers or other textile products but for making paper (Dadd 2004). Textile fiber can be classified into natural fibers such as cotton and wool and synthetic fibers. By recycling cotton wastes, we not only conserve landfill space and impacts but also, we reduce the amount of land, water, energy, pesticides and human labor that goes into cotton production.

The textile recycling industry represent one of the most important recycling activities from solid wastes because it is a labor intensive activity and can provide a lot of job opportunities as well as the availability of textile waste every where. Most of the textile recycling firms are small, family-owned businesses with 5–20 semi-skilled and marginally employable workers at the primary processing level including used clothing dealers, rag graders, and fiber recyclers.

Textile waste can be classified into two categories pre-consumer category and post-consumer category (Dadd 2004). Pre-consumer textile waste category consists of by-product materials from the textile, fiber and cotton industries. The pre-consumer textile waste can be recycled into new raw materials for the automotive, furniture, mattress, coarse yarn, female accessories, home furnishings, paper and other industries. Post-consumer textile waste category consists of any type of garments or household article, made of some manufactured textile, that the owner no longer needs and decides to discard. These items are discarded either because they are worn out, damaged, or have gone out of fashion. Many items made from fabric items recut to make new items, such as t-shirts cut to make cleaning cloths.

Food Waste Recycling, Composting

Food waste recycling can take place through aerobic fermentation; composting or anaerobic fermentation (biogas). Composting is the most recommended method for recycling food wastes. Composting is a process that involves biological decomposition of organic matter, under controlled conditions, into soil conditioner (El Haggar et al. 1998). Aerobic fermentation is the decomposition of organic material in presence of air. During the composting process, microorganisms consume oxygen, while CO_2, water, and heat are released as result of microbial activity as shown in Figure 13.15.

Factors Affect Composting Process

Four main factors control the composting process: moisture content, nutrition (carbon:nitrogen ratio of the material), temperature and oxygen (aeration).

Moisture Content

The ideal percentage of the moisture content is 60% (El Haggar et al. 1998). The initial moisture content should range from 40% to 60% depending on the components of the mixture. If the moisture content decreases less than 40%, microbial activity slows down and became dormant. If the moisture content increases above 60%, decomposition slows down and odor from anaerobic decomposition is emitted.

Carbon to Nitrogen Ratio

Micro organisms responsible for the decomposition of organic matter require carbon and nitrogen as a nutrient to grow and reproduce. Microbes work actively if carbon:nitrogen ratio is 30:1. if carbon ratio exceeded 30, the rate of composting decreases. Decomposition of the organic waste material will slow down if C:N ratios are as low as 10:1 or as high as 50:1.

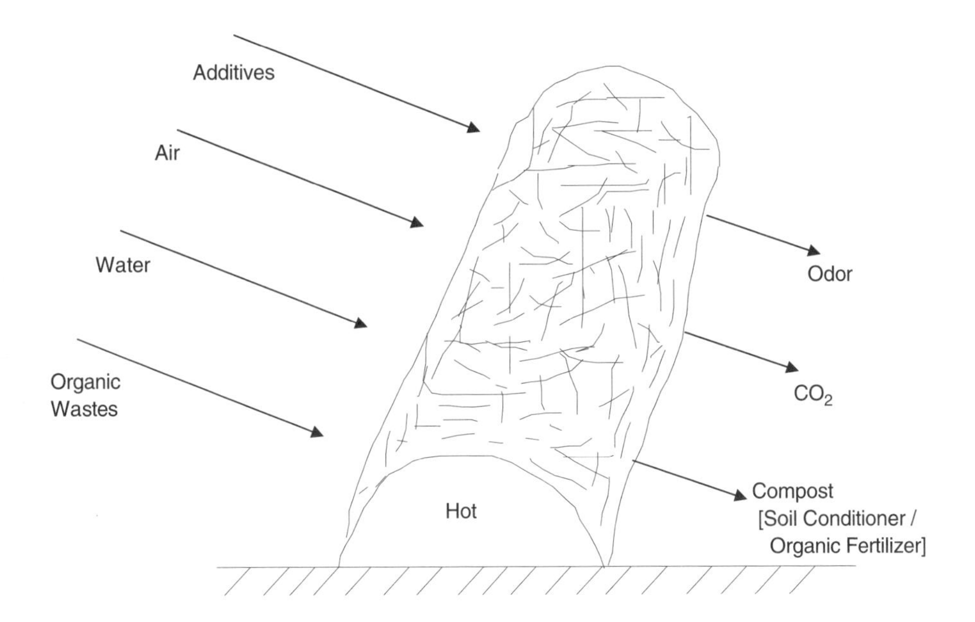

FIGURE 13.15. Composting process.

Temperature

The activity of bacteria and other micro-organisms produce heat while decomposing (oxidize) organic material. The ideal temperature range within the compost to be efficient varies from 32°C to 60°C. If the temperature outside this range, the activity of the micro-organisms slow down, or might be destroyed.

The increase of temperature while composting above 55°C, kill the weeds, ailing microbes and diseases including *Shigella* and *Salmonella*; this help to reduce the risk of diseases' transmission from infected, and contaminated materials. The outside temperature has an effect on the composting process. In winter, the composting process is slower than spring and summer.

Oxygen (Aeration)

A continuous supply of oxygen through aeration is a must to guarantee aerobic fermentation (decomposition). Proper aeration is needed to control the environment required for biological reactions and achieve the optimum efficiency. Different techniques can be used to perform the required aeration according to the composting techniques. The most common types of composting techniques are: natural composting, forced composting, passive composting and vermin-composting.

Natural Composting
Piles of compost are formed along parallel rows as shown in Figure 13.15 and continuously moisturized and turned. The distance between rows can be determined according to the type and dimension of the turning machine (Tchobanoglous et al. 1993). Piles should be turned about three times a week at summer and once a week at winter to aerate the pile and achieve homogenous temperature and aeration throughout the pile (Fig. 13.16). This method needs large surfaces of land, many workers, and running cost.

Advantages

- Natural Aeration
- Low cost

Disadvantages

- Odor emission
- Needs long time for maturation
- Needs labor for turning

Passive Composting
Parallel rows of perforated high-pressure PVC piping are placed at the bottom, on which compost is added above it. The pipes are perforated with 10 cm holes to allow air to enter the composting piles as shown in Figure 13.17. The pipe manifold help in distributing the air uniformly. Air flows through the ends of the pipes to the compost. This system is better than the natural system because of the limited flow rates induced by the natural ventilation. This method needs limited surfaces of lands, less running cost, and does not need skilled workers. This method is recommended for cost-effective and the most economic aeration method. Therefore, it is the most suitable method for the developing communities that want to achieve maximum benefit from the food recycling with the minimum capital

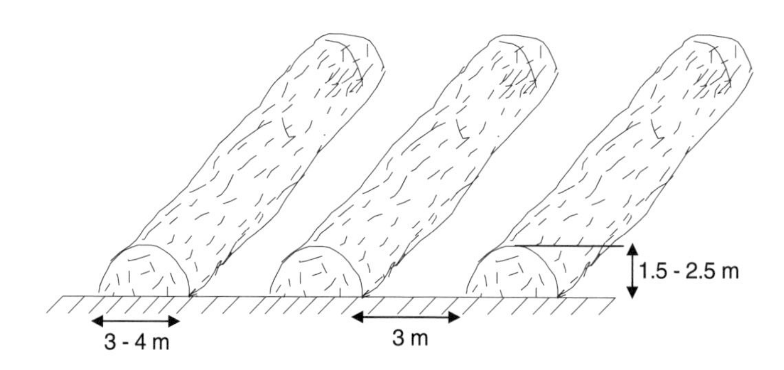

FIGURE 13.16. Natural composting process.

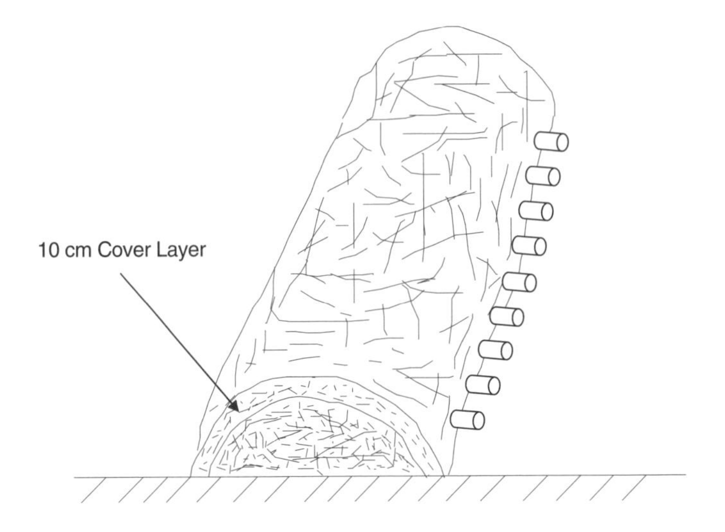

FIGURE 13.17. Passive aeration process.

investment and good quality soil conditioner or organic fertilizer if NPK (nitrogen:phosphorous:potassium) will be adjusted with natural rocks (El Haggar el al. 2004).

Advantages

- Natural convection
- Low running cost
- Less maturation time
- Odor, can be controlled by adding a top layer of finished compost

Disadvantages

- Relatively high cost compared with natural aeration

Forced Aeration

It works like the previous system except that the ends of plastic pipes are connected to blowers that force (or suck) the air through the compost with a specific rate and velocity. Otherwise if the air rate exceeded a certain limit, the temperature inside the compost pile decreases affecting the microbial activity. Also, the air velocity during day should always be higher compared with the air velocity at night. This system needs higher technology with air velocity control and more energy consumption. That is why, it is less economic compared to the other two systems and it is not recommended for rural or developing countries that want to make profit out of all recycling processes. This method needs capital investment, skilled workers, and running cost.

Advantages

- Odor, can be controlled by adding a top layer of finished compost
- Less maturation time compared with natural aeration

Disadvantages

- Needs electrical source near compost
- High capital cost and velocity control
- Require skilled workers

Vermi Composting

It is an ecologically safe and economic method that depends on the worms' characteristic of transforming the organic wastes to fertilizers that are extremely beneficial to earth. There are two types of earthworms that are used due to their insensitivity to environmental changes:

- The red wiggler (*Eisensia foetida*)
- The red worm (*Lumbricus rebellus*)

Under suitable aeration, humidity, and temperature, worms feed on organic wastes and expel their manure (worm castings) that break up soil providing it with aeration and drainage. It also creates an organic soil conditioner as well as a natural fertilizer. Worm casting has more nutrients than soil conditioner in terms of nitrogen, phosphorous, etc.

A mature worm will produce a cocoon every 7 to 10 days that contains an average of seven baby worms that mature in approximately 60 to 90 days, and in 1 year each 1000 worms produce 1,000,000 worms (El Haggar 2003a).

Vermi composting can be used in houses easily by using a special container (worm bin) that can be placed in any place that is not subjected to light such as kitchen, garage, and basement. The organic waste is put in this container and the worms with them. The worms are odorless and free from disease.

2.4 Rejects

The rejects problem starts at the sorting stage. The plastic bags that are contaminated with the organic material are not collected because cleaning them is very costly. Also items such as small pieces of glass can hurt a worker's hand so they are not collected. In a typical municipal solid waste composting plant, the raw materials coming to the plant are food waste mixed with contaminated plastic bags and other rejects. The compost piles, after reaching maturation stage, are driven by a conveyor belt to a drum-like machine that tends to separate the organic material from the rejects as shown in Figure 13.18. This drum rotates with the compost inside and the

FIGURE 13.18. Composting plant.

cutting tools located at the inner surface decompose further the organic material. As the drum rotates the organic material penetrates through the openings along the body of the drum while the rejects that rotate and move along the axis of the drum pass by an exit opening at the end of the drum (Sawiries et al. 2001). Figure 13.18 shows a drum in a typical composting plant in Egypt, with the piles of compost in front of it and rejects on the other end. These rejects represent 20%–25% of the compost pile (El Haggar and Toivola 2001). Rejects can be gotten rid of using incineration, landfills, or recycling.

Incineration

Rejects can be incinerated to convert it into ash. Incineration is the process of thermally combusting solid waste. There are various types of incinerators and the type used depends on the type of waste to be burnt. Conceptually, incinerators can be classified into the following common types of systems (Dasgupta and El Haggar 2003).

- Liquid feed incinerator
- Rotary kiln incinerator
- Grate-type incinerator
- Fluidized bed incinerator

Liquid feed incinerators can handle liquid waste while rotary kiln incinerators can handle both liquid wastes as well as solid wastes. Grate-type incinerators are used for large irregular shaped solid waste to allow air to pass through the grate from below into the wastes. The fluidized-bed type incinerator is used for liquid, sludge and/or uniformly sized solid waste.

Liquid Feed Incinerator

A large number of hazardous liquid waste incinerators used today are of this type. The waste is burned directly in a burner or injected into a flame zone or combustion zone of the incinerator chamber through atomizing nozzles. The heating value of the waste is the primary determining factor for the location of the injection point of the liquid waste. Liquid injection-type incinerators are usually refractory-lined chambers (horizontal or vertical flow, with up or down options), generally cylindrical in cross-section, and equipped with a primary burner (waste and/or auxiliary fuel fired). Often secondary combustors or injection nozzles are required where low heating value waste liquids are to be incinerated. Liquid incinerators operate generally at temperatures from 1000°C to 1700°C. Residence time in the incinerator may vary from milliseconds to as much as 2.5 seconds. The atomizing nozzle in the burner is a critical part of the system because it converts the liquid waste into fine droplets. The viscosity of the waste determines whether good atomization of the liquid is possible. Two-fluid atomizers, using compressed air or steam as an atomizing fluid, are capable of atomizing liquids with viscosities up to 70 centistokes. The physical, chemical, and thermodynamic properties of the waste must be considered in the design of the incinerator.

The method of injection of the liquid waste is one of the critical factors in the design and performance of these incinerators. The atomizer design is therefore an important design parameter of this system. The reasons for injecting the liquid waste as a fine spray are: (1) to break up the liquid into fine droplets, (2) to develop the desired pattern for the liquid droplets in the combustion zone with sufficient penetration and kinetic energy, and (3) to control the rate of flow into the combustion zone. Organic liquids pass through three phases before oxidation takes place. The liquid droplets are heated, vaporized, and superheated to ignition temperature. In a good atomizer, the droplet size will be small providing greater surface area and resulting in rapid vaporization. For example, the burning time for a 300 micron droplet is 150ms while only 30 milliseconds for a 125 micron droplet. Depending on the type of liquid waste and the combustion conditions required, there are generally three types of atomizers. Proper mixing of air with the atomized droplet is very important for complete oxidation. If the liquids contain fine solids, the design must allow the particles to be carried to the gas stream without agglomeration to affect proper combustion. Sufficient time must be provided to permit complete burnout of the solid particles in the liquid suspension. Inorganic particles carried in the liquid waste stream may become molten and agglomerate into molten ash. The combustor must be designed to collect the molten ash without plugging the flow passages of the incinerator. Primary and secondary combustion chambers are used in liquid feed incinerators. Primary units are used to burn wastes, which have sufficient heating value to burn without auxiliary fuel. Secondary units require auxiliary fuel. Sufficient air must be provided at all times to oxidize the organics in the combustion chamber. Incinerators can produce soot when burning under insufficient oxygen and poor air mixing conditions. Soot can clog up nozzles, and accumulate in the chamber, impairing burning conditions.

Rotary Kiln Incinerator

The rotary kiln is often used in solid/liquid waste incineration because of its versatility in processing solid, liquid, and containerized wastes. The kiln is refractory-lined. The shell is mounted at a slight incline (about 5 degrees) from the horizontal plane to facilitate mixing the waste materials. A conveyor system or a ram usually feeds solid wastes and drummed wastes. Liquid hazardous wastes are injected through a nozzle(s). Noncombustible metal and other residues are discharged as ash at the end of the kiln. Rotary kilns are also frequently used to burn hazardous wastes.

Rotary kiln incinerators are cylindrical, refractory-lined steel shells supported by two or more steel trundles that ride on rollers, allowing the kiln to rotate on its horizontal axis. The refractory lining is resistant to corrosion from the acid gases generated during the incineration process. Rotary kiln incinerators usually have a length-to-diameter ratio (L/D) between 2 and 8. Rotational speeds range from 0.5–2.5 cm/s, depending on the kiln periphery. High L/D ratios and slower rotational speeds are used for wastes requiring longer residence times. The kilns ranged from 2–5 meters in diameter and 8–40 meters in length. The burners for the kilns ranged from 10 million British Thermal Units (BTU) per hour to 100 million BTU per hour.

Rotation rate of the kiln and residence time for solids are inversely related; as the rotation rate increases, residence time for solids decreases. Residence time for the waste feeds varied from 30 to 80 minutes, and the kiln rotation rate ranged from 30 to 120 revolutions per hour. Another factor that has an effect on residence time is the orientation of the kiln. Kilns are oriented on a slight incline, a position referred to as the rake. The rake typically is inclined 5 degrees from the horizontal.

Rotary kiln incinerators are designed with either a co-current or a counter-current chamber. In the counter-current design, waste is introduced at the end opposite the burner and flows down the rake toward the burner, while combustion gases are drawn up the rake. In a co-current design, the waste feed is introduced at the burner end and flows down the rake, while the combustion gases are also drawn down the rake. Most rotary kiln incinerators were of the co-current design, which provides for more rapid ignition of the waste feed and greater gas residence time for combustion than does the counter-current design.

Hazardous or nonhazardous wastes are fed directly into the rotary kiln, either continuously or semi-continuously. Devices such as arm feeders, auger screw feeders, or belt feeders can be used to feed solid wastes. Hazardous liquid wastes can also be injected by a waste lance or mixed with solid wastes. Drums and cartons of hazardous waste may be fed directly into the kiln but are often shredded first. Rotary kiln systems typically include secondary combustion chambers of afterburners to ensure complete destruction of the hazardous waste. Operating kiln temperatures rang from 800°C to 1300°C in the secondary combustion chamber or afterburner depends on the type of wastes. Liquid wastes are often injected into the kiln combustion chamber.

The advantages of the rotary kiln include the ability to handle a variety of wastes, high operating temperature, and continuous mixing of incoming wastes. The disadvantages are high capital and operating costs and the need for trained personnel. Maintenance costs can also be high because of the abrasive characteristics of the waste and exposure of moving parts to high incineration temperatures.

Cement kiln incinerator is an option that can be used to incinerate most hazardous and non-hazardous wastes. The rotary kiln type is the typical furnace used in all cement factories. Rotary kilns used in cement industry are much larger in diameter and longer in length than the previously discussed incinerator.

The manufacture of cement from limestone requires high kiln temperatures (1400°C) and long residence times, creating an excellent opportunity for hazardous waste destruction. Further, the lime can neutralize the hydrogen chloride generated from chlorinated wastes without adversely affecting the properties of the cement. Liquid hazardous wastes with high heat contents are an ideal supplemental fuel for cement kilns and promote the concept of recycling and recovery. As much as 40% of the fuel requirement of a well-operated cement kiln can be supplied by hazardous wastes such as solvents, paint thinners, and dry cleaning fluids. The selection of hazardous wastes to be used in cement kiln incinerators are very important not only to treat the hazardous wastes but also to get some benefits out of it as alternative fuel and alternative raw material without affecting both the product properties as well as gas emissions. However, if hazardous waste is burned in a cement kiln, attention has to be given to determine the compounds that may be released as air emissions because of the combustion of the hazardous waste. The savings in fuel cost due to use of hazardous waste as a fuel, may offset the cost of additional air emission control systems in a cement kiln. Therefore with proper emission control systems, cement kilns may be an economical option for incineration of hazardous waste.

Grate-type Incinerator

Grate-type incinerators are used for incineration of solid wastes. These types of incinerators are suitable for large irregular shaped wastes, which can be supported on a stationary or moving alloy grate, which allows air to pass through the grate from below into the waste. Generally, grate-type incinerators have limited application for hazardous waste incineration because the high temperature required in the chamber may affect the material of the grate. The primary furnace is followed by a secondary chamber, where additional air and fuel are added, to complete destruction of all toxic emissions.

Fluidized-Bed Type Incinerator

This type of incinerator is used for liquid, sludge, or uniformly sized solid waste. The fluidized-bed incinerator utilizes a fluidized bed consisting of sand or alumina on which combustion occurs. Air flow is applied from below which has sufficient pressure, that it will fluidize the bed of sand, or hold it in suspension, as long as the velocity of the air is not so great that it transports the sand out of the system. This is a fluidized bed in which the particles of the bed are in suspension, but not in flow. Waste is added/injected into this fluidized bed. The air with which the bed is fluidized is heated to at least ignition temperature of the waste, and the waste begins to burn (oxidize) within the bed. Most of the ash remains in the bed, and some exits the incinerator into the air pollution control system. Heat also exits with the flue gases and can be captured in a boiler or used to preheat combustion air.

In almost all types of waste incinerators, primary and secondary combustion chambers are necessary to complete the combustion and oxidation to achieve the required Destruction and removal efficiency (DRE). The primary chamber's function is to volatilize the organic fraction of the waste. The secondary chamber's function is to heat the vaporized organics to a temperature where they will be completely oxidized.

Advantages of incineration

- It is applicable to all kinds of waste.
- Incinerators are made to avoid air pollution through air pollution control units.
- The ash resulting from the combustion only occupies around 10% of the solid waste volume.
- Energy may be recovered from incineration through many ways like gas to water heat exchange. The water is converted to steam, which may then be used to generate electricity through steam turbines.

Disadvantages of incineration

- Incinerator construction requires high capital cost.
- Incineration operation and management require high cost and skilled workers
- Wastes require energy to be burnt.
- The air pollution control systems are very expensive. On the other hand, the emissions and the ash resulting from incineration are

extremely dangerous. If not properly controlled, they cause air pollution that can have dangerous effects on human health.

Landfill

Rejects and/or ash produced from the incineration process should be landfilled. Most people think of landfills as dumpsites, just an open hole in the ground where solid waste is put with all kinds of animals and insects wondering around. This is not true. A landfill is a very complicated structure, very carefully designed either into or on top of the ground in which garbage is isolated from the surrounding environment. Unlike compost pile, landfill keeps the trash under conditions that minimizes its decomposition. There are many steps before starting to construct a landfill. The first step is to choose a site. After choosing the appropriate location for a landfill, the designing process starts. Beside the design of the lining and coverage of the landfill, a leachate collection system, biogas collection system and a storm water drainage system should also be designed and planned for implementation during operation.

The bottom liner isolates the solid waste from the soil preventing the groundwater contamination. The liner is usually some type of durable, puncture-resistant synthetic plastic (polyethylene, high-density polyethylene, polyvinyl chloride). In landfills constructed below surface level, a side liner system is used in mechanical resistance to water pressure, drainage of leachate and prevention of lateral migration of biogas.

The amount of space is directly related to the capacity and the lifetime of the landfill. To maximize the landfill's lifetime, trash is compacted into areas, called cells. Each cell contains only one day's trash. Each cell is covered daily with six inches of compacted soil. This covering seals the compacted trash from the air and prevents pests (birds, rats, mice, flying insects, etc.) from getting into the trash.

It is impossible to totally exclude water from the landfill. The water percolates through the cells and soil in the landfill. As the water percolates through the trash, it picks up contaminants (organic and inorganic chemicals, metals, biological waste products of decomposition). This water with the dissolved contaminants is called leachate and is typically acidic. It is then collected into a pond by means of perforated pipes.

Bacteria in the landfill break down the trash in the absence of oxygen (anaerobic) because the landfill is airtight. A byproduct of this anaerobic breakdown is landfill gas, which contains approximately 60% methane and 40% carbon dioxide with small amounts of nitrogen and oxygen. This presents a hazard because the methane can explode and/or burn. So, the landfill gas must be removed or collected for utilization. To do this, a series of pipes are embedded within the landfill to collect the gas. After the landfill has been closed a layer of soil is put above it to prepare it for landscaping.

Disadvantages of landfill

- Landfill construction requires high capital cost.
- Mismanagement of landfill may cause soil contamination, water contamination as well as air pollution.
- Landfill operation require high capital cost.
- The leatchate collection and treatment systems and gas collection system might require huge capital.

From the above analysis, it is clear that incineration and landfilling processes for solid waste rejects require a huge capital and might cause environmental problems if it is not managed and operated properly.

In conclusion, recycling has proved to be the most suitable solution to the problem of rejects in developing countries compared with incineration and/or landfill according to the last 2-Rs of the 7-Rs golden rule discussed before.

Rejects Recycling

Recycling of solid waste rejects was developed using three steps approach. First, trying to innovate new techniques for un-recyclable raw materials like rejects. Second, to develop different valuable and economical products from recycled rejects. Market study is very important step in this approach to guarantee the sustainability of the project.

Advantages of recycling

- Using a resource that would otherwise be wasted
- Reducing or prevent the amount of waste going to landfill
- Reducing the costs involved in the disposal of waste, which ultimately leads to savings for the community
- Providing employment
- Protect natural resources
- Reduce pollution

Reject's Technologies

The rejects recycling system (El Haggar 2004c) consists of an indirect flame furnace with a mixer and a hydraulic press that is based on molding technology. After the rejects have been agglomeratedd, they are mixed with sand and plastic additives to adjust the properties and appearance. The mixture is then entered into a mixer to be heated to 180°C. The hot paste is then transferred to the hydraulic press to be pressed into bricks, interlocks, table toppings, wheels, manholes, road ramps and other products. The manufacturing process of rejects (silica-plast) products consist of the following steps.

1. Sieving the rejects to remove organic wastes and return the organic waste back to the composting process

2. Agglomerating the rejects (contaminated plastic waste)
3. Sieving of sand to remove oversize grain
4. Mixing the agglomerated rejects with sieved sand and heating the mix indirectly at 1809 C
5. Pouring the hot mix into moulds and press it in a hydraulic press to reach required density and shape
6. Cooling the product

The hot mix can be easily molded into any forms and any decorative shape according to the shape of the mold. The key issue behind this technology is the continuous mixing of sand and plastics to guarantee homogeneous distribution of materials and good quality products. The schematic diagram of this process is shown in Figure 13.19.

Product Development

Product development from recycled materials is a must to maximize the benefits. This leads to a number of economic opportunities to remanufacture products with the recovered material. Just as market forces cannot be ignored when introducing a new product, they must also be taken into account when introducing remanufactured products. The market analysis for these products may be more difficult than new ones because customers are more concerned with product durability and cost of maintenance.

BRICKS

The idea of bricks was the first application of the rejects because it was the easiest to be produced from rejects. The major problem with bricks is the adhesion problem since it is made out of plastics and sand. Adhesive materials are very expensive which add to the cost of the brick. There was a cheap way to bind them together using pins as shown in Figure 13.20 for

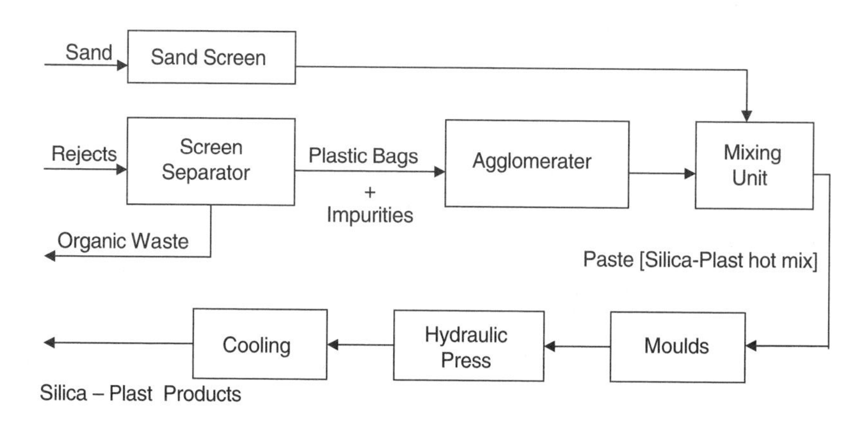

FIGURE 13.19. Schematic diagram of silica-plast products.

FIGURE 13.20. Development of bricks.

easy assembly and disassembly, but the density of the brick is very high so the weight of the brick from rejects was heavy compared with ordinary bricks. After producing the bricks it was not economic, as its price was not profitable when compared to the ordinary bricks including adhesives. There was only one useful application for these bricks in military purposes because it resists the effect of bullets.

Interlocks

The idea of bricks developed to produce interlocks from rejects by changing the shape of the mold. There were several applications for interlocks. It is used in pavements, gardens, factory floors, backyards, etc. The production of interlocks was much more profitable than brick, it does not need adhesives. The cost of silica-plast interlocks is 30%–50% of the traditional types of interlock depend on top layer coating. The appearance of the interlocks can be improved by coating the top layer of the interlock. The top layer of plastic waste shown in Figure 13.21 is a cheap coating layer since it is made out of plastic waste. The plastic waste that was used is of clothes hanger type that was crushed and melted in the heater and put in the press with the reject paste to produce this top layer.

Table Toppings

After the improvement of interlocks the idea of table topping emerged. A mould was manufactured with the required dimensions to produce the topping of the table as shown in Figure 13.22. Flute is one of the cheap coatings that could be used to improve the appearance and protect the environment from any emissions as shown in Figure 13.23.

Grani top layer by a mixture of rocks and granite granules from industrial wastes of rocks can be used to improve the quality and the appearance

FIGURE 13.21. Development of interlocks.

FIGURE 13.22. The table topping from rejects.

of the top layer as shown in Figure 13.23. Crushing of rocks is done by ball crushers followed by a process of sieving mixed with organic polymer and poured in another mold with 1 cm space all around (Figure 13.24).

Bricks proved to be incompetent due to the difficulty in adhesion. Interlocks and table toppings proved to be very competitive. For the interlock the plastic coat was the optimum coating because it is made out of clean waste plastic. This plastic coat covered completely the black color of the rejects and it was proved to be durable. The table topping had many

FIGURE 13.23. Different top layers using Flott or Grani industrial waste. **A**: Flott top. **B**: Grani industrial waste.

FIGURE 13.24. Final shape of the table.

coatings, each coating was used in a different application and the granis layer was proved to be the optimum due to its durability and appearance.

WHEELS

Wheels are another application of the rejects as shown in Figure 13.25. It is used instead of ordinary wheels used in slow moving cars. The cost of these

FIGURE 13.25. Wheels with different sizes.

wheels is 10% of ordinary wheels. It has the disadvantage of dipping in the sandy roads or farms when carrying high loads in the unpaved roads. So, practically they can be used in paved roads.

Manholes

Manholes are another application of the rejects. It can be used instead of ordinary cast iron manholes with all different geometry and dimensions as shown in Figure 13.26. Manholes proved to be an excellent product out of rejects from social, economical, and technical point of views.

Road Ramps

Road ramps are used to control car speed in residential areas, schools and hospital. Traditionally, road ramps are made from cast iron, rubber or asphalt mix. The new material made out of sand-plast mix can be used to replace the cast iron or the rubber papers as shown in Figure 13.27. It provides an excellent alternative from the technical, economical, and environmental point of views.

Water Wave Breakers

Water waves are one of the main problems facing coastal zone and shore protection. Reinforced concrete blocks are always used as a water wave breaker to protect the shores. It costs, a lot of money and should be replaced

A

B

FIGURE 13.26. Sand-plast manholes.

FIGURE 13.27. Road ramp.

frequently depends on wave velocity and wave height. The alternative material is sand-plast with high percentage of plastics to increase the life time of the sand-plast blocks. This might require further analysis for leachate to make sure there is no water contamination.

Conclusion

Getting rid of MSW constituted a problem worldwide. All the components of municipal solid waste could be recycled. Several ways of getting rid of rejects, like incineration and landfill, were discussed and proven unfeasible for usage in developing countries and also worldwide. To reach 100% recycling, rejects of MSW which constitute 20%–25% of total MSW should be recycled.

Rejects recycling proved to be the most feasible for the developing countries since it is the most economic choice. The products that were developed were: bricks, interlocks, table toppings, wheels, manholes, road ramps, etc. The bricks proved to be uncompetitive because it is expensive to be bound. Also the bricks had high density. The interlock is a more competent application compared with bricks. However, the main problem was the bad appearance of the interlock, so a protective and good-looking coating were developed to improve the appearance. The plastic waste layer was the best alternative. The plastic waste layer is made of clean plastic waste so the cost of the material is still low. Also it is durable and good-looking. The table toppings are another competitive product. Many coatings were developed, the grani layer proved to be the best coating because it is more durable and good looking. The most competitive products from rejects are manholes and road ramps since they are used a lot in different applications and can replace another product with much less cost and same technical specification.

Agricultural Waste

Rural communities have a historical agricultural tradition for thousands of years and future plans for expansions. In order to combine the old traditions with modern technologies to achieve sustainable development, wastes should be treated as a by-product. The main problems facing rural areas nowadays are agricultural wastes, sewage and municipal solid waste as explained before. They represent crisis for sustainable development in rural village and to the national economy. However, few studies have been conducted on the utilization of agricultural waste for composting and/or animal fodder but none of them has been implemented in a sustainable form. This section combines all major sources of pollution/wastes generated in a rural area in one complex called environmentally balanced rural waste complex (EBRWC) to produce fertilizer, energy, and animal fodder according to market and need.

The amount of agricultural waste in developing countries is huge and causes a lot of environmental pollution if it is not properly utilized. It is also represent a very important natural resource which might provide job opportunities and valuable products. Some of the agricultural waste is used as animal fodder, others used as a fuel in very primitive ovens causing a lot of health problems and damaging the environment. The rest are burning in the field causing air pollution problems. The type and quantity of agricultural waste in developing countries changes from one country to another, from one village to another and from one year to another because farmers always look for the most profitable crops suitable for the land and the market. The main crops with the highest amount of waste are the rice straws, corn stalks, wheat/barley straws and cotton stalks, baggase, etc.

Rural villages in developing countries were connected with drinking water supply without a sewer system. Other places in urban and semi urban communities have no sewage treatment networks. Instead under each house there is a constructed septic tank where sewage is collected or connected directly to the nearest canal through a PVC pipe. Some houses pump their sewage from the septic tank to sewer car once or twice a week and dump it some where else far from the house.

In general, a huge amount of sewage and solid waste are generated in the village. Because of the lack of sewer system and garbage collection system in the village, sewage is drained in the water canals as well as garbage is thrown in the nearest canal causing soil, water and visual pollution as well as health problems. Some canals are used for irrigation, other canals used as a source of water for drinking.

Agriculture biomass resources in developing countries are huge. Fifty percent of the biomass is used as a fuel in rural areas by direct combustion in low efficiency traditional furnaces. The traditional furnaces are primitive mud stoves and ovens that are extremely air polluting and high energy inefficient. The agriculture biomass waste (resources) are primary consist of cotton stalks, rice straw, etc.

One of the main agricultural wastes is a cotton stalk. The available amount of cotton stalks in Egypt is estimated as 1.6 million tons/year (corresponding to 740,000 TOE/year). According to the ministry of agriculture regulations and resolutions that commit the farmers to dispose the cotton plant residues through environmentally safe disposal method immediately after harvesting (within 15 days). The easiest and cheapest method is to burn the cotton stalks as soon as possible in the field. The reason behind this regulation is to kill insects and organs carrying plant diseases. The cotton stalks will be stored for a long time giving chance for the cotton warms to complete the worm live cycle and attach the cotton crop in the next season. Such process leads to a total energy loss estimated as 0.74 MTOE/year that count for a high money value ($35 million/year) in addition to the negative environmental impacts due to releasing vast amount of green house gases.

Moreover, the traditional storage systems for plant residues in the farms and roofs of the buildings gives a chance for insects and diseases to grow and reproduce as well as fire hazardous.

ABBC Technologies

The four corner stone technologies for agricultural waste are animal fodder, briquetting, biogas, and composting (ABBC technologies). These technologies can be developed based on demand and need. In principal three agricultural waste recycling techniques can be selected to be the most suitable for the developing communities. These are animal fodder and energy in a solid form (briquetting) or gaseous form (biogas) and composting for land reclamation. There are some other techniques, which might be suitable for different countries according to the needs such as gasification, fiber boards, pyrolysis, etc. These techniques might be integrated into a complex that combine them altogether to allow 100% recycling for the agricultural waste. Such a complex can be part of the infrastructure of every village or community. Not only does it allow to get rid of the harms of the current practice of agricultural waste, but also of great economical benefit.

The amount of agricultural waste varies from one country to another according to type of crops and farming land. These waste occupies the agricultural lands for days and weeks until the simple farmers get rid of these waste by either burning it in the fields or storing it in the roofs of their houses; the thing that affects the environment and allows fire villages and spread of diseases. The main crops responsible for most of these agricultural wastes are the rice, wheat, cotton, corn, etc. These crops were studied and three agricultural waste recycling techniques were set to be the most suitable for these crops. The first technology is animal fodder that allows the transformation of agricultural waste into animal food by increasing the digestibility and the nutritional value. The second technology is energy, which converts agricultural wastes into energy in a solid form (briquetting) or gaseous form (biogas). The briquetting technology that allows the transformation of agricultural waste into briquettes that can be used as useful fuel for local or industrial stoves. The biogas technology can combine both agricultural waste and municipal waste water (sewage) in producing biogas that can be used in generating electricity, as well as organic fertilizer. The last technology is composting, that uses aerobic fermentation methods to change agricultural waste or any organic waste into soil conditioner. The soil conditioner can be converted into organic fertilizer by adding natural rocks to control N : P : K ratio, as explained before. Agricultural waste varies in type, characteristics and shape, thus for each type of agricultural waste there is the most suitable technique as shown in Figure 13.28.

A complex combining these four techniques is very important to guarantee each waste has been most efficiently utilized in producing beneficial outputs like compost, animal food, briquettes and electricity. Having this complex will not only help the utilization of agricultural waste, it will help solving the sewage problem as well that face most of the developing countries, as a certain percentage of the sewage will be used in the biogas production and composting techniques to adjust carbon to nitrogen ratio. An efficient collection system should be well designed to collect the agricultural waste from the lands to the complex in the least time possible to avoid having these wastes occupying agricultural land. These wastes are to be shredded and stored in the complex to maintain continuous supply of agricultural waste to the system and in turns continuous outputs.

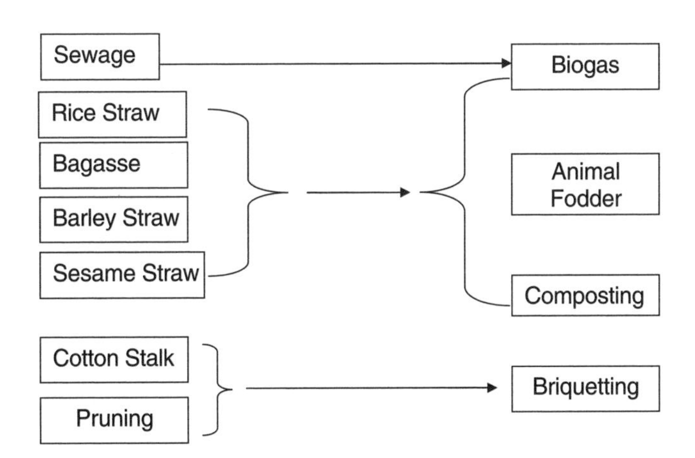

FIGURE 13.28. Matching diagram between output technology and agricultural waste.

Animal Fodder

The deficiency of animal foodstuff in developing countries caused importing raw material with high cost or reducing the animal production. Transforming some of these wastes into animal foodstuffs will help a great deal in overcoming this deficiency. These wastes have a high content of fiber that makes them uneasily digestible. The size of the waste in its natural form might be too big or tough for the animals to eat. To overcome these two problems several methods were used to transform the agricultural waste to a more edible form with a higher nutritional value and more digestibility.

Mechanical and chemical treatment methods were used to transform the shape of the roughage (waste) in an edible form. The further addition of supplements can enrich the foodstuffs with the missing nutritional contents. The mechanical treatment method consists of chopping, shredding, grinding, moistening, soaking in water and steaming under pressure. The mechanical method has been proved to give good results with high digestion by animals but they were never widespread on the market level because of high cost and unfeasible for small farms.

The chemical treatment method with urea or ammonia is more feasible than the mechanical treatment method. The best results were obtained by adding 3% of ammonia (or urea) to the total mass of the waste. It is recommended to cover the treated waste with a wrapping material made usually out of polyethylene 2mm thick. After 2 weeks (summer), 3 weeks (winter), the treated waste is uncovered and left for 2–3 days to release all the remains ammonia before use as an animal food.

Briquetting

Briquetting process is the conversion of agricultural waste into uniformly shaped briquettes that is easy to use, transport and store. The idea of briquetting is using materials that are not usable due to a lack of density, compressing them into a solid fuel of a convenient shape that can be burned like wood or charcoal. The briquettes have better physical and combustion characteristics than the initial waste. Briquettes will improve the combustion efficiency using the existing traditional furnaces. In addition to killing all insects and diseases as well as reducing the destructive fires risk in countryside.

The advantages of briquetting

- Get rid of insects
- Decrease the volume of waste
- Efficient solid fuel of high thermal value
- Low energy consumption for production
- Protect the environment

- Provide job opportunities
- Less risk hazardous

The raw materials suitable for briquetting are rice straws, wheat straws, cotton stalks, corn stalks, sugar cane waste (baggase), fruit branches, etc. However, in the suggested complex cotton stalks and fruit branches are best utilized by briquetting. The briquetting process starts with collection of wastes followed by size reduction, drying, and compaction by extruder or press.

Biogas

Biogas is the anaerobic fermentation of organic materials by microorganisms under controlled conditions. Biogas is a mixture of gasses mainly methane and carbon dioxide gases that results from anaerobic fermentation of organic matter by the action of bacteria. Biogas is ranked low in priority in some developing countries because of lack of energy policy and there is no plan of the share of biogas of the total biomass potential.

Huge amounts of organic wastes are generated in rural communities such as agriculture waste, sludge from municipal treatment plants, and organic waste from garbage, food processing plants as well as animal manure and dead animals. Table 13.6 shows a sample of types and quantities of organic wastes generated in Egypt. All these can be considered as a biomass that is organic carbon-based material, which could be an excellent source for biogas and fertilizer.

Biogas activities in Egypt until now have focused mainly on small scale plants with a digester volume of 5–50 m^3 with a few exceptions such as the Gabel Al-Asfar Plant located near Cairo, with a total capacity of 1 million cubic meter. The total energy potential in Egypt of centralized biogas plants with 50 to 500 tons/day input was estimated to be about 1 million TOE.

Composting

Composting is the aerobic decomposition of organic materials by microorganisms under controlled conditions as discussed above. In 1876 Justus von

TABLE 13.6
Sources and Quantities of Organic Wastes in Egypt

Organic waste Source	*Quantity*
Agricultural waste	25 million tons of dry material per year
Municipal solid waste	6.6 million tons of dry organic waste/year
Sewage treatment plants	4.3 million tons of dry sludge/year

Liebig (Epstein, 1997), a German chemist, had figured out that northern African lands that were supplying two-thirds of the grains consumed in Rome were becoming less fertile, losing their quality and productivity. After doing some brief research, he knew the reason behind such a phenomenon. It was due to the fact that when crops were exported from North Africa, their waste never came back—on the contrary they were flushed into the Mediterranean. The agricultural waste that comes out of rice, cotton, corns, etc. are rich in organic matter. This matter was given by the soil and now the soil wanted it back to continue producing healthy crops. However, that was never the case. This, in his opinion, was a breaking of the natural loop that should give the land back its nutrients. He named such a phenomena the "direct flow." Accordingly, this German scientist came out with artificial fertilizers. Although the artificial fertilizer was meant to compensate the soil for its loss of organic matter, it was never the same as natural fertilizers.

Composting is one of the famous recycling processes for organic waste to close the natural loop. The major factors affecting the decomposition of organic matter by microorganisms are oxygen and moisture. Temperature, which is a result of microbial activity, is an important factor too. The other variables affecting the process of composting are nutrients (carbon and nitrogen), pH, time and physical characteristics of raw material (porosity, structure, texture, and particle size). The quality and decomposition rate depends on the selection and mixing of raw material.

Aeration is required to recharge the oxygen supply for the microorganisms. Passive composting method (El Haggar 2003a) is the recommended technique for developing community from the technical and economical factors as explained before. The main advantages of composting is the improvement of soil structure by adding the organic matter and pathogens structure as well as utilizing the agricultural wastes that can cause high pollution if burned.

Because compost materials usually contain some biological resistant compounds, a complete stabilization (maturation) during composting may not be achieved. The time required for maturation depends on the environmental factors within and around the composting pile. Some traditional indicators can be used to measure the degree of stabilization such as decline in temperature, absence of odor, and lake of attraction of insects in the final products.

Compost can be adjusted by adding natural rocks, such as phosphate rocks (source of phosphorous), feldspar (source of potassium), dolomite (source of magnesium), to produce organic fertilizer for organic farming (El Haggar et al. 2004). Organic farming means better taste, no effect on people's health, and it is less harmful to environment. Organic farming seeks to reduce external cost, produce good yields, save energy, maintain biodiversity, and keep soil healthy.

Integrated Complex

The idea of the integrated complex is to combine the above-mentioned technologies under one roof, the thing that will help utilization of each agricultural waste with the most suitable technique that suits the characteristics and shape of the waste. The main point in this complex is the distribution of the wastes among the four techniques, as this can vary from one village to the other according to the need and market for the outputs. The complex is flexible and the amount of the outputs from soil conditioner, briquettes, and animal food can be controlled every year according to resources and need.

The distribution of these wastes on the four techniques, animal fodder, briquetting, biogas, composting, should be based on:

- The need to utilize all the sewage (0.5%–1.0% solid content) using the biogas technique.
- Some agricultural waste will be added to the sewage to adjust the solid contents to 10% in the biogas system.
- Biogas generated will be used to operate the briquetting machine and other electrical equipment.
- Fertilizer from biogas unit (degraded organic content) will be mixed with the compost to enrich the nutritional value.
- Cotton stalks will be utilized using briquetting technique because cotton stakes are hard and bulky for all other techniques and has high heating value.

Based on the above criteria, EBRWC will combine all wastes generated in rural areas in one complex to produce valuable products such as briquettes, biogas, composting, animal fodder and other recycling techniques for solid wastes. The main outputs of EBRWC are fertilizer, energy, animal fodder, and other recycling materials depending on the availability of wastes and according to demand and need.

The flow diagram describing the flow of materials from waste to product is shown in Figure 13.29. First the agricultural waste are collected, shredded and stored to guarantee continuous supply of waste into the complex. Then according to the desired outputs the agricultural wastes are distributed among the four techniques. The biogas should be designed to produce enough electrical energy for the complex, then the secondary output of biogas that is slurry is mixed with the composting pile to add some humidity and improve the quality of the compost. And finally briquettes, animal food, and compost are main outputs of the complex.

Environmentally Balanced Rural Waste Complex

The environmentally balanced rural waste complex (EBRWC) shown in Figure 13.30 can be defined as a selective collection of a compatible activ-

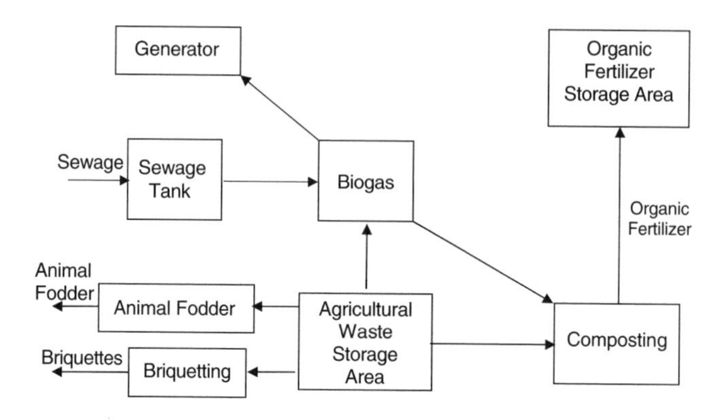

FIGURE 13.29. Material flow diagram.

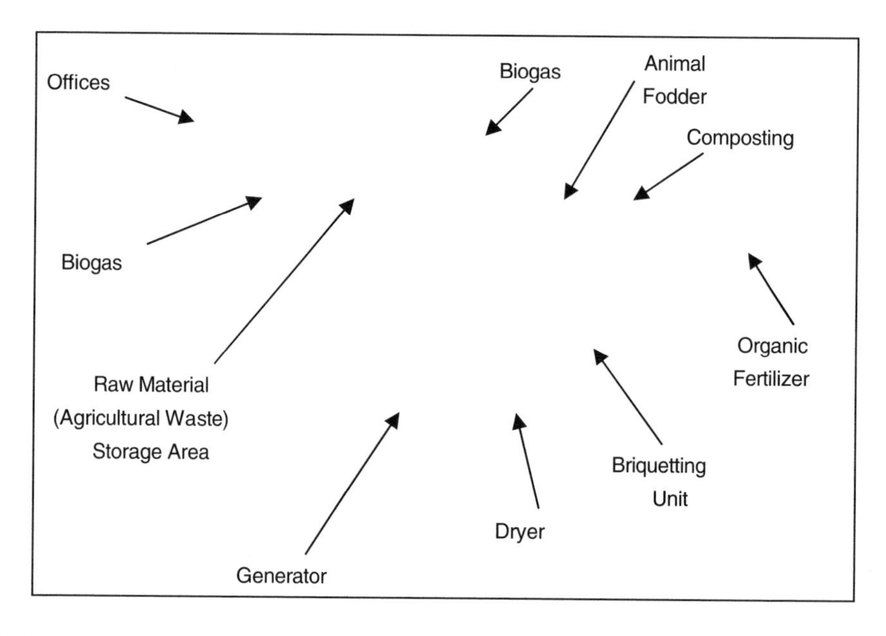

FIGURE 13.30. Environmentally balanced rural waste complex flow diagram.

ities located together in one area (complex) to minimize (or prevent) environmental impacts and treatment cost for sewage, municipal solid waste and agricultural waste. A typical example of such a rural waste complex consists of several compatible techniques such as animal fodder, briquetting, anaerobic digestion (biogas), composting, and other recycling techniques for solid wastes that will be located together within the rural waste complex. Thus, EBRWC is a self-sustained unit that draws all its inputs from within the rural wastes achieving zero waste and pollution. However, some emission might be released to the atmosphere, but this emission level would be significantly much less than the emission from the raw waste coming to the rural waste complex.

The core of EBRWC is material recovery through recycling. A typical rural waste complex would operate to utilize all agricultural waste, sewage and municipal solid waste as sources of energy, fertilizer, animal fodder and other products depend on the constituent of municipal solid waste. In other words, all the unusable wastes will be used as a raw material for a valuable product according to demand and need within the rural waste complex. Thus, a rural waste complex will consist of a number of such compatible activities, the waste of one being used as raw materials for the others generating no external waste from the complex. This technique will produce different products as well as keep the rural environment free of pollution from the agricultural waste, sewage, and solid waste. The main advantage of the complex is to help the national economy for sustainable development in rural areas. The reader is referred to Chapter 11 for many examples of industrial complexes.

Collection and transportation system is the most important component in the integrated complex of agricultural waste and sewage utilization. This is due to the uneven distribution of agricultural waste that depends on the harvesting season. This waste need to be collected, shredded and stored in the shortest period of time to avoid occupying agricultural lands, spread of disease and fire villages.

Sewage does not cause problems in transportation as it is transported through underground pipes from the main sewage pipe of the village to the system. Sewage can also be transported by sewage car which is most common in rural areas since pipelines might cost a lot of money.

Household Municipal Solid Waste

Household municipal solid waste represents a crisis for rural areas where people dump their waste in the water canals causing water pollution as well as visual pollution. The household municipal solid waste consists of organic materials, paper and cardboard, plastic waste, tin cans, aluminum cans, textile, glass, and dust. The quantity changes from one rural community to another. It is very difficult to establish recycling facilities in rural areas where the quantities are small and changes from one place to another. It is recommended to have a transfer station(s) located in each community to separate the wastes, compacted and transferred to the nearest recycling centers as explained before. The transfer station consists of a sorting conveyer belt to sort all valuable wastes from the organic waste; compacted through a hydraulic press. The collected organic waste can be mixed with other rural waste to composting or biogas as explained before.

Conclusion

Several compatible techniques will be located together within the agricultural complex such as animal fodder, briquetting, biogas, composting, and other recycling techniques for municipal solid wastes (transfer station).

Converting rural waste into energy sources such as briquettes (solid fuel) and/or biogas (gaseous fuel) that is equivalent to millions of TOE/year will save a lot of money and avoid environmental negative impacts and health hazardous as a result of field burning process.

Soil conditioner or organic fertilizer which will be produced from composting and biogas units will enrich the soil, increase the crop production, decrease the chemical fertilizer and close the natural loop.

Converting the agricultural waste into animal foodstuff will help a great deal in overcoming the deficiency of animal foodstuff in developing countries.

The EBRWC approach will combine all wastes generated in rural areas in one complex to produce valuable products. An EBRWC complex will consist of a number of compatible activities to produce different products as well as keep the rural environment free of pollution generated from burning agricultural wastes, disposal of sewage in water canals and disposal of municipal solid wastes. EBRWC will provide job opportunities for young people in developing countries

The outputs of the EBRWC are valuable and needed goods to be sold. EBRWC is flexible and can be adjusted with proper calculations to suit every village; moreover, the complex can be adjusted every year according to the main crops cultivated in the village that usually varies from year to year. The key element to the success of this solution lies in the integration of these ABBC technologies to guarantee that each type of waste is most efficiently utilized.

Industrial Waste

Since the beginning of human history, industry has been an open system of materials flow. People transformed natural materials: plants, animals, and minerals into tools, clothing, and other products. When these materials were worn out they were discarded or dumped in the same condition as they are, and when the refuse buildup bothered them, the habitants changed their location, which was easy to do at that time due to the small number of habitants and the vast areas of land.

Among the goals of industry must be the preservation and improvement of the environment. With increasing industrial activity all over the world new ways have to be developed to make large improvements in the whole industrial interactions with the environment.

An open industrial system—one that takes in materials and energy, creates products and waste materials and then throws most of these—will probably not continue indefinitely and will have to be replaced by a different system. This system would involve, among other things paying more attention to where materials end up, choosing materials and manufacturing processes to generate a more circular flow. Until quite recently, industrial societies have attempted to deal with pollution and other forms of waste largely through regulation. Although this strategy has been partially successful, it has not really gotten to the root of the problem. To do so will require a new paradigm for our industrial system—an industrial ecology whose processes resemble those of a natural ecosystem (Frosch 1994).

Industrial Ecology

Industrial ecology (IE) is the study of industrial systems that operate more like natural ecosystems. A natural ecosystem tends to evolve in such a way that any available source of useful material or energy will be used by some organism in the system. Animals and plants live on each other's waste matter. Materials and energy tend to circulate in a complex web of interactions: Animal wastes and dead plant material are metabolized by microorganisms and turned into forms that are useful nutrients for plants. The plants in turn may be eaten by animals or die, decay and go around the cycle again. These systems do, of course, leave some waste materials; otherwise we would have no fossil fuels. But on the whole the system regulates itself and consumes what it produces (Frosch 1994).

Emerging during about the past 10 years, industrial ecology is a new approach to the analysis and design of sustainable political economies (Frosch 1995). Allenby (1999) calls industrial ecology the science of sustainability. Several other characteristics of stable ecosystems also suggest new norms to pursue in thinking about sustainability. Prigogine (1955) observed several very interesting features about steady state biological systems. One is that they are in a state of minimum entropy production, that is, the system is functioning with the least degree of dissipation of energy (and materials) thermodynamically possible in a real situation. These systems also exhibit a high degree of material loop closing. Materials are circulated through a web of interconnection with scavengers located at the bottom of the food web turning wastes into food. Even long-lived biological systems eventually succumb to environmental and internal stresses. They are not ideal models for a concept that implies flourishing forever. Ayres (1989) coined the term "industrial metabolism" as the web of flows of energy and material. When modeling an industrial economy consisting of an interconnected system of energy, material, and money flows such a system will supply an analytic means to repair the break in both the economic and environmental sciences. Daly (1977) and others have stressed the importance of including material flows in economic flows analysis, noting the fundamental connections of economics to natural resources. Daly (1977) and earlier Georgescu-Roegen (1971) developed a steady state framework for describing modern economic systems and for designing policy, invoking basic laws of thermodynamics and ecological systems behavior as part of the grounding. Expanding the typically sectoral or firm-level models used by policy analysts and corporate planners to material and energy flows during the entire life cycles of economic goods should, in theory, reduce the probability of suboptimal solutions and of the appearance of unintended consequences. To convert part of these ideas into an industrial design context, a set of design rules have to be established for the innovation of more environment friendly and sustainable products and services. A few of these rules were developed by Ehrenfeld (1997).

- Close material loops
- Use energy in a thermodynamically efficient manner; employ energy cascades
- Avoid upsetting the systems metabolism; eliminate materials or wastes that upset living or inanimate components of the system
- Dematerialize; deliver the function with fewer materials

Industrial ecology as the "normal" science of sustainability (modifying slightly the phrase) as used by Allenby (1999) promises much in improving the efficiency of humans' use of the ecosystem. Technological improvements are not always better in the full sense of sustainability without taking environment into consideration, where zero pollution is a must for industrial ecology. Cooperation and community are also important parts of the ecological metaphor of sustainability. Industrial ecology is the net resultant of interactions among zero pollution, eco-industrial parks, and life cycle analysis.

Eco-Industrial Parks

One of the most important goals of industrial ecology (Frosch 1994)—making one industry's waste another's raw materials—can be accomplished in different ways. The most ideal way for IE is the eco-industrial park (EIP). They are industrial facilities clustered to minimize both energy and material wastes through the internal bartering and external sales of wastes. One industrial park located in Kalundborg, Denmark has established a prototype for efficient reuse of bulk materials and energy wastes among industrial facilities. The park houses a petroleum refinery, power plant, pharmaceutical plant, wallboard manufacturer, and fish farm that have established dedicated streams of processing wastes (including heat) between facilities in the park. The gypsum from neutralization ("scrubbing") of the sulfuric acid produced by a power plant is used by a wallboard manufacturer, spent fermentation mash from a biological plant is being used as a fertilizer, and so on. The success of the EIP depends on the ability to innovate, access to talent, markets, and the ability to meet profit conditions or cost constraints and on achieving close cooperation between different companies and industrial facilities.

Industrial Ecology Barriers

Even though the industrial ecology concept has a lot of advantages from economical, environmental, and social point of views, there are still some barriers for implementations.

The barriers to industrial ecology fall into five categories (Wernick and Ausubel 1997), namely technical, market and information, business and financial, regulatory and regional strategies.

Technical Barriers

Technical issues are one of the main challenges for IE to approach zero pollution. It requires a lot of innovation to convert waste into money or prevent it at the source. Overcoming the technical barriers associated with recovering materials from waste streams is necessary but an insufficient step for stimulating the greater use of wastes in the economy. Technology making recovery cheap and assuring high quality input streams must be followed by encouraging regulations and easy informational access. Finally a ready market must be present.

Market and Informational Barriers

These are inseparable from institutional and social strategies. Due to absent direct governmental interference, the markets for waste materials will ultimately rise or fall based on their economic vitality. Markets are sophisticated information processing machines whose strength resides in a large part on the richness of the informational feedback available. One option for waste markets are dedicated "Waste Exchanges" where brokers trade industrial wastes like other commodities. By using "Internet Technology" to facilitate the flow of information, the need for centralized physical locations for either the stuff or for the traders in the stuff may be minimal. Research is needed on waste information systems that would form the basis for waste exchanges. A stock market can be developed based on waste material. Systems would need to list available industrial wastes as well as the means for buyers and sellers to access the information and conduct transactions. The degree to which such arrangements would allow direct trading or rely on the brokers to mediate transactions presents a further question. As part of the market analysis for waste materials, research is needed to understand past trends regarding the effect of price disparities between virgin and recovered materials, and to assess the effect of other economic factors associated with waste markets, such as additional processing and transportation costs. A further matter for investigation concerns whether some threshold level of industrial agglomeration is necessary to make such markets economically viable. Progress is already being made on this front.

The Chicago Board of Trade (CBOT), working with several government agencies and trade associations, has begun a financial exchange for trading scrap materials. Other exchanges such as the National Materials Exchange Network (NMEN) and the Global Recycling Network (GRN) facilitate the exchange of both materials recovered from municipal waste streams and of industrial wastes. Analysts might propose ideas for improving or facilitating the development of these exchanges. The value of such exchanges as a means of improving the flow of information depends on the deficiency of the current information flow, and how much this particular aspect of recycling plays in recycling's success or failure. The CBOT is different from the other exchanges in that it is a financial market—starting now as a cash exchange with hopes that it will evolve into a forward and/or futures market.

A simple waste exchange is premised on the notion that opportunities for exchange are going unrealized. A cash exchange has a related premise that there is a need for what economists call price discovery. Finally, a futures or forward market exists to allow the risk associated with price volatility to be traded independent of the commodity.

Business and Financial Barriers

The private firm is the basic economic unit and collectively constitutes the mechanism for reducing inventions and innovations to practice, in service of environmental quality or other goals. Corporations employ a spectrum of organizational approaches to handle environmental matters. In some cases the environment division of a corporation concerns itself exclusively with regulatory compliance and the avoidance of civil liability for environmental matters. For other firms the environment plays a more strategic role in corporate decision-making. Decisions made at the executive level strongly determine whether or not companies adopt new technologies and practices that will affect their environmental performance. Also, the manner in which corporations integrate environmental costs into their accounting systems, for instance how to assign disposal costs, bears heavily on its ability to make both short and long term environmentally responsible decisions.

Research is needed to better understand the role of corporate organization and accounting practices in improving environmental performance and the incentives to which corporations respond for adopting new practices and technologies. Such studies would examine the learning process in corporate environments as well as investigate how corporate culture influences the ultimate adoption or rejection of environmentally innovative practices.

Regulatory

Environmental regulation strongly induces companies to appreciate the environmental dimensions of their operations. Businesses must respond to local, national, and international regulatory structures established to protect environmental quality.

Although few question that regulations have helped to improve environmental quality, many argue that wiser, less commanding regulations would improve quality further at less cost. Agreements on hazardous waste tightly regulate the transport of these wastes across state and national boundaries, perhaps reducing opportunities for re-use and encouraging greater extraction of virgin stocks. Elements of the national regulatory apparatus for wastes, heavily regulate the storage and transport of wastes and dictate waste treatment methods that also serve to dissuade later efforts at materials recovery.

Regional Strategies

Often geographic regions may provide a sensible basis for implementing IE. Industries tend to form spatial clusters in specific geographic regions based on factors such as access to raw materials, convenient transportation, technical expertise, and markets. This is particularly true for "heavy" industries requiring large resource inputs and generating extensive waste quantities. Furthermore, the industries supporting large industrial complexes tend to be located within reasonable proximity to their principal customers. Due to the unique character of different regions this work could proceed in the form of case studies of regions containing a concentration of industries in a particular sector.

Case Studies

Two case studies will be discussed in this section related to cleaner production to approach zero pollution for sustainable development. These case studies are related to the cement industry (El Haggar 2000) and iron and steel industries (Korany and El Haggar 2000).

Cement Industry

The cement industry is one of the main industries necessary for sustainable development. It can be considered the backbone for development. The main pollution source generated from cement industry is the solid waste called cement by-pass dust, which is collected from the bottom of the dust filter. It represents a major pollution problem in Egypt where around 2.4 million tons per year of cement dust is diffused into the atmosphere causing air pollution problems because of its size (1–10 microns) and alkalinity (pH 11.5).

Cement by-pass dust is naturally alkaline with a high pH value and represents a major pollution problem. The safe disposal of cement dust costs a lot of money and still pollutes the environment. The chemical analysis for the by-pass dust is shown in Table 13.7.

TABLE 13.7
Chemical Analysis of Cement By-Pass Dust

Chemical formula	*Percentage*
SIO_2	9.0–13.0
Al_2O_3	3.0–4.0
Fe_2O_3	2.0–2.5
CaO	45.0–48.0
MgO	1.7–1.9
SO_3	4.0–11.0
Na_2O	3.0–8.0
K_2O	2.0–6.0
Cl	4.0–13.0

Because of the high alkalinity of the cement by-pass dust, it can be used in the treatment of the municipal sewage sludge, which is considered another environmental problem in developing countries since it contains parasites such as *Ascaris* and heavy metals from industrial waste in the city. Although sludge has a very high nutritional value for land reclamation, it might contaminate the land. The safe disposal of sludge costs a lot of money and direct application of sludge for land reclamation has a lot of negative environmental impacts and is very hazardous to health.

Mixing the hazardous waste of cement by-pass dust with the environmentally unsafe sewage sludge will produce a good quality fertilizer. Cement by-pass dust will enhance the fermentation process of the organic waste and kill all microbes and parasites. The high alkalinity cement by-pass dust fixes the heavy metals present in the product and converts them into insoluble metal hydroxide. Hence preventing metal release in the leachate. Agricultural wastes must be added to the mix to adjust the carbon to nitrogen ratio as well as the pH value for better composting (El Haggar 2000). The produced fertilizer from composting is safe for land reclamation and free from any parasites or microbes that might exist in raw sludge.

Utilization of Cement By-pass Dust

Two types of primary sludge from sewage treatment plants were used: the first one from a rural area where no heavy metals were included and the second from an urban area where heavy metals exist.

The uniqueness of this process is related to the treatment of primary sludge which is heavily polluted with *Ascaris* eggs (most persistent species of parasites) using passive composting technique. This technique is very powerful, very efficient with much less cost (capital cost and running cost) than other techniques as explained before. First, primary sludge will be mixed with 5% cement dust for 24 hours. Second, agricultural waste as a bulking agent will be mixed for passive composting treatment. Passive composting piles will be formed from sludge mixed with agricultural waste (bulking agent) and cement dust with continuous monitoring of the temperatures and CO_2 generated within the pile. Both parameters are good indicators of the performance and digestion process undertaken within the pile.

Passive composting technology has shown very promising results, especially by adding cement dust and agricultural wastes. Results show that *Ascaris* has not been detected after 24 hours of composting mainly due to the high temperature elevations reaching 70°C to 75°C for prolonged periods, as well as the high pH from cement dust. Also, the heavy metal contents were way beyond the allowable limits for both urban sludge as well as rural sludge.

Formulation of Objectives and Strategies

The main objective of this case study is to treat the wastes (cement by-pass dust, primary sludge from sewage treatment plants as well as agricultural waste) as a by-product to produce a valuable material instead of dumping them in the desert or burning them in the field. This technique will protect the environment and establish a new business in developing countries where cement by-pass dust exists. If cement by-pass dust does not exist, quick lime can be used to treat the MWW sludge. Sludge has a very high nutritional value but is heavily polluted with *Ascaris* and sometimes with heavy metals (treatment plants located in urban areas). Direct application of sludge for land reclamation has negative environmental impacts and health hazardous.

Cement by-pass dust is alkaline with high pH value and represents a major pollution problem throughout developed and developing countries. The safe disposal of cement dust costs a lot of money and still pollutes the environment because it is very fine dust with high pH (>11) and has no cementing action.

Agricultural waste has no heavy metals and contains some nutrients, which will be used as a bulking agent. The bulking agent can influence the physical and chemical characteristics of the final product. It will also reduce the heavy metal content of the sludge and control C:N ratio for composting.

Environmental and Social Indicators

- Preventing biomass field burning process
- Reducing the greenhouse gas emissions
- Reducing the total suspended particulate (TSP) as well as PM_{10} and $PM_{2.5}$ as a result of cement by-pass dust diffused to the air and burning the organic waste

- Killing of parasites, insects and diseases carried live on the plant residues and sludge during the aerobic fermentation process that will be accelerated by the by-pass cement dust
- Improving the land reclamation and public health not only for rural areas but also for remote areas
- Develop a new business opportunity and increase in investment and demand for biofertilizer
- Providing job opportunities

Sustainability

Our generations have converted more of earth's resources into waste and pollution. The question is what can we do to convert this waste into a valuable product and to be self-sustained. There are millions of tons of dry agriculture waste generated every year and millions of tons of dry sludge from municipal treatment plants as well as millions of tons cement by-pass dust diffused in to the atmosphere in all developing countries. Table 13.8 shows waste sources and quantities in Egypt.

The Agricultural waste can be collected in each community to one or more centralized units mixed with cement by-pass dust as well as sludge from treatment plants to produce a good quality soil conditioner or fertilizer. The produced fertilizer is free from any parasites or microbes, which exist in raw sludge. Collecting such waste and selecting the suitable places for waste mixing utilization to be centralized in different locations might be a challenging.

Iron and Steel Industry

Steel industry is one of the essential industries for the development of any community. In fact, it is really the base for numerous industries that could not have been established without steel industry. The European industrial revolution at the beginning of this century was actually founded on this industry. There are three basic routes to obtain finished steel products: (1) integrated steel production, (2) secondary processing, and (3) direct reduction. Integrated steel production involves transforming coal to coke in coke ovens, while iron ore is sintered or belletized prior to being fed into the blast furnace (BF). The ore is reduced in the blast furnace to obtain hot metal containing some 4% carbon and smaller quantities of other alloying ele-

TABLE 13.8
Wastes Sources and Quantities in Egypt [El Haggar, 2004]

Quantity	*Source*
18.36 million tons of dry material per year	Agricultural waste
2.4 million tons cement by-pass dust	Cement factories
4.3 million tons of dry sludge/year	Municipal treatment plants

ments. Next the hot metal is converted to steel in the basic oxygen furnace (BOF). Then, it is continuously cast to obtain semi-finished products, such as blooms, bars or slabs. These semi-finished products are rolled to the finished shapes of bars, sheet, rail, H or I beams.

The secondary processing, often called minimills, starts with steel scrap which is melted in an electrical arc furnace (EAF). The molten steel produced is possibly treated in a ladle furnace and then continuously cast and finished in a rolling operation. Originally, minimills provided only lower grade products, especially reinforcing bars. But, they recently have been able to capture a growing segment of the steel market.

An alternative mode of steel production is the direct reduction method. In this method, production starts with high grade Iron ore pellets which are reduced with natural gas to sponge pellets. Then, the sponge iron pellets are fed into an electrical arc furnace. The resulting steel is continuously cast and rolled into a final shape.

The problem of the solid waste generated from iron and steel industry is not only hindering the use of millions square meters of land for more useful purposes but also contaminating it. Many of these waste materials contain some heavy metals such as barium, titanium, and lead. Also, it is well known that toxic substances tend to concentrate in slag. Health hazardous of heavy metals and toxic substances are well known. Based on the concentration levels, some slags may be classified as hazardous waste materials. Furthermore, ground water is susceptible to serious pollution problems due to the likely leaching of these waste materials. This section describes the different types of solid waste materials generated from the iron and steel industry and the associated environmental problems. Different techniques of managing these waste materials are presented with a focus on utilizing slag and dust in civil engineering applications. Test results of many research efforts in this area are summarized. In addition, numerous ideas to mitigate the environmental impacts of this problem are suggested.

Environmental Problems

Iron making in the BF produces a slag that amounts to 20%–40% of hot metal production. BF slag is considered environmentally unfriendly when fresh because it gives off sulfur dioxide and, in the presence of water, hydrogen sulfide and sulphoric acid are generated. These are at least a nuisance and at worst potentially dangerous. Fortunately, the material stabilizes rapidly when cooled and the potential for obnoxious leachate diminishes rapidly. However, the generation of sulphoric acid causes considerable corrosion damage in the vicinity of blast furnaces. In Western Europe and in Japan, virtually all slag produced is utilized either in cement production or as road filling. In Egypt, almost two-thirds of the BF slag generated is utilized in cement production. Some 50 to 220 kg of BOF slag is produced for every metric ton of steel made in the basic oxygen furnace, with an average value of 120kg/metric ton (Szekely 1995). At present, about 50% of BOF slag is utilized worldwide, particularly for road construction and as an addition to cement kilns (Szekely 1995).

Recycling of slags has become common only since the early 1900s. The first documented use of BF slag was in England in 1903 (Featherstone and Holliday 1998); slag aggregates were used in making asphalt concrete. Today, almost all BF slag is used either as aggregate or in cement production. Steel-making slag is generally considered unstable for use in concrete but has been commercially used as road aggregates for over 90 years and as asphalt aggregates since 1937. Steel-making slag can contain valuable metal and typical processing plants are designed to recover this metal electromagnetically. These plants often include crushing units that can increase the metal recovery yield and also produce materials suitable as construction aggregates. Although BF slag is known to be widely used in different civil engineering purposes, the use of steel slag has been given much less encouragement.

BF dust and sludge are generated as the offgases from the blast furnace are cleaned, either by wet or dry means. The dust and sludge typically are 1 to 4% of hot metal production (Lankford et al. 1985). These materials are less effectively utilized than BF slag. In some cases, they are recycled through the sinter plant, but, in most cases, they are dumped and landfilled. Finding better solutions for the effective utilization of BF dust and sludge is an important problem that has not yet been fully solved. BOF dust and sludge are generated during the cleaning of gases emitted from the basic oxygen furnace. The actual production rate depends on the operation circumstances. It may range from about 4 to 31kg/metric ton of steel produced, and has a mean value of about 18kg/metric ton (Szekely 1995). The disposal or utilization of BOF dust and sludge is one of the critical environmental problems needing solution.

Electrical arc furnace produces about 116kg of slag for every metric ton of molten steel. Worldwide, about 77% of the slag produced in EAFs is reused or recycled (Szekely 1995). The remainder is landfilled or dumped. Due to the relatively high Iron content in EAF slag, screens and electromagnetic conveyors are used to separate the Iron to be reused as raw material. The EAF slag remained is normally aged for at least 6 months before being reused or recycled in different applications such as road building. All efforts in Egypt have focused on separating the Iron from EAF slag without paying enough attention to the slag itself. However, pilot research conducted at Alexandria University has investigated the possibility of utilizing such slag (El-Raey, 1997). The test results proved that slag asphalt concrete could in general fulfill the requirements of the road-paving design criteria.

EAF dust contains appreciable quantities of zinc, typically 10% to 36%. In addition, EAF dust holds much smaller quantities of lead, cadmium, and chromium. EAF dust has been classified as a hazardous waste

(K061) by the EPA, and therefore its safe disposal represents a major problem. Although there are several technologies available for processing this dust, they are all quite expensive, on the order of $150 to $250 per metric ton of dust.

The Problem in Egypt

Until recently, there was a lack of consciousness in Egypt either to the environmental impacts or to the economic importance of waste materials generated from iron and steel industry. Apart from the granulated BF slag used in producing slag cement, all types of waste from any steel plant were simply dumped in the neighboring desert. Based on the production figures of the major steel plants in Egypt and the generation rates of different waste materials per each ton of steel produced, the annual waste materials generated in Egypt was estimated and over and above the stockpile (Table 13.9). There are stockpiles of 10 million tons of air-cooled BF slag and another 10 million tons of BOF slag laying in the desert.

Obviously, waste generators will be required to pay a certain fee per ton disposed. In 1993, the cost of disposing one ton of steel plant wastes in the United States was $15 (Foster 1996). Large tonnage of iron and steel slags are increasingly produced in Egypt and the huge space needed to dump them has become a real challenge. To have an idea about the considerable area needed for disposal, it is quite enough to know that the 20 million tons of BF and BOF slags currently available are estimated to occupy 2.5 million square meters.

Potential Utilization

Many of the environmental problems of solid waste materials generated from the iron and steel industry have been known for some time and attempts have been made to tackle them with varying degrees of success. During the past few years, the iron and steel industry has been able to produce some creative solutions to some of these environmental problems. It is highly probable that many other creative solutions also could emerge as a result of a well thought out and well supported research programs.

TABLE 13.9
Waste Materials Generated from Iron and Steel Industry in Egypt

Type of Waste	*Annual Amount Generated (tons)*
Blast Furnace Slag	600,000
Basic Oxygen Furnace Slag	200,000
Electrical Arc Furnace Slag	300,000
Blast Furnace Dust	20,000
Basic Oxygen Furnace Dust	Not Collected
Electrical Arc Furnace Dust	15,000
Rolling Mill Scales and Sludge	25,000

Processing of slag is a very important step in managing such waste material. Proper processing can provide slag with high market value and open new fields of application. Cooper et al. (1986) discussed the recent technologies of slag granulation. The main steps of the granulation process were addressed with schematic drawings including verification, filtering, and denaturing systems. The most recent continuous granulation technology at that time was introduced in detail with the help of many illustrative figures as shown in Figure 13.31.

Foster (1996) addressed the high cost of disposing wastes generated from the steel industry and discusses an innovative idea from South Africa to manage BOF slag, which has a limited usage. He came up with a new idea for processing BOF slag. This process starts with preparing the slag by grinding it, mixing it with a reductant such as sawdust or charcoal and feeding it into a modified cyclone type preheater. This reduction process removes iron oxide from iron. The slag is then passed over a magnet which removes the iron particles. The low-iron slag is then mixed with other materials, such as clay, to produce an acceptable type of cement kiln feed.

Featherstone and Holliday (1998) introduced the idea of dry slag granulation (Fig. 13.32). The existing slag treatment methods, the new dry granulation method, and the value of granulated slag products were reviewed. The development, application and advantages of the dry method of granulating molten slags were described. The dry granulated slag was proved to have many environmental advantages over conventional processes while generating a product of equal quality in addition to its low-cost and simplicity.

Swamy (1993) presented an extensive and critical examination of the use of ground granulated BF slag in concrete. It was shown that the use of BF slag in concrete can lead to concrete combines high strength and excellent durability. Apart from its ability to reduce the temperature rise due to hydration, test results showed that BF slag has a hidden potential to

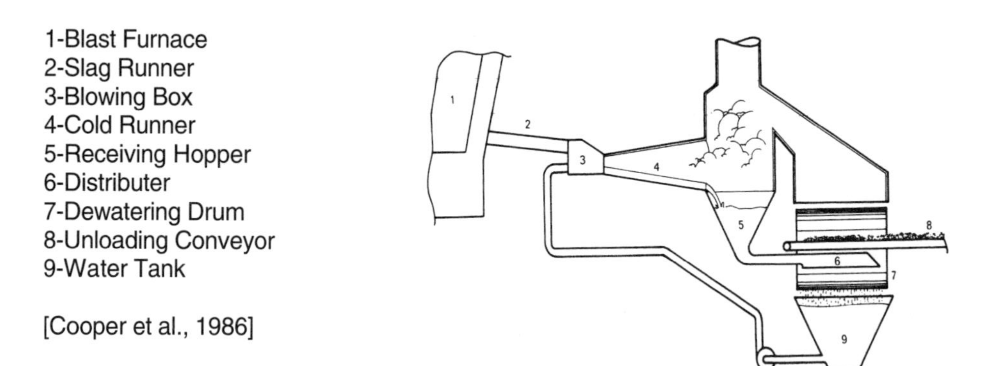

FIGURE 13.31. Continuous slag granulation system.

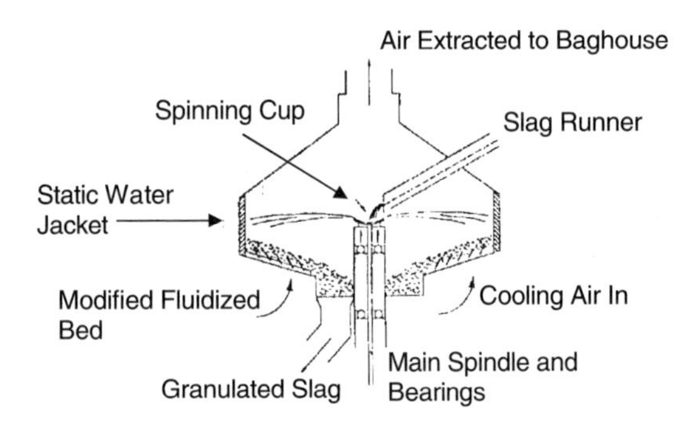

FIGURE 13.32. Dry slag granulation (Featherstone and Holliday, 1998).

TABLE 13.10
Compressive Strength Development of Slag Concrete [Swamy, 1993]

Mix	*Age, days*	*Compressive Strength, (MPa)*
A	1	7.20
B	1	4.10
A	3	28.90
B	3	19.00
A	7	39.00
B	7	27.40
A	28	46.80
B	28	34.20
A	90	54.90
B	90	38.20
A	180	57.10
B	180	36.47

contribute high early age strength, excellent durability and very good chemical resistance. A mix proportioning method was advanced which assured the development of early strength for slags of normal surface area of 350–450 m^2/kg. Table 13.10 summarizes the compressive strength development up to 180 days age for mixes with 50% (A) and 65% (B) slag replacement of coarse aggregates. Curing was shown to be a critical factor which affects early age strength, continued strength development and fine pore structure responsible for durability. It was also shown that with a well-defined curing period, the mineralogy and chemistry of slag could be mobilized to develop a very fine pore structure which is far superior to that of Portland cement concrete. Such a fine pore structure can impart a very high resistance to concrete to the transport of sulfate and chloride ions and water.

Nagao et al. (1989) proposed a new composite pavement base material made of steel-making slag and BF slag. When the new composite base material was prepared by mixing steel-making slag, air-cooled slag, and granulated blast furnace slag in proportions of 65%, 20%, and 15%, respectively, it was found to have material properties and placeability similar to those of conventional hydraulic and mechanically stabilized slags. Also, it was found feasible to quickly and economically suppress the swelling of steel-making slag by the hot water immersion that involves hydration reaction at 70°–90°C, under which slag can be stabilized in 24 hours at an expansion coefficient of 1.5% or less, as proposed by Japan Iron and Steel Federation.

One of the interesting research efforts in Egypt to find fields of application for iron and steel slags among other waste materials is the project carried out by Morsy and Saleh (1996) from 1994 to 1996. This project was funded by the Scientific Research and Technology Academy to investigate the technically sound and feasible utilization of two solid waste materials, iron and steel slags and cement dust, in addition to some other different liquid wastes. The study dealt with BF slag, both air-cooled and water-cooled, and BOF slag. The use of such slags in road paving as a base, subbase, and surface layers was examined through laboratory and pilot field tests. The results proved that these slags are suitable for use in all paving layers. Better performance and higher California bearing ratio were obtained for slags compared to conventional stones. The Egyptian standards for ballast require that the weight of cubic meter of ballast not be less than 1.1 ton and that Los Angeles abrasion ratio does not exceed 30%. Test results showed that the properties of Iron and steel slags surpassed the requirements of these standards and can be used as ballast provided that suitable grading is maintained.

Another study performed in Egypt in 1997 (El-Raey) by the Institute of Graduate Studies and Research at Alexandria University in conjunction with Alexandria National Iron and Steel company was aimed at investigating the use of EAF slag as road-paving base material and as coarse aggregate for producing concrete suitable for applications such as wave breakers, sidewalk blocks and profiles, and manhole covers. The physical and chemical properties of EAF slags of different ages were determined. EAF slag was crushed to the desired size and the applicability of slag in producing asphalt concrete was tested laboratory by Marshall test. Table 13.11 shows some of the results obtained for asphalt concrete containing slag. The test results proved that slag asphalt concrete could in general fulfill the requirements of the road-paving design criteria. EAF slag was successfully used as a coarse material for the base layer in a field-scale test.

The study covered also the use of EAF slag as coarse aggregate for concrete and the obtained results revealed that slump values for slag concrete were lower than those of gravel concrete by 33% for the same water/cement ratio. The unit weight of slag concrete was found to be 2.6–2.7 t/m^3 while it ranged from 2.35–2.38 t/m^3 for gravel concrete. The higher unit weight is attributed to the higher specific gravity of slag compared to gravel, 3.5 and

TABLE 13.11
Marshall Test for Asphalt Concrete Containing Slag

Asphalt Content, %	*Unit Weight (Ib/ft^3)*	*Stability (Ib)*	*Flow/ inch*	*Voids in Agg., %*	*Voids in Total Mix*
3.5	2.62	2279	10.2	14.8	9.2
4.5	2.72	2726	11.7	14.3	5.9
5.5	2.77	2550	12.4	14.5	3.5
6.5	2.73	2100	15.9	15.6	1.5
Design Criteria	–	>1800	8–18	>13	3–8

2.65, respectively. For the same cement content and water/cement ratio, and at both early and later ages, slag concrete exhibited higher compressive strength than gravel concrete, with an average increase of 20% at 7 days age and 10% at 28 days age. The same classical effects of water/cement ratio, aggregate/cement ratio, and curing on gravel concrete were observed for slag concrete. From a durability point of view, no sign of self-deterioration was noticed for slag concrete. Slag concrete has been successfully applied in manufacturing sidewalk blocks and profiles, manhole covers, and balance-weights.

Korany and El Haggar (2000) investigated the utilization of different slag types as coarse aggregate replacements in producing building materials such as cement masonry units and paving stone interlock. Cement masonry specimens were tested for density, water absorption, and compression, and flexural strengths. While the paving stone interlocks were tested for bulk density, water absorption, compressive strength, and abrasion resistance. They also studied the likely health hazards of the proposed applications. The test results proved in general the technical soundness and suitability of the introduced ideas. Most of the slag solid brick units showed lower bulk density values than the commercial bricks used for comparison. All slag units exhibited absorption percentages well below the ASTM limit of 13%. A substantially higher compressive strength results were reached for all masonry groups at 28-day age compared to the control and commercial bricks as seen in Figure 13.33. All test groups showed higher compressive strength than the ASTM limit of 4.14 MPa for nonload-bearing units. At slag replacement levels higher than 67%, all groups resulted in compressive strength higher than the ASTM requirement of 13.1 MPa for load-bearing units. All slag types resulted in paving stone interlocks having water absorption values far below the ASTM limit as shown in Figure 13.34. All slag paving stone interlocks showed higher compressive strength and abrasion resistance than the control specimens made of dolomite. Moreover, the proposed fields of application were found to be safe to the environment and have no drawbacks based on the heavy metals content and water leaching test results.

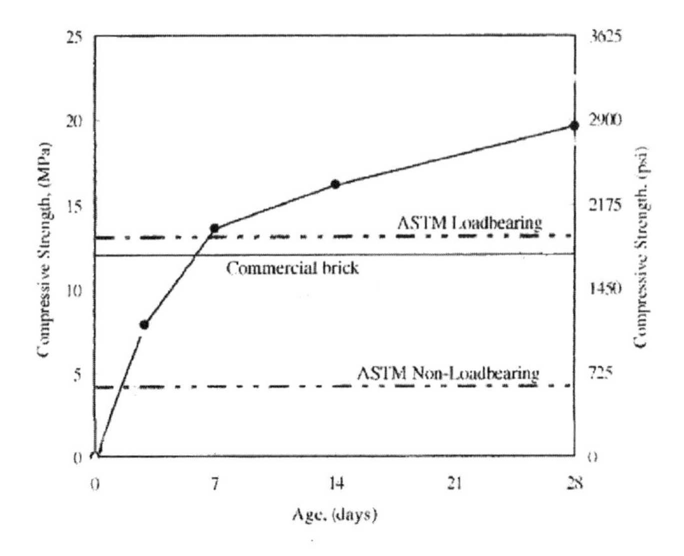

FIGURE 13.33. Development of compressive strength with age for solid brick groups at 100% slag replacement levels (Korany and El Haggar, 2000).

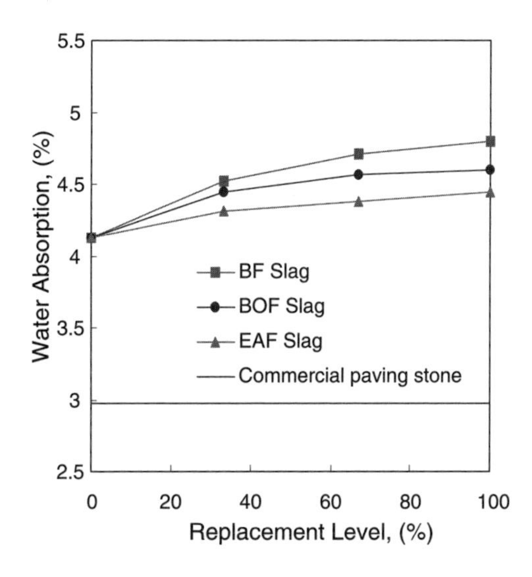

FIGURE 13.34. Comparison between absorption percentages of slags used for paving stone units at different replacement levels (Korany and El Haggar, 2000).

Szekely (1995) proposed a comprehensive research program to reduce fume formation in the BOF and EAF, find an effective approach to reduce and utilize steel-making slag, and to effectively use the oily sludge produced in rolling mills. Related environmental problem areas were discussed and preliminary solutions were identified. From Szekely's viewpoint, although several technologies are available for treating EAF dust, they are quite expensive and satisfactory solutions for the EAF dust problem have not yet been produced. He suggested some possible solutions worthy exploration such as modifying the charging, blowing and waste gas exhaustion system to minimize dust formation. Another proposed solution is to examine the composition of the dust produced during different phases of furnace operation and, if appropriate, segregates the recovered dust.

Some of the methods used to turn steel plant dust into a valuable raw material were described by one of the solid waste processing companies in its article published in 1997 (Heckett MultiServ 1997). One of the commonly used methods is micro-pelletizing where dust is mixed with lime as a binder and pelletized to produce a fine granular form, the major proportion being in the size range of 2–3 mm with a total size range of 1–10 mm. The water content is adjusted to 12% during mixing and the pellets are air-cured for a minimum of 3 days before charging to the sinter plant where they account for 3% of the total charge. The article also addressed the direct injection process currently in trials in Germany and the UK, where injection is used to pass fine dust into the liquid metal in the furnace. Fine dust is blended with hydrating dusts such as burnt lime and carbon. The metal oxide content of the material is about 70% and has a particle size range of 0–8 mm making it ideal for direct injection into a range of furnaces.

Conclusions

Besides the economic and technical importance of utilizing waste materials generated from iron and steel industry, this activity is of great importance from the environmental protection point of view. The first environmental impact is the useful consumption of the huge stockpiles of these waste materials. When these waste materials are used as replacements for other products such as cement, the natural resources which serve as raw materials are preserved. Also, air pollution levels will be reduced due to the reduction in fuel consumption. Another important impact that has been almost completely overlooked is the considerable energy loss as contained in these waste materials especially slags. The temperature of the molten slag at the furnace outlet is 1500°C and drops to 1300°C when slag reaches the disposal yard. Recent dry granulation processes give the chance to recover a great part of this heat.

All slag types should be treated as by-products rather than waste materials. All existing and new steel plants should have a slag processing unit within the factory property to extract steel from slag and to crush the slag and sieve it to the desired grading for ease of promotion.

The chemical analysis of the EAF dust showed that it consists mainly of iron with a percentage higher than 75% by weight. This iron content is higher than that of the Egyptian ore currently used. Therefore, steel makers generating EAF dust can sell it to the iron and steel factories working by blast furnaces where dust can be blended with the ore as a rich source of iron. Cement factories may also use the EAF dust as a flux material to accelerate reactions and reduce energy consumption.

Sustainable Development Tools

Concerning environmental issues, there is a common misconception that environmental protection comes at the expense of economic development or vice versa. This is clearly portrayed when communities faced with economic crises settle for alternatives that sacrifice environmental integrity such as incineration, treatment, or the construction of landfills even though these solutions in fact are extremely important but are not sustainable and economically expensive as explained before.

Sustainable development is formally defined by the World Commission on Environment and Development (WECD) as "development that meets the needs of the people today without compromising the ability of future generations to meet their own needs." Therefore, sustainable development refers to a shared commitment towards steady economic growth, given that this economic growth does not compromise the satisfactory management of available environmental resources. Resource allocation, financial investments, and social change are directed in a sound manner that guarantees their sustainability or continuation with time and thus they are made consistent with both future and present needs. Another notable definition for sustainable development is "economic and social changes that promote human prosperity and quality of life without causing ecological or social damage" (Nantucket Sustainable Development Corporation 2003). Industries are therefore encouraged to flourish but also to realize their impacts on the environment and society around them. Thus, it can be concluded that sustainable development is a concept that is not only exclusive to policy makers and environmentalists, it should be a matter of concern to industries, businesses, and society.

The practice of sustainable development is not a new one. This is a concept that has been repeatedly used in an effort to sustain and/or preserve resources of any type. However, formal attention and labeling of this concept began during the 1970s. In 1972, the global community came together in Stockholm to discuss international environmental and development issues for the first time in the United Nation Conference on Human Environment. This conference was the first significant link between business and environment to take responsibility for the environmental problem that uncontrolled industrial development was causing. The conference resulted in the creation of United Nations Environment Programme (UNEP) to adopt a global action plan for protecting the environment. From its creation until now it tried to develop guidelines and tools for the above cause.

In 1986, the World Commission on Environment and Development (WCED) was established. This commission's report is what first spread the term "sustainable development" and it became the benchmark for thinking about global environmental and development issues. The crest of global attention towards sustainability was during the United Nations Conference on Environment and Development in 1992 held in Rio de Janeiro. During this conference an action agenda was produced. This Agenda 21 was a comprehensive global plan of action for local, national and global sustainable development. An equally comprehensive summit was the Johannesburg Earth Summit in 2002, which was more focused on eradication of poverty as it also revived the commitment towards global sustainable development. These summits augmented with vast global efforts have aided in increasing awareness as well as multilateral agreements concerning various sustainability issues and critical environments.

The concept of sustainable development is a methodology that attempts to encompass social, technological, economic, and environmental aspects. Thus, the interactions and impacts of these four factors on each other are focused upon rather than the fallacy that they are independent of one another. Environmental, technological, social, and economic growth is seen to reinforce each other thus attaining win-win solutions that do not compromise any aspect. In order to develop a methodology, a number of tools are required. The main tools necessary for implementing sustainable development are cleaner production, environmental management system, 7-Rs golden rule, industrial ecology, environmental impact assessment, and information technologies.

Cleaner Production

The UNEP launched the Cleaner Production Programme in 1989 in response to the need to reduce worldwide industrial pollution and waste. Positive future expectations exist for the spread of cleaner production concept, as it combines maximum effect for the environment with significant economic savings for the industry incorporating the economic perspectives that concerns any business as discussed before.

Although it is only up to the industry to adapt cleaner production concept, but the role of the government is required to provide the environment that will encourage industries to begin their own CP programs. The tools that the government uses as regard to their country's needs and circumstances include:

Adopting regulations: The regulations specify the environmental goals, methodology of achieving them, and technology used.

Economic instruments: The economic instruments are used to make the costs of pollution more expensive than the costs of cleaner production. Two forms of these instruments exist: rewards and penalties.

Provide support measures: Government can provide support in four main areas
Providing information about CP
Assisting in the development of management tools in the industry
Organizing training workshops
Promoting the concept in engineering schools and universities

Provide external assistance: Assistance to developing countries can take several forms: financial aid, development of case studies in different sectors, technology transfer, and exchange of expertise.

Provide guidelines for implementing CP: To initiate cleaner production it requires three main steps: structured methodology, management commitment, operator's involvement.

Methodology for CP Assessment

The main goal of undertaking a CP assessment or CP audit in the industry is to able to identify opportunities for cleaner production. The methodology for the cleaner production opportunity assessment is as follows:

- Forming cleaner production team from CP expert and industrial expert(s)
- List process steps with process flow diagram(s) indicating inputs, outputs as well as wastes and emissions
- Identify wasteful process with cause analysis
- Analyze process steps with material and energy balance including cost of waste streams
- Develop CP opportunities and select workable opportunities
- Select CP solutions according to
  Technical feasibilities
  Economic feasibility
  Environmental aspects
  Social aspects
- Implementation: prepare for implementation through operator's involvement and management commitment
- Assessment: monitor and evaluate results
- Sustainability: maintaining CP
- Go back to re-identify another wasteful process and continue

CP Techniques

The key difference between cleaner production and other methods like pollution control is the choice of the timing, cost, and sustainability. Pollution control follows "react and treat" rule, while cleaner production adopts "prevent better than cure." Such that, cleaner production focuses on before-the-event techniques that can be categorized as follows.

- Source reduction
  - Good housekeeping
  - Process changes
    - Better process control
  - Equipment modification
    - Technology change
    - Input material change
- Recycling
  - Onsite recycling
  - Useful by-products through offsite recycling
- Product modification

Keeping in mind these techniques while performing the audit will generate options for cleaner production in the industry.

One element of sustainable development that is extremely important is the change in attitudes concerning the way development and the environment are seen. There are many prerequisites that must be fulfilled so that sustainable development can be practiced correctly. There must be willingness to try new methods, sincere commitment to the cause, an open mind towards new ideas and tools, good team dynamics, as well as a structured methodology for the work. Current attitudes that act as barriers preventing the implementation of practices of sustainable development are fear and unawareness. Fear of the new, of criticism, and of breaking a dog-eared system is a common attitude. Also, basic unawareness coupled with this fear causes individuals to refuse ideas even before fully understanding them. The many stereotypes and incorrect information that continue to mislead many decision makers are a major barrier to sustainable development. However, this problem can be and is in the process of being remedied through the use of deliberate awareness campaigns that target various members of the community, primarily decision makers.

Environmental Management System

To attain sustainable development many elements have been outlined and researched. One of the elements widely discussed is integrating EMS within industries. An EMS consists of a systematic process that allows an organization to "assess, manage, and reduce environmental hazards." Thus the continuous monitoring of environmental impacts concerning that organization is integrated into the actual management system guaranteeing its continuation as well as commitment to its success.

Environmental management system is a part of the overall management system of an organization, which consists of organizational structure, planning, activities, responsibilities, practices, procedures, processes, and resources for developing, implementing, achieving, reviewing, and maintaining the environmental policy.

The EMS provides several benefits through continual implementation and development that include:

- Financial benefits: through cost savings as well as increasing local and international market competitiveness
- Improve company's performance and image
- Reducing business risks

In general an EMS should include the following elements as shown in Figure 13.35:

- Management commitment and environmental policy
- Planning for the environmental policy
- Implementing the environmental planning
- Evaluation and corrective/preventive actions
- Management review

CP-EMS Model

Cleaner production and environmental management systems are located at the top of sustainable development tools. Huge efforts in spreading these concepts worldwide are dedicated especially to developing country due to the immediate environmental and financial benefits they generate if properly applied as explained before.

The EMS can provide a company with a decision-making structure and action plan to bring cleaner production into the company's strategy, management and day-to-day operations. As a result, EMS will provide a tool for cleaner production implementation and pave the road toward it. Thus, integrating cleaner production technologies with EMS as shown in Figure 13.36 will help the system to approach zero pollution and maximize the benefits where both CP benefits and EMS benefits will be integrated together.

7-Rs Golden Rule

A new hierarchy for waste management to approach zero pollution was developed at The American University in Cairo by El Haggar (2001c, 2003c). This approach is based on the concept of adapting the best practicable environmental option for individual waste streams and dealing with waste as a by-product. This 7-R golden rule for zero pollution (regulation, reduce, reuse, recycle, recovery, rethinking, and renovation) is the basis for industrial ecology. This theory will band the disposal and treatment facilities and develop renewable resources. This theory was implemented in different industrial sectors as well as different projects with full success as explained

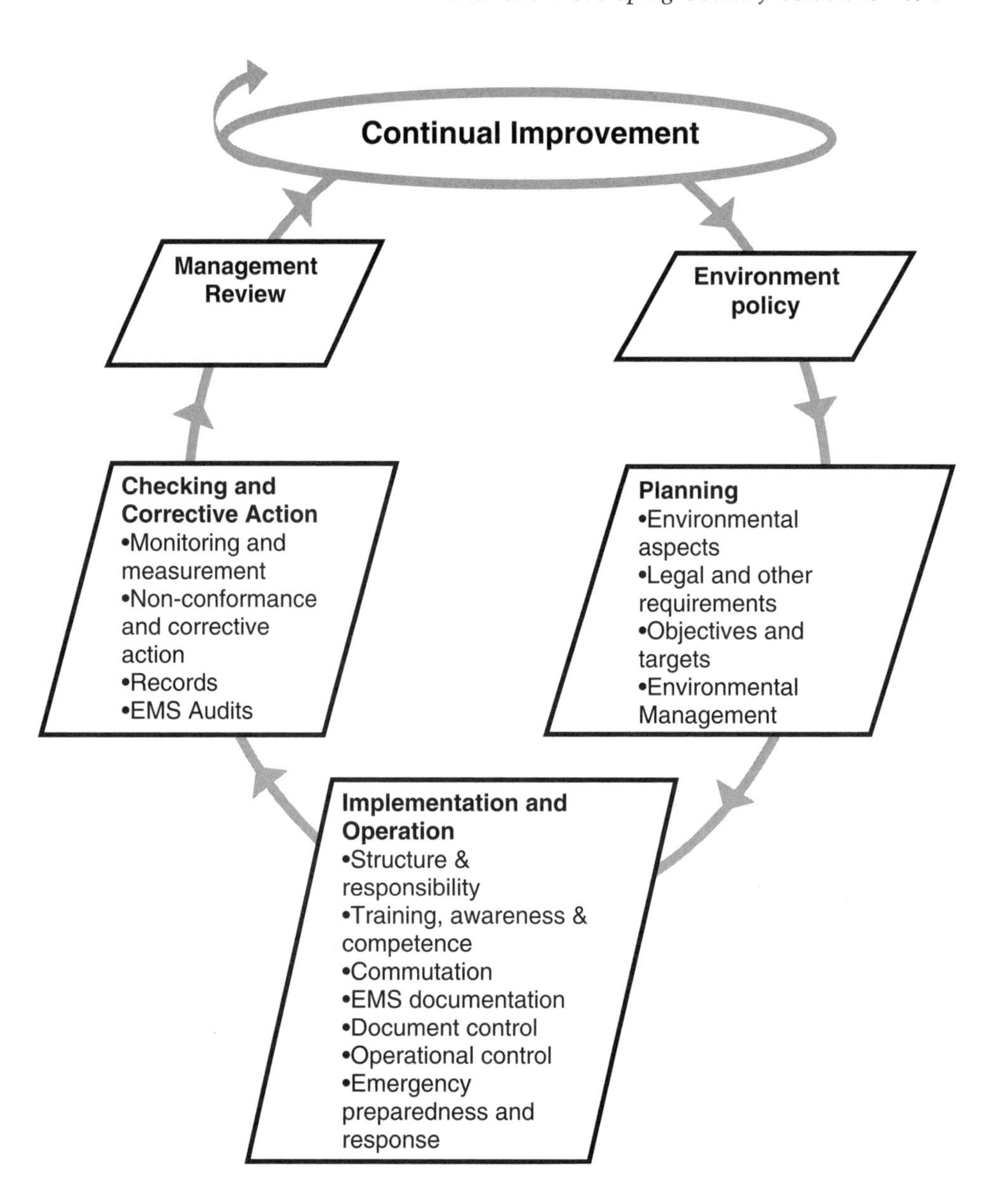

FIGURE 13.35. Environmental management system (EMS) model.

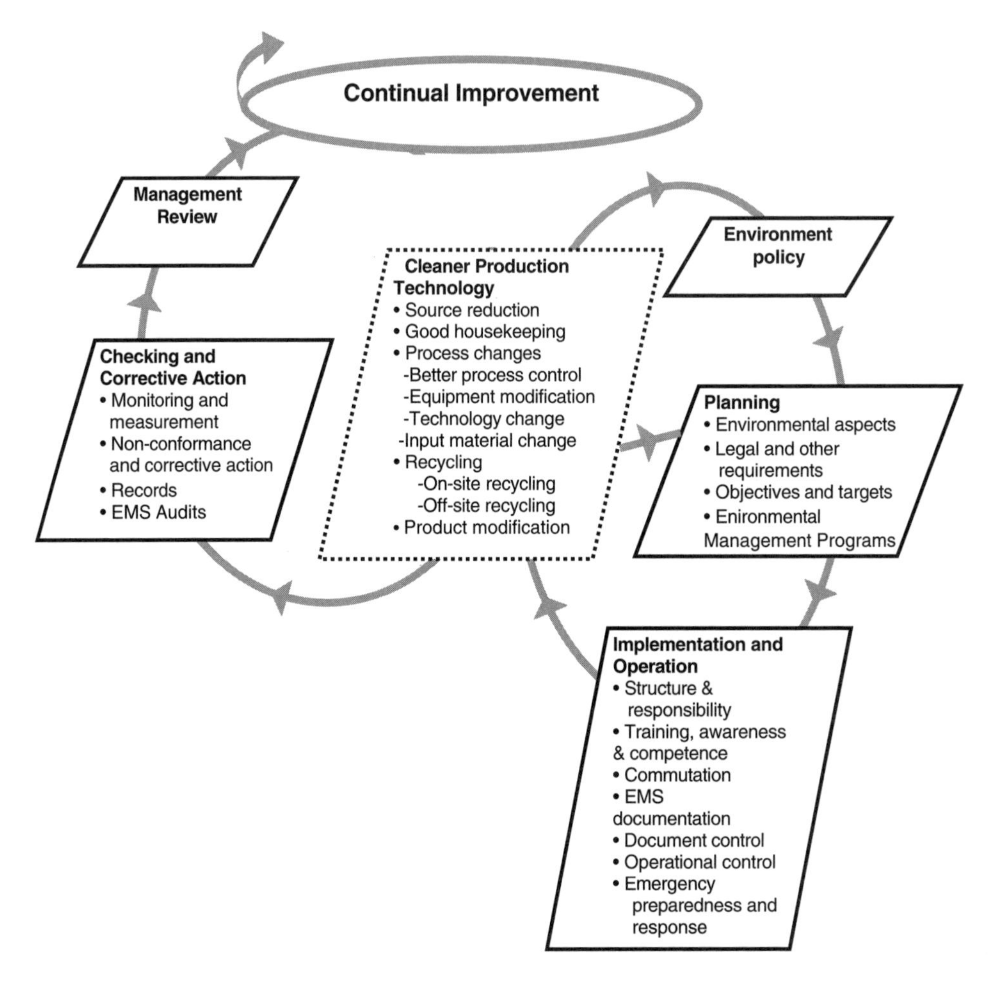

FIGURE 13.36. Cleaner production (CP) and environmental management systems (EMS) model.

before. The theory can be implemented in small industries as well as large industries.

Industrial Ecology

We are talking about sustainable development all the time, but we have lost direction to our goal because we need to ask first: How can we develop sustainability? Our real environmental and economical problem in this century is that development of science and technology have increased human capacity to extract resources from nature, then process it, use it, but finally it is not returned back to the environment to regenerate it.

Unsustainable human activities are creating an open loop that can not continue and has to reach one day to a dead end. On the contrary, sustainability is the rule that governs natural economics. Natural economics regulates every transaction involving the exchange of any of its resources among the members of the system. Natural economics never pushes nature's inventory of resources beyond critical limits for sustainability. Human economic systems should learn from natural economic systems that there is a need to attain cooperation among individuals to enhance the economic efficiency in the use of resources for each member of the community that will in return maximize economic effectiveness for the community as a whole.

The tool for sustainable human economic systems is industrial ecology, and the means to achieve industrial ecology are cleaner production concepts and environmental management system to be able to reach zero pollution. Industrial ecology is the study of industrial systems that operate like natural systems. In a natural system there is a closed materials and energy loop, where animal and plant wastes are decomposed by microorganisms into useful nutrients to the plants, then plants produce food to the animals to consume it, and finally the loop is closed again when animals die and are either converted to fossil fuels or nutrients for the plants. Industries can follow the same system through making one industry's waste another's raw materials, where materials and energy are circulated in a complex web of interactions.

Now, IE can be defined as the study of industrial systems that operate more like natural ecosystems. It can be achieved by the proper implementation of cleaner production technologies in order to approach zero pollution. The proper implementation of CP can lead to environmental benefits as well as economical and social benefits, if a good management system exists. Adapting EMS or ISO 14001 can develop the foundation of a good management system. Therefore the integration of CP with EMS as explained before is a must in order to reach zero pollution/waste. In other words, industrial ecology can be implemented within the activity through the implementation of the proposed CP-EMS model shown in Figure 13.36. The proposed CP-EMS model or a modified EMS model to integrate cleaner production technologies with environmental management system as shown in Figure 13.36 will promote the concept of CP technologies throughout the EMS steps starting from environmental policy—planning—implementation (especially the training component) all the way to checking and corrective action for sustainable development. Top management should be aware of the benefits of cleaner production to their business and the environment.

Another tool for industrial ecology is 7-Rs golden rule for zero pollution. As the basis for industrial ecology, it can be considered as a methodology for approaching industrial ecology. This theory will help band the disposal facilities (landfills) as well as the treatment facilities (such as thermal treatment or incineration) and avoid the environmental impacts and develop renewable resources.

Industrial Ecology Barriers

Even though the industrial ecology concept has a lot of advantages from economical, environmental, and social points of view, there are still some barriers for implementations.

The barriers to industrial ecology fall into six categories: technical, market and information, business and financial, regulatory, legal, and regional strategies.

Technical issues require a lot of innovation to convert waste into money or prevent it at the source. The markets for waste materials will ultimately rise or fall based on their economic vitality and can be enhanced through information technology tools. One option for waste markets are dedicated "Waste Exchanges" where brokers trade industrial wastes like other commodities. Developing business plans and providing financial support will help promoting the concept of industrial ecology. The private firm is the basic economic unit and collectively constitutes the mechanism for producing inventions and innovations to practice. The government should regulate planning for industrial ecology and sustainable development. Industries tend to form spatial clusters in specific geographic regions based on factors such as access to raw materials, convenient transportation, technical expertise, and markets. This requires regional strategies provided by local and federal governments.

Environmental Impact Assessment

Environmental impact assessment (EIA) can be defined as "the systematic examination of unintended consequences of a development project or program, with the view to reduce or mitigate negative impacts and maximize on positive ones" (EEAA 1996; El Haggar and El-Azizy 2003). In practice, this means studying and analyzing the environmental feasibility of any proposed project (under construction) because the implementation or operation of the project may affect the environment, natural resources and/or human health.

EIA is a tool that ensures the protection and conservation of the environment and natural resources and ensures a sustainable economic development for any project under construction. EIA studies and analyzes the environmental feasibility of any proposed project that could be a new one or an existing project that needs expansion or renovation. Through this analysis, the negative impacts of the project on the environment are studied as well as the mitigating measures to comply with environmental regulations and accordingly proposed projects that guarantee the protection of the environment are implemented. Alternative solutions for mitigating measures are provided with complete analysis from technical, economical, environmental and social aspects to ensure the sustainability of the project.

EIA must be performed for new establishments or projects and for expansion or renovation of existing establishments. EIA studies the effect of the surrounding environment on the project as well as the effect of the project on the surrounding environment. It also looks at the different processes involved in product production, including inputs and outputs. EIA tries also to find ways of minimizing the environment impacts of the project. This study if implemented properly will ensure sustainability for the project especially that now it has become necessary to provide this assessment before starting any project.

The purpose of EIA is to ensure the protection and conservation of the environment and natural resources including human health aspects against uncontrolled development. The long term objective is to ensure a sustainable economic development that meets present needs without compromising future generations ability to meet their own needs. EIA is an important tool in the integrating environmental management approach.

Information Technology

Since the 1980s, computer-based IT has become the focus of the global world information. The environmental situation is continually changing as a result of human activities, so it is very important to obtain accurately and timely the information on various environmental changes, as well as provides an excellent information process to global knowledge (UNEP, 1997).

Information technology can act as a facilitator who will ignite the participatory process and to assist people in spelling out their ideas, perceptions, attitudes, and knowledge so as to attain self-development. One of the main tools for information technology is the Internet. Internet is a two-way medium to communicate, give info, as well as receive that matches bottom-up environmental activities. It will improve interaction between people and institutions, better access to government by those governed, and ease in the general access to information. IT also faces challenges especially in the developing countries, such as:

- If new technology may not be available for all, in return it may create new divisions between the "haves" and the "have nots." Developing countries need equal access to the new world of possibilities through technology.
- It can face barriers not directly related to the technology itself, such as:
  Illiteracy in general
  Computer illiteracy
  Technical and education variation of users

Information technology can help promoting the concept of environmental awareness (EA), which might increase the people interest on environment. People's response to improving their environment depends on the depth of their perception on the environmental problems and their willingness to act in favor of that. Participatory environmental awareness, such as bottom-up approaches, helps them to identify problems priority to them, understand how to solve these problems, and encourage them to be involved in the planning and implementation stages. To make participatory EA a success, an efficient information process is required as a tool. Information technology is the tool that matches the needs of modern times, however, it is faced with several challenges in developing countries.

Public environmental awareness will not be created overnight, but it grows gradually and slowly. The government cannot take the major role in this process, as it is more controlled and directed by the public. The required role of the government, NGOs, and educated individuals is to find efficient and creative means to reach to the public through information technology.

Conclusion

Industrial development and environmental protection cannot be achieved without establishing the concept of industrial ecology. The main tools necessary for establishing industrial ecology are cleaner production, environmental management systems, and zero pollution. The concept of *industrial ecology* will help the industrial system to be managed and operated more or less like a natural ecosystem hence causing as less damage as possible to the surrounding environment. The 7-Rs golden rule encompasses regulation, reducing, reusing, recycling, recovering, rethinking, and renovation, and is the basic tool for industrial ecology.

Establishing industrial ecology within the industrial activities will avoid landfill, incineration and/or treatment. The cost of treatment and safe disposal of waste through incineration or landfills is escalating exponentially. Locating waste disposal sites (landfills) are becoming more difficult and expensive. The environmental and health impacts of landfills and incinerations are becoming more dangerous and disaster for the community and national economy.

Industrial ecology can be considered as science of sustainability promises much in improving the efficiency of human use of the ecosystem. Technological improvements are not always better in the full sense of sustainability without taking environment into consideration, where zero pollution is a must for industrial ecology. Cooperation and community are also important parts of the ecological metaphor of sustainability.

Review Questions

1. State the quantity and types of agricultural wastes produced in your community or nearby community. Develop an environmentally bal-

anced rural waste complex (EBRWC) for such waste as well as a risk analysis according to demand and need. Compare your complex with Figure 13.30.
2. "EMS can be used as a tool to promote the cleaner production concept." Explain this statement and develop a methodology to integrate cleaner production techniques and environment management systems. Compare your methodology with Figure 13.36.
3. Identify the quantity and types of municipal solid waste (MSW) generated in your community and develop an environmentally balanced municipal solid waste complex (EBMSWC) using the guidelines given in Figure 13.4.
4. Identify the wastes that are now recycled from MSW in your community. What is the percentage of waste that is not being recycled? What is the current practice for such waste or rejects? Develop a technology that can be used to recycle the unrecyclable material (rejects) according to Figure 13.19. State the products that can be produced according to the demand and need concept in your community.
5. Identify the quantity and quality of industrial solid waste generated in your community or nearby community. What is the current practice of such waste? Develop a methodology to utilize such waste using the industrial ecology concept and 7-Rs golden rule.

References

Allenby, B. R. 1999. *Industrial Ecology, Policy Framework and Implementation.* Englewood Cliffs, NJ: Prentice Hall.

Ayres, R. U. 1989. Industrial metabolism. In: J. H. Ausubel and H. E. Sladovich, eds., *Technology and Environment* (pp.23–49). Washington, DC: National Academy Press.

CAPMAS, Central Agency for Public Mobilization and Statistics. 1999. *Statistical Yearbook.* Cairo, Egypt.

CEPI, The Confederation of European Paper Industry. 1999. *Annual Statistics.*

Cichonski, T. J., K. Hill. (eds). 1993. *Recycling Sourcebook.* Washington, DC.

Cooper, A. W., M. Solvi, M. Calmes. 1986. Blast furnace slag granulation. *Iron and Steel Engineer* July, 63:46–52.

Dadd, D. L. 2004. Textile Recycling by Worldwise, Council for Textile Recycling, Secondary Materials and Recycled Textiles. SMART, United States (www.worldwise.com).

Daly, H. E. 1977. *Steady State Economics.* San Francisco: Freeman

Dasgupta, A., S. M. El Haggar. 2003. Industrial Hazardous Waste Treatment and Disposal Guidelines. Report submitted to USAID, EEPP, June.

Ehrenfeld, S. 1997. Industrial ecology: A new framework for product and process design.*Journal of Cleaner Production* 5(1–2):87–95.

EEAA, Egyptian Environmental Affairs Agency. 1994. Comparing environmental health risks in Cairo. *Project in Development and Environment* 2, C.

EEAA, Egyptian Environmental Affairs Agency. 1996. EIA Guidelines, Egypt, October.

El Haggar, S. M. 1996. Recycling of solid waste. In: 3rd Annual AUC Research Conference, Environmental Protection for Sustainable Development, April 21–22.

El Haggar, S. M. 1998. Recycling of solid waste. In: Grater Cairo, Ain Shams First Conference on Engineering and Environment, Cairo, Egypt, May 9–10.

El Haggar, S. M. 2000. The use of cement by-pass dust for sewage sludge treatment. In: The International Conference for Environmental Hazardous Mitigation, ICEHM, Cairo University, Egypt, September 9–12.

El Haggar, S. M. 2001a. Paper recycling in Egypt. In: International Symposium for Recovery and Recycling of Paper, University of Dundee, UK, March 19–20.

El Haggar, S. M. 2001b. Environmentally balanced rural waste complex for zero pollution. In: Enviro 2001, The Third International Conference for Environmental Management and Technologies, Cairo, Egypt, October 29–31.

El Haggar, S. M. 2001c. New cleaner production hierarchy for zero pollution. In: Enviro 2001, The Third International Conference for Environmental Management and Technologies, Cairo, Egypt, October 29–31,

El Haggar, S. M. 2002. Zero pollution for sustainable development in Egyptian industries. In: International Congress on Sugar and Sugar Cane By-products, Diversification, Havana, Cuba, June 17–22.

El Haggar, S. M. 2003a. Solid waste management using cleaner production technologies. In: *Environmental Balance and Industrial Modernization*, (Chapter 7). Cairo, Egypt: Dar El-Fikr El-Araby Book Co.

El Haggar, S. M. 2003b. Reaching 100% recycling of municipal solid waste generated in Egypt. In: The International Symposium for Advances in Waste Management and Recycling, Dundee, UK, September 9–11.

El Haggar, S. M. 2003c. 6-R Golden Rule for Industrial Ecology. In: Enviro 2003, The Fourth International Conference on Environmental Technologies, September 30–October 2.

El Haggar, S. M. 2004a. Industrial Ecology for Renewable Resources, International Conference for Renewable Resources and Renewable Energy: A Global Challenge, Terista, Italy, June 10–12.

El Haggar, S. M. 2004b. Cradle-to-cradle for industrial ecology. *Advances in Science and Technology of Treatment and Utilization of Industrial Waste*, CMRDI and University of Florida, US-Egypt Joint Fund, Cairo, Egypt, June 6–10.

El Haggar, S. M. 2004c. *Solid Waste Management: Alternatives, Innovations and Solutions, Fundamentals and Mechanisms for Sustainable Development* (Series 3). Cairo, Egypt: Dar El-Fikr El-Araby Book Co.

El Haggar, S. M., R. Baher. 1999. Design, manufacturing and testing of a waste paper recycling system. *International Journal of Environment and Pollution* 11(2):211–227.

El Haggar, S. M., E. M. El-Azizy. 2003. *Environmental Impact Assessment, Basis and Mechanisms of Sustainable Development* (Series 1). Cairo, Egypt: Dar El-Fikr El-Araby Book Co.

El Haggar, S. M., M. Toivola. 2001. Mixing plastic waste with sand as raw material for eco-building products. WASTE 2001, The Middle East Congress and Exhibition of Recycling and Waste Management, February 28–March 2.

El Haggar, S. M., M. F. Hamouda, M. A. Elbieh. 1998. Composting of vegetable waste in subtropical climates. *International Journal of Environment and Pollution* 9(4).

El Haggar, S. M., S. El-Attawi, A. Mazen. 2001. Recycling and deinking of newspaper mixed with magazine. International Symposium for Recovery and Recycling of Paper, University of Dundee, UK, March 19–20.

El Haggar, S. M., B. E. Ali, S. M. Ahmed, M. M. Hamdy. 2004. Solubility of natural rocks during composting. Second International Conference of Organic Agriculture, Cairo, Egypt, March 25–27.

El-Raey, M. 1997. Utilization of Slag Produced by Electric Arc Furnace at Alexandria National Iron and Steel Company. Institute of Graduate Studies and Research, Alexandria University (final report).

Epstein, E. 1997. *The Science of Composting*. Lancaster, PA: Technomic Publishing Co., Inc.

Erkman, S. 1997. Industrial ecology, a historical view. *Journal of Cleaner Production* 5(1–2):1–10.

Featherstone, W. B., K. A. Holliday. 1998. Slag treatment improvement by dry granulation. *Iron and Steel Engineer* July:42–46.

Foster, A. 1996. How to make a heap from slag. *The Engineer* April, 282:22.

Frosch, R. A. 1994. Industrial ecology: minimizing the impact of industrial waste. *Physics Today* Nov., 47(11):63–68.

Frosch, R. A. 1995. Industrial ecology: adapting technology for a sustainable world. *Environment* Dec., 37(10).

Georgescu-Roegen, N. 1971. *The Entropy Law and the Economic Process*. Cambridge, MA: Harvard University Press.

Heckett MultiServ. 1997. Reprocessing steel plant fines. *Steel Times* Jan., 225(1): 19–24.

Kilby, E. 2001. Current statistics on recovered paper. International Symposium for Recovery and Recycling of Ppaper, University of Dundee, UK, March 19–20.

Korany, Y., S. M. El Haggar. 2000. Utilizing slag generated from iron and steel industry in producing masonry units and paving interlocks. 28th CSCE Annual Conference, London, Ontario, Canada, June 7–10.

Lankford, W. T., N. L. Samways, R. F. Craven, H. E. McGaannon. 1985. *The Making, Shaping and Treating of Steel*. Pittsburgh, PA: Herbrick and Held.

Lardinos, I., A. Van de Klundert. 1995. *Plastic Waste, Options for Small-Scale Resource Recovery, Urban Solid Waste* (Series 2). Amsterdam: The TOOL Publications.

McKinney, R. 1991. *Recycling Fiber—The State of the Art in Europe*. World Pulp and Paper Technology.

Morsy, M. M., M. S. Saleh. 1996. Industrial and non-industrial solid waste utilization in road building and as ballast for railroads. Transportation and Communications Research Council Report, Scientific Research and Technology Academy, Egypt.

Nantucket Sustainable Development Corporation. 2003. Sustainable Nantucket—A compass for the future. Summer.

Newel, J. 1990. Recycling Britain. *New Scientist* Sept., 46.

Nemerow, N. L. 1995. *Zero Pollution for Industry*. New York: Wiley.

Nagao Y., et al. 1989. Development of new pavement base course material using high proportion of steel making slag properly combined with air-cooled and granulated blast furnace slags. Nippon Steel Technical Report No. 43.

Prigogine, I. 1955. *Thermodynamics of Irreversible Processes*. Springfield, IL: Charles Thomas.

Sawiries, Y. L., S. M. El Haggar, A. M. Ghanem. 2001. Environmentally balanced municipal solid waste complex for zero pollution. In: Enviro 2001, The Third International Conference for Environmental Management and Technologies, Cairo, Egypt, October 29–31.

Swamy, R. N. 1993. Concrete with slag: High performance and durability without tears. The Fourth International Conference on Structural Failure, Durability and Retrofitting, Singapore, pp. 206–236.

Szekely, J. 1995. A research program for the minimization and effective utilization of steel plant wastes. *Iron and Steelmaker* January:25–29.

Tchobanoglous, G., H. Theisen, S. Vigil. 1993. *Integrated Solid Waste Management* New York: McGraw-Hill.

UNEP, United Nations Environment Programme. 1997. The role of information technology in environmental awareness-raising, policy-making, decision-making, and development aid. Papers from an executive seminar, September 3.

UNEP, United Nations Environment Programme. 1997. Cleaner production at pulp and paper mills: A guidance manual publication in cooperation with the National Productivity Council, India.

Vogler, J. 1984. *Small Scale Recycling of Plastics*. Rugby, UK: Intermediate Technology Publications.

Wernick, I. K., J. H. Ausubel. 1997. Industrial ecology: Some directions for research. Program for the Human Environment (pre-publication draft).

Wogrolly, E., M. Hofstatter, E. Langshwert. 1995. Recycling of plastic waste. *Advances in Material Technology Monitor*, Vienna: United Nations industrial Development Organization UNIDO, 2(4):1–6.

CHAPTER 14

International Frameworks for Environmental Solutions

Fritz Balkau

1. Introduction

Many environmental problems now extend well beyond the local level, sometimes even threatening the stability of our planet's life-support systems. Greenhouse gas emissions, long-range air pollution, dispersion of toxic chemicals, and transport of unwanted species by ships and air all have global scope, as do the extinction of species and the unwanted spread of genetically modified organisms.

Recent reviews of the global environmental condition (Tellus Institute 2002; UNEP 2003; Worldwatch Institute 2004) provide us with an eloquent summary of the environmental challenges facing society. They tell us that our world, home to an increasing population with rising living standards, has a global system of production and trade that is putting enormous pressure on our living space, that is to say on the environment which provides our resources and receives our wastes. If we do not act decisively and soon, we face a bleak future of ecological collapse and social chaos.

These same reviews advocate solutions—more or less radical and at various levels of detail depending on the authors—that will prevent the chaos from happening. These involve a range of policy, regulatory, and voluntary options to be implemented by various social partners. All of them, however, will only be successful if there is greater commitment to act by politicians, a deeper level of engagement of civil society and a more widespread application of appropriate technologies and techniques. As is appropriate for a global dilemma, many of the proposed solutions have a global reach, and most require a concerted, collective effort by nations and individuals.

Our environmental "toolbox" already contains a whole range of solutions. Depending on circumstances, we can consider technology-based solutions, voluntary instruments for business, tighter regulations, more stringent use of assessment procedures, and so on. A recent feature of environmental solutions is their international harmonization, relying on common methodologies and protocols, and global databases of basic knowledge. Standardization is partly driven by globalized business that wants the same approach wherever it operates. Environmental specialists at local level also find it useful to share experience and research through common networks, and action frameworks.

Faced with so many potential solutions, how are we to select the most appropriate instrument for the task ahead? Answering this question requires a sense of strategy if we are to choose approaches that perform well at the local level but also give good results in the larger global framework of environmental action. In many instances it is actually the framework which will determine the solution.

This chapter shows how approaches and initiatives are applied. However, to get a full picture we first need to understand their context, and to do that we must return to the roots of environmental work and trace its evolution to the present day.

The History and Context of Environmental Solutions

Environmental management is still a young profession. Its first challenge, during the middle part of the 20th century, was to deal with impacts that were obvious to the eye and serious enough to demand urgent action. The immediate focus was on what we now call point-source pollution, often originating from heavy industry. But it was also necessary to deal with problems such as the destructive London smog, the release of untreated sewage into inland waterways, and the uncontrolled dumping of chemical wastes.

Successes in dealing with these problems were undeniable. However, more dilute or dispersed waste streams came into sharper focus once the major point sources were under control, and there was a gradual realization that polluters also included average citizens driving their cars or farmers spraying their fields. There was a realization that many of the early, and costly, "solutions" simply moved pollutants from one environmental compartment to another, and the work had to begin all over again. In fact, economic concerns quickly came to the fore. Cost-effectiveness fell rapidly when it came to addressing the more dilute or dispersed streams. Where the polluter could not pay but had political muscle, exemptions to environmental regulations were often made and pollution continued unabated.

Conventional waste treatment solutions soon hit the steep part of the curve of diminishing returns and the search began for more sophisticated approaches based on other concepts, and also involving other actors. There was more emphasis on changing polluting practices, on partnerships and on incentive-based approaches to supplement the existing arsenal of point-source regulations and standards.

The scope of the problems was also found to be broader than had been expected. Pollution does not respect frontiers, and emissions released in one country often impact its neighbors or, in some cases, the entire planet. Transboundary problems such as long-range air pollution and the still-current practice of exporting wastes to other countries began to feature in our daily news.

Additional control programs were proposed, rooted in a more life-cycle view of materials and products and in a consensus that it is more efficient to tackle the driving forces of a problem than to deal with the final impacts. There was more talk about "preventive" or "upstream" action, where the key actors are often mainstream business or development professionals but may also be ordinary citizens in the form of consumers. Given that many of the driving forces have their origins in globalized trade and consumption patterns, trade measures became a more visible part of environmental policy.

As the costs of environmental programs increased it became necessary to reconsider the objectives. Is zero pollution a realistic goal? We had long been accustomed to avoid this question, sheltering behind the convenient standards promulgated by government authorities. These, however, deal principally with the environmental quality endpoints and so gave little guidance when work started on the upstream driving forces. Additional management criteria, objectives and goals were needed to guide the future programs under construction. Cost-benefit calculations became an integral part of environmental policy.

The environmental professional thus became increasingly confronted by competing or even conflicting environmental objectives, made even more complicated by the multifunctional character of the ultimate goal, sustainable development. Here, global guidance was sadly lacking; our objectives have evolved faster than the development of new methodologies and policies. Even nearly 15 years after the Earth Summit in Rio de Janeiro there are still few guidelines on how to reconcile environmental quality with economic performance and social welfare. The internationally agreed on Millennium Development Goals, the multiple objectives of the World Summit on Sustainable Development (WSSD) in Johannesburg 2002, and even the comprehensive Agenda 21 all suggest useful goals. But they deal with lists of single issues and give no hint of how these can be integrated during the implementation process. We found ourselves, quite suddenly, with a host of goals and a full toolbox, but with no guidance manual to tell us how to put them all together in a coherent manner.

Innovative solution frameworks tried to fill the void. Holistic models based on life-cycle material-flow concepts seemed to offer the necessary framework within which individual solutions could be applied. Rather than focus on problems per se, the new approaches would deal with the driving forces. Instead of dealing with waste disposal issues, there would be an attempt to optimize global resource flows in line with the capacity of the planet's ability to provide. There was much excitement at the prospect of these holistic approaches finally providing the global solution called for by the doomsday reports.

The search for more effective approaches did not necessarily obviate the need for simple emission standards and prescriptive regulations. Their relatively limited field of application was counterbalanced by their relative simplicity of operation. They have also remained in service for other reasons. The new alternative approaches were not well understood by established experts and environmental officials, many of whom lacked experience and contacts in the mainstream sectors where the new instruments are applied. The fact that traditional solutions are strongly technology-based while the newer approaches look for organizational and management solutions is also an important factor.

At this stage in the new century the environmental professional is indeed still only "half-way to the future" (Tellus Institute 2002), selecting from a host of methodologies and tools, placing them in one or other of a variety of competing holistic frameworks, and often proceeding with inadequate data. Solutions must solve today's problems at the local level but must also make sense in a larger international framework that considers future generations and global issues as well. The following sections look at each of these issues in turn.

Solution Concepts

The environmental professional's tool box today contains an impressive array of methodologies, techniques and instruments, with an astounding level of similarity across regions and sectors. We have seen that countries also begin to experiment with new approaches once the limitations of the traditional "command and control" approach become more evident.

The multitude of environmental solutions is nevertheless based on a relatively small number of concepts that are combined and synthesized in different ways and often showing a high degree of synergy. Although we still lack a "Grand Unified Theory" of environmental action, custom-made solutions can now be created for most situations.

Some of the important concepts which underpin contemporary environmental solutions are summarized below:

Reducing pollution and conserving natural resources by means of a **recycling economy** where wastes are systematically collected, processed and re-used at the national level. Design for recycling improves the possibilities of products to be disassembled and converted. Reverse distribution improves collection. Manufacturing based on recovered materials improves the market for recovered materials. Japan is one country that has adopted a comprehensive 3Rs policy of recover, recycle, re-use to bring waste materials back into useful service (Inami 2001). While bringing undoubted benefits at a local level, such a policy deals less well with the impacts from resource production in far-away places where resources are mined or with the return of end-of-life products from the global consumers. Trade in waste across national boundaries, even for recycling, is politically unpopular. And the trend to smaller and lighter products, and greater use of composites, remains an obstacle for future recycling economics. Simple recycling will always be an important component of environmental policy, but needs to incorporate other elements before it can address environmental impacts in a holistic fashion.

Industrial ecology (Frosch 1989) is a more holistic approach to materials management (including waste recycling) than we have at present. Its goal of providing synergies among various production units—industrial symbiosis—leads to an optimization of the overall system rather than simply improving each individual process stage. It requires an appropriate management framework for such collective solutions to be applied.

See International Society for Industrial Ecology (www.is4ie.org). For a complex system of mutual resource exchange in a heavy industry complex in Australia (see Figure 14.1) (http://cleanerproduction.curtin.edu.au/pub/2004/ieinpracticeinkwinana.pdf). The other classic example of this type of industrial symbiosis is the industrial complex of Kalundborg in Denmark (http://www.indigoev.com/Kal.html).

The fact that the solutions based on industrial ecology generally involve more than one company is a significant barrier in our competitive society. The increasing practice of corporate supply-chain management is helping to overcome this barrier insofar as it is applied to noncompeting segments of the life-cycle. Collective action by potential competitors, i.e., horizontally across a particular sector, is rather less evident although some examples are known in co-operation as in sector-wide R&D for example. The "zero emissions initiative" of Guenter Pauli[1] is an excellent example of applied industrial ecology category. The cradle-to-cradle idea of Michael Braungart (http://www.mcdonough.com/cradle_to_cradle.htm) similarly depends on collective action to achieve higher degrees of resource sustainability than we normally see. China has incorporated some of the above ideas into its Circular Economy concept (http://www.chinadaily.com.cn/english/doc/2004–10/02/content_379464.htm) that has now been adopted as official government policy. The city of Kawasaki, in Japan, has combined the ideas of industrial ecology and recycling economy to form a new approach to regional redevelopment (www.city.kawasaki.jp/28/28sangyo/home/ecotown/ecoen.htm).

Dematerialization (Weizsäcker 1997) refers to a deliberate attempt to reduce the energy and materials intensity of our lifestyle to relieve the environmental stress from our production system. The term *factor 10* arises

1. http://www.zeri.org/

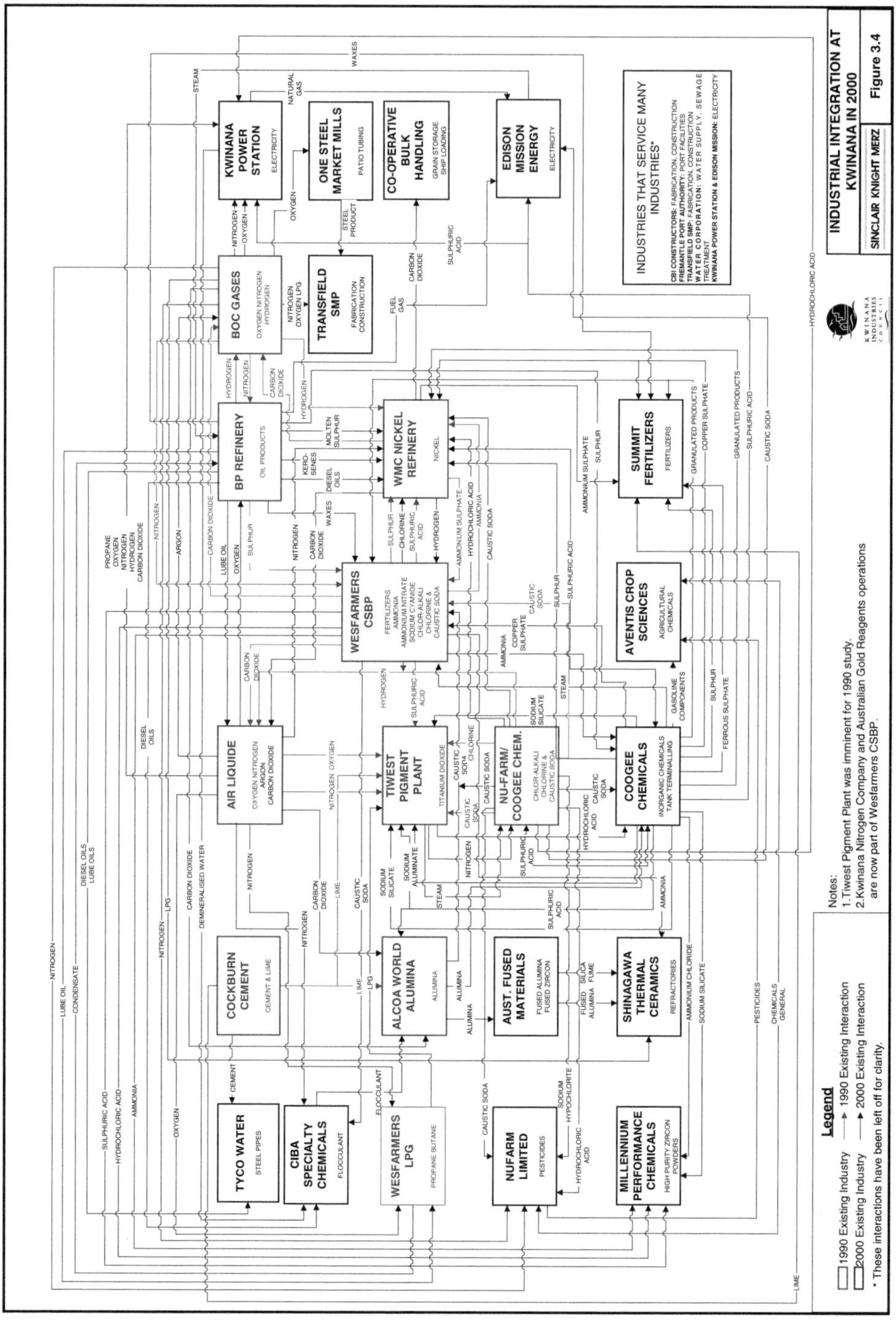

TYCO WATER
STEEL PIPES

COCKBURN CEMENT
CEMENT & LIME

AIR LIQUIDE
OXYGEN NITROGEN ARGON CARBON DIOXIDE

BP REFINERY
OIL PRODUCTS

BOC GASES
OXYGEN NITROGEN HYDROGEN

KWINANA POWER STATION
ELECTRICITY

CIBA SPECIALTY CHEMICALS
FLOCCULANT

TRANSFIELD SMP
FABRICATION CONSTRUCTION

ONE STEEL MARKET MILLS
PATIO TUBING

WESFARMERS CSBP
FERTILIZERS AMMONIA AMMONIUM NITRATE SODIUM CYANIDE CHLOR-ALKALI CHLORINE & CAUSTIC SODA

WESFARMERS LPG
PROPANE BUTANE

ALCOA WORLD ALUMINA
ALUMINA

TIWEST PIGMENT PLANT
TITANIUM DIOXIDE

WMC NICKEL REFINERY
NICKEL

CO-OPERATIVE BULK HANDLING
GRAIN STORAGE SHIP LOADING

NUFARM LIMITED
PESTICIDES

AUST. FUSED MATERIALS
FUSED ALUMINA FUSED ZIRCON

NU-FARM/ COOGEE CHEM.
CHLOR-ALKALI CHLORINE & CAUSTIC SODA

EDISON MISSION ENERGY
ELECTRICITY

MILLENNIUM PERFORMANCE CHEMICALS
HIGH PURITY ZIRCON POWDERS

SHINAGAWA THERMAL CERAMICS
REFRACTORIES

COOGEE CHEMICALS
INORGANIC CHEMICALS TANK TERMINALLING

AVENTIS CROP SCIENCES
AGRICULTURAL CHEMICALS

SUMMIT FERTILIZERS
FERTILIZERS

INDUSTRIES THAT SERVICE MANY INDUSTRIES*

CBI CONSTRUCTORS: FABRICATION, CONSTRUCTION
FREMANTLE PORT AUTHORITY: PORT FACILITIES
TRANSFIELD SMP: FABRICATION, CONSTRUCTION
WATER CORPORATION: WATER SUPPLY, SEWAGE TREATMENT
KWINANA POWER STATION & EDISON MISSION: ELECTRICITY

Legend

1990 Existing Industry — 1990 Existing Interaction
2000 Existing Industry — 2000 Existing Interaction
* These interactions have been left off for clarity.

Notes:
1. Tiwest Pigment Plant was imminent for 1990 study.
2. Kwinana Nitrogen Company and Australian Gold Reagents operations are now part of Wesfarmers CSBP.

KWINANA INDUSTRIES COUNCIL

INDUSTRIAL INTEGRATION AT KWINANA IN 2000

SINCLAIR KNIGHT MERZ | Figure 3.4

FIGURE 14.1

from simple observation that the earth's resources could only sustain a high-consumption lifestyle for about 10% of the world's population. Major dematerialization of our products and services will be necessary if sustainable development is to reach all parts of the globe. As a first approximation a factor of 4 improvements in industrial resource use could be achieved by technology alone—if extensively and universally applied. The remaining improvements would require changes also in consumption patterns, social organization, personal values, and so on. Dematerialization is a conclusion of the study of a situation—it makes no proposals how solutions should be achieved, and it is unfortunate that some skeptics have imagined authoritarian coercive approaches behind the calculations. Major dematerialization results are already being achieved by industry in many products as the business benefits of this approach become evident.

Upstream action through **cleaner production**,[2] pollution prevention, waste minimization and in-plant recycling, i.e., preventing wastes from being produced in the first place seems a logical approach, so logical that we can ask why it is not the basis of mainstream action on environment.

> Cleaner Production is the continuous application of an integrated preventive environmental strategy to processes, products, and services to increase overall efficiency, and reduce risks to humans and the environment. Cleaner Production can be applied to the processes used in any industry, to products themselves and to various services provided in society. (UNEP's Cleaner Production, http://www.uneptie.org/pc/cp/understanding_cp/home.htm).

From a process point of view the main elements of cleaner production can be seen in the diagram below. But as the focus of cleaner production has now broadened beyond industrial processing wastes to include also other elements of the product cycle, it is necessary to also look at the consumption agenda (see Figure 14.2).

Despite the inherent reasons why we continue to adopt "cure rather than prevention" action is amply explained in the publications emanating from such programs. An important contributing factor is that prevention action is best taken by "mainstream" production managers and ordinary consumers, and is less effective when led by environmental specialists and agencies. There is thus resistance in both the production and the environmental constituencies to such an approach, albeit for different reasons. The major challenge is organizational and managerial, not technical.

The term **sustainable consumption** rose out of the debate on Agenda 21 and subsequently became an integral part of WSSD. It is based on a con-

2. Industry and Environment, Vol. 25, No. 3–4: Cleaner Production—Seventh International High-level Seminar, Prague.

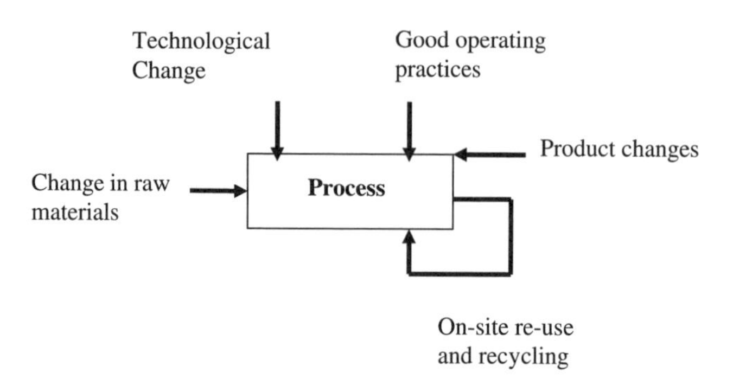

FIGURE 14.2. Process elements for cleaner production options.

sideration of global carrying capacity and our currently excessive demand on resources. It articulates a need for different patterns of consumption (both qualitatively and quantitatively) if sustainable development is to be achieved. It is important to stress that sustainable consumption does not demand a reduction in living standards, often the reverse is true. Rather it suggests a clearer reflection on what we need to consume to achieve our personal aspirations and environmental ambitions. Reducing waste is an important first step, and many small changes in personal behavior can achieve significant economic and environmental gains, e.g., temperature regulation of homes, more sedate driving habits, avoiding overpurchasing, and so on. There is also a component of product choice. In many situations the major environmental impact occurs during the use phase rather than manufacturing, as for example motor vehicles, buildings, many appliances. Finally, end-of-life equipment is becoming an increasing environmental issue in our society, with the benefits and costs of short-life products unequally shared. Sustainable consumption is still a young concept, and will evolve further as its relevance becomes recognized by a wider stakeholder base. Some components of sustainable consumption action include

- Reduce easily avoidable waste and resource consumption
- Eco-design of products and processes
- Green procurement by organizations and individuals
- Product-service systems
- Green consumer guidelines
- Collection and reverse distribution of end-of-life products

Already we are seeing some visible action in for example corporate and government green purchasing (procurement), eco-design of products, sustainability performance criteria, product service systems and take-back schemes. Consumers and companies are also becoming more conscious of waste reduction opportunities in day to day activities.

Life-cycle management (LCM) is based on the observation that various environmental impacts are expressed differently along the materials or product chain, and that the most effective action is often not the point at which the impact occurs. In many cases various actors along the life-chain need to jointly agree on a solution, as for example to control potentially hazardous exposures, to ensure return of end-of-life or off-specification materials, and to influence the production chain to avoid emissions and wastes later on. Some individual producers are already exercising extended producer responsibility along the life-chain (www.oecd.org) as far as their products and services are concerned. Environmental supply-chain management (www.capresearch.org/publications) for incoming raw materials and subcontractor services is also becoming more common.[3]

More comprehensive LCM is an exercise in collective (not to say cooperative) management by a group of stakeholders. In a globalized society LCM will often involve action across national boundaries. Some of the known examples are industry-led as in the recently created international management code for cyanide that links manufacturers, suppliers, transporters and end-users (see www.cyanidecode.org). Such codes are intended to slow the accelerating trend to the adoption of substance bans along the supply chain. For example the Tiffany corporation announced (http://www.professionaljeweler.com/archives/news/2004/110804story.html) that it will not sell gold metal that was produced by cyanide-based processes, nor to market "conflict diamonds." The Kimberley Process (http://www.kimberleyprocess.com:8080/site/) to track the origin and distribution of gem-quality diamonds was created to avoid the terrible reputation of conflict diamonds in tarnishing the image of ornamental gems. The Green Lead Initiative (www.greenlead.com) is a further example of industry itself adopting LCM to try to ensure a long-term future for its products (Fig. 14.3).

There are other examples in the palm oil industry (www.sustainable-palmoil.org) and other commodity sectors. LCM is a multi-stakeholder management solution where technology plays a part but where organizational commitment is the first stage of action.

Most of the solution concepts above can still be defined in some way as problem solving since the objectives are simply the negation of a known undesirable impact. The Natural Step (TNS) (www.naturalstep.org) is one of the few approaches that starts from a fixed desirable environmental end-point and then uses "back-casting" to tell us what we need to do to get

3. Many chemicals added during the growing of foods or the processing of materials are expelled during subsequent refining stages elsewhere, for example, pesticides used in hide preservation of raw hides being washed out at the tanneries.

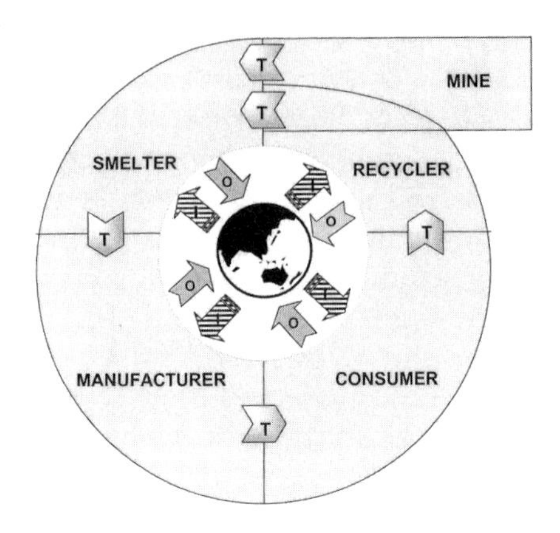

FIGURE 14.3. The Green Lead cycle.

there. The objectives of TNS are to ensure environmental health and well being of people by ensuring that our activities remain within the carrying capacity of the ecosystem. It advocates that we should shape our society by taking from the earth only what it can sustainably provide, and not to put back substances that it cannot assimilate. Implementation of TNS uses some of the holistic life-cycle approaches outlined above.

Functional approach is a non-branded term proposed by UNEP as a response to the WSSD agenda on sustainable consumption and production. This is another endpoint-based approach that uses as its starting point the basic needs of people, and then looks at the best (and most pollution-free) way of fulfilling them. Basic needs are generally agreed to include food, shelter, clothing, health, mobility, leisure, social interaction, but the exact mix of needs and wants is a function of contemporary culture. It remains to see how to fulfill them with the optimum mix of technologies, products and services (i.e., product-service systems, PSS) that respects the environment as well as social constraints along the supply-chain. The functional approach has already been used by some leading companies to define their future product profiles, but has not been applied at the level of entire societies or social groups. Experience will eventually show whether the functional approach concept can be converted into a mainstream decision-making instrument.

The various solution concepts above are not independent, and can be combined in different ways to suit the needs of governments and business, and to reflect national circumstances and cultures. We should nevertheless be mindful of "partial solutions" that make a certain contribution but when institutionalized may actually block the path to more holistic and integrated approaches. For example, recycling often mesmerizes its proponents to such an extent that they ignore to the need to also implement other components of preventive environmental action.

In implementing the above approaches there is a need to break away from the past pre-occupation with wastes and to focus more on management of materials and resources (Hawken et al. 2003). Several prominent international organizations—including the International Solid Waste Association (ISWA) (www.iswa.org) and the International Council on Mining and Metals (ICMM) (www.icmm.com) have already adopted a position on "integrated materials management" rather than dealing exclusively with wastes. Increasingly the debate in international conferences is shifting from residue treatment towards more holistic resource management. A life-cycle approach can give more visibility to environmental issues related to resource extraction to balance the traditional focus on waste disposal.

The greater focus on upstream management leads us to a redefinition of the traditional waste action hierarchy to develop a new pyramid (Figure 14.4) based on product considerations:

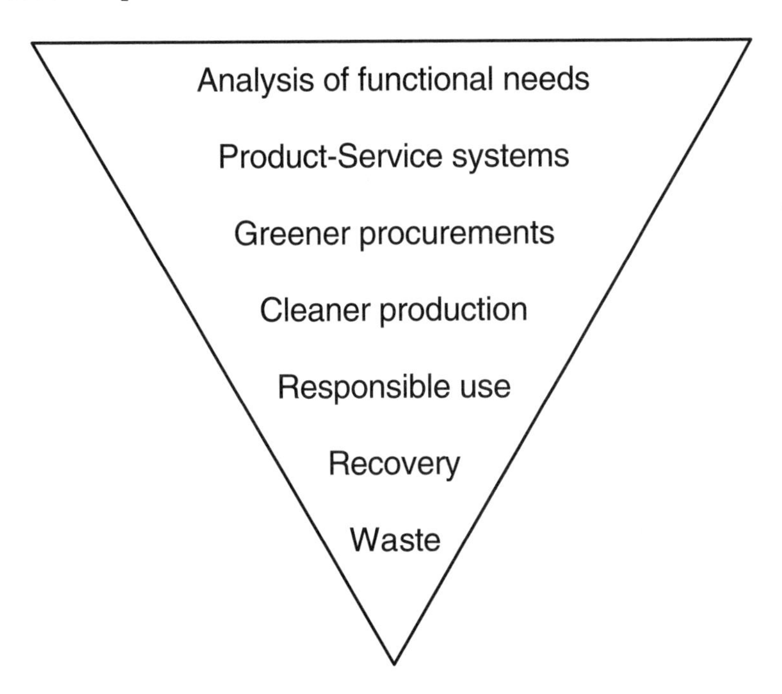

FIGURE 14.4. Product Pyramid.

Such a hierarchy defines clearly the contributions of various mainstream actors and professions to sustainable development, and puts the needs of the future consumer rather than the waste producer at the top. In this model the environmental profession plays the role of co-ordination and information, and in the provision of sophisticated data, assessment and evaluation services. Traditional waste management is an element of the above, with the usual waste hierarchy a valuable guide when dealing with the lower half of the product pyramid.

The next sections look at some practical instruments that can help to put these ideas into practice.

Solution Tools

The previous section described how it is important to choose an appropriate environmental strategy for environmental solutions; this section examines some of the tools that can be used in the framework of such strategies.

The Role of Technology and Techniques

In the consumption-production pyramid above it is especially the production, use and residue management stages where technologies play in a major role. Some of the most important applications are discussed below, following the more traditional order of the waste management hierarchy which finds a useful application at this level in the pyramid.

Clean Technologies and Technology Assessment

Insofar as it is better not to pollute in the first place than to efficiently deal with waste, cleaner production technologies (CTs) have a high priority in environmental technology development. Many modern production processes are already inherently cleaner than their traditional precursors, and naturally find their way into use whenever new manufacturing plant is installed. While some newer technologies are more expensive, they soon pay for themselves in terms of higher output, reliability and product quality, and more efficient use of expensive raw materials. More efficient technology is an important component of the cleaner production approach.

A change to clean technologies is still largely a business affair, motivated more by fiscal instruments like taxation, write-off times, removal of subsidies, and investment promotion than by environmental legislation, although tighter standard do play some role in deciding the moment of new technology investment. Any instrument that encourages a modernization in manufacturing plant will also increase the uptake of cleaner technologies.

Much of the current industrial pollution problem is due to old equipment still in use. A rapid industrial turnover is thus beneficial from an environmental point of view. More attention also has still to be paid to the practice of "recycling" older equipment to poorer countries when facilities in the West are modernized.

Their obvious benefits notwithstanding, the complexity of the environmental agenda now makes it important to understand the full dimension of all technologies. A new process will not necessarily have the same high performance on all environmental criteria. The Environmental Technology Assessment (EnTA) (http://www.unep.or.jp/ietc/Publications/

Integrative/EnTA/AEET/index.asp) was designed as an evaluation procedure to allow a fuller understanding of the implications of a technology, just as LCA studies the complete impact of a product or material. Some degree of environmental evaluation is often already implicitly carried out for specific equipment choices by both the provider and the purchaser, however, a formal assessment procedure ensures that all aspects are systematically considered. EnTA also has a useful application in the study of generic technology types in order to understand the long-term implications of radically new product systems brought to market, e.g., hydrogen fuels, GMOs, and so forth.

Intrinsically Safe Processes (ISP) and Risk Reduction
Post-event (i.e., retrospective) risk reduction is expensive, and an upstream prevention approach through "intrinsically safe technologies" and "green chemistry" concepts are increasingly being applied to new investments. The dramatic disasters of Bhopal and Exxon Valdez among others demonstrate the limits of add-on safety technologies which can eventually become as much as an encumbrance as asset. More effort is now going into risk reduction as industry and governments ask themselves serious questions about liabilities, legacies and risk management. ISP (see Zwetsloot in http://www.uneptie.org/media/review/vol25no3-4/I&E25_34.pdf) is a natural companion of cleaner technology, as upstream hazard reduction tasks a similar path to that of earlier environmental pollution considerations.

The features that lead to intrinsically safe technologies include:

- Reduced use of unstable, toxic, or highly hazardous chemicals
- Avoidance of high pressure/high temperature processes
- Stable reactions
- Autonomous shut-down in safe mode
- Reduced storage, handling, and transfer of reagents inside and outside the plant
- Fewer, simpler materials transfers, reduced need for pipes, pumps, and valves

ISP goes hand in hand with the recent concept of green chemistry (http://www.rsc.org/is/journals/current/green/greenpub.htm) and its preference for safer, lower impact chemicals, processes, and products.

Add-on safety retains an essential place in a larger policy of intrinsic safety and preventive mindsets. It is still an important professional area for industry and government professionals, relying on sophisticated monitoring, and engineering devices. In addition, planning to cope with irregular events eventual accidents remains a "must" and well-rehearsed **emergency response** systems at industrial plants continue to be needed. This is more an organizational and communications challenge than a technological one.

Health and safety (H&S) systems are notoriously difficult to make work in all countries, the more so in poorer countries where risk is omnipresent and a safety mentality has no deep roots in local culture (how long did it take in the West?) In the implementation of an industrial safety culture, the work of the DuPont stands out, and this company has a thriving consulting business to share its learning experience with other industries and sectors (www.dupont.com).

It can be observed that in initiatives to achieve higher in-plant safety, **community safety** is often overlooked. A number of serious industrial accidents in the chemicals, mining, and transport sectors have led to impacts also in the public domain with occasional serious loss of life offsite. The Stava mine collapse in 1985 killed 268 villagers in Italy; the Bhopal chemical plant in India over 3000; the fertilizer explosion in Toulouse, France, over 30. A recent gas-well blowout in China killed several hundred nearby people. Risk reduction in the process industries is important, but so is a higher level of *a*wareness and *p*reparedness for *e*mergencies at *l*ocal *l*evel (APELL). In 1988 an international consortium led by UNEP prepared written guidelines on how to involve the community in the preparation of official emergency response plans. (See APELL: A Process for Responding to Technological Accidents; available at http://www.uneptie.org/pc/apell/publications/apellmanual.html.) Like H&S, APELL requires attention to communication, information, trust-building and risk monitoring but this time focused on civil society rather than workers. The tools are transparency, patience, attentiveness rather than public relations and sophisticated data. Here the environmental scientist turns sociologist and communicator.

Waste Treatment Technologies, Site Remediation

Often simply termed "environmental technologies,"[4] **waste treatment processes** are a group of industrial technologies employed to ensure that wastes that are already produced do not become (or remain) a nuisance. Like most production technologies, they also generally produce a further secondary residue stream which must be treated in turn, and so down the line.

The desirability of cleaner production notwithstanding, waste treatment is still a high priority in environmental programs, and will remain so until the backlog of inefficient processes and the legacy of widespread polluted environments is behind us. Waste treatment has the inconvenience that it, unlike cleaner production, it incurs a cost with little economic return. Their high capital cost and need for ongoing operation ensures that

4. We would like to think that the best environmental technologies are those production processes that produce no waste.

treatment plants are little used in developing countries, and even when installed, they may not actually be functioning or perhaps not even switched on. There is also the problem of what to do with treatment residues which are costly to dispose of, provided always that disposal facilities actually exist. There is no free lunch when it comes to waste treatment.

The situation concerning **remediation of polluted land** is even more serious. Unless required by legislation, few sites would be remediated today as the cost of clean-up often far exceeds the value of the land. This applies even to Western countries which have taken a slow and expensive path to dealing with their pollution legacy. Remediation techniques are invariably expensive, complicated and protracted, and have inherent risks to personnel working on such sites and even the outside community.[5] Few countries have adequate facilities to receive the large volumes of contaminated soil that many such sites produce. Development of simpler, safer and more cost-effective remediation technologies is one of the outstanding challenges of the waste management world. A review of some current technologies is available at Environment Canada (www.on.ec.gc.ca/pollution/ecnpd/tabs/tab22-e.html).

Environmental Techniques

Adequate **environmental data** is essential to allow the above tools and techniques to be employed. At the site level, **environmental monitoring** and audits will bring up the data needed for local decision-making. Some data may be made available through corporate environmental reports or via mandatory emissions reporting. Extensive monitoring of environmental quality on wider national or regional scale through defined campaigns allows "state of environment" reports to be prepared that can give action programs a scientific basis. A number of countries and regions now publish regularly such reports to inform all stakeholders of the progress in environmental work. At the global level such data is compiled in the Global Environment Outlook (GEO) (http://www.unep.org/geo/geo3), which is published periodically by the United Nations Environment Programme. A number of other data compilation mechanisms at the national level rely on self-monitoring data (usually required under regulation), such as the Toxic Release Inventory (TRI) Program (http://www.epa.gov/tri) in the U.S. TRI has encouraged similar programs in other countries (see http://www.

5. No injuries were reported during the dramatic spill of mine waste from the Aznalcollar mine in Spain in 1998. However 5 fatalities occurred from road accidents during the cleanup phase which involved increased traffic of heavy machinery on rural roads. www.boliden.com

mapcruzin.com/globalchem.htm#country) and several international programs have drawn inspiration from it (see http://www.chem.unep.ch/prtr/default.htm).

Monitoring data are relatively easily compiled although quality assurance remains a challenge. Conversely, data requirements for **life-cycle assessment** are more demanding in view of their complexity and diverse origins. The SETAC/UNEP Life-cycle Initiative (LCI) has been established to improve the data sets and assessment procedures underlying international LCA practice (http://www.uneptie.org/pc/sustain/lcinitiative/home.htm). With better data, the growing practice of life-cycle management will become easier.

As not everything can be measured, **environmental modeling** has assumed an increasing importance in environmental work, helping to focus attention on the most important features of a particular environmental situation. Climate change, long-range atmospheric pollution and oceanographic modeling have captured much public attention, but at a smaller scale many environmental professionals rely on models of various levels of sophistication to give day to day advice on how to proceed in their work. Clean-up programs for contaminated sites often rely on chemical dispersion models to determine the best way to proceed. **Scenario development** has become popular to investigate sustainable futures and the policies and actions needed to attain them. The Tellus Institute examined models labeled "Conventional Worlds," "Fortress Worlds," and "Great Transitions" (see http://www.tellus.org/index.asp). Some aspects of these were eventually also incorporated also in UNEP's work. Many other organizations and companies have engaged in scenario development to help orient their future priorities and programs.

A variety of specialized techniques has been developed over the past few decades to assist professionals to analyse and evaluate the environmental problems revealed by the data compilations. **Environmental assessments** in particular have attained a degree of sophistication that now requires specialized training and guidelines for effective use. To the earlier Environment Impact Assessment (EIA) (http://www.iaia.org/eialist.html) have been added life-cycle assessments of products and materials (LCA), Environmental Technology assessments (EnTA) (see earlier), site assessments and audits, and a number of other more specialized tools. Training material on EIA has been prepared by UNEP for worldwide use (http://www.unep.ch/etu/etp/index.htm).

Of particular importance now are also procedures to help **prioritize environmental problems** to allow a focused cost-effective approach to dealing with them. The identification, evaluation, and prioritization of contaminated sites is a particular example of this. Similar prioritization techniques have also been applied by large mining companies to identify which of their many sites are in most urgent need of environmental and security upgrading.

Management Tools for Environmental Solutions

Upstream preventive work relies heavily on influencing corporate and consumer behavior. Environmental management tools are thus designed to exert an influence on organizational structures, procedures and practices. Typical element are shown in Table 14.1.

As the management of complex environmental affairs requires the same systematic, all-encompassing approach as any other organizational functions, the 1990s saw the development of formal **environmental management systems** (EMS). To ensure uniform application these were soon standardized, first in the UK, and then through the International Organization for Standardization (ISO) as standard no. ISO 14 001. The European Union has extended this approach through its EMAS guideline (http://europa.eu.int/comm/environment/emas/index_en.htm).

EMS and EMAS are not stand-alone exercises—they represent that part of the normal management system of an organization that intends to deal with environmental issues. They draw their inspiration from the better known ISO 9 000, and can in many ways be regarded as a type of quality assurance instrument, i.e., quality of environmental management.

In order to implement an EMS, subsidiary instruments are necessary. For instance environmental audit and assessment procedures are designed to provide managers with the information needed for quality decision-making. Several such instruments have themselves been standardized under the 14000 family of standards of ISO. Protocols for other instruments such as reporting (a requirement in most EMS) were also developed under other frameworks such as GRI (www.globalreporting.org).

The range of environmental management tools is now quite large, and often demands specialist skills and working experience in its application. The group of *assessment tools* was mentioned earlier, and is most closely associated with traditional views of environmental action because they deal explicitly with environmental information. Several upstream *action tools* are much more "management" focused, such as environmental (life-cycle) supply-chain management, product service systems, green procurement and so on. A third group of instruments can be loosely termed *communication tools* since they aim to manage a two-way flow of information between decision-makers, partners, stakeholders, clients and by-standers. A review of environmental management tools was published by UNEP in 1995 (Hawken 2003), and while no longer complete in some respects, it still provides a good background overview. A basic list of useful instruments is shown in the table below. Since 1995 many more guideline documents have been published and there is now a tendency to put such information online to improve access and updating. See for example:

1. UNEP Environmental Management Tools (http://www.uneptie.org/pc/pc/tools/index.htm)
2. The Environmental Management Navigator (www.em-navigator.net)

3. A managers guide for the oil industry (http://www.oilandgasforum.net/oefonline)
4. A range of corporate tools and approaches (http://www.gemi.org/docs/PubTools.htm)
5. A recent and forward-looking compilation (Füssler 2004) was prepared by the Global Compact and WBCSD; however, this is not available online at present

The greater sophistication of environmental management instruments has resulted in further reflection on their relationship with environmental regulations, both traditional and the more recent performance-based variants. The industry now believes that it is better equipped to itself deal with certain aspects of the environmental solution, most obviously those that affect management. This combined with the re-thinking of regulatory approaches in recent years has led inevitably to the drive for more effective self-governing *codes by industry*, see for example "Voluntary Industry Codes of Conduct for the Environment" (www.uneptie.org/Outreach/vi/pub_codes.htm). The best-known (and still arguably the most effective) is Responsible Care (www.americanchemistry.com) in the chemicals industry.

Such codes are not "self-regulation" as sometimes claimed by both proponents and detractors. The codes are a supplement, not a replacement, to legal obligations, and are unworkable in the absence of clear policy frameworks and targets by the authorities. Thus a number of codes go considerably further than simply advocating legal compliance by advocating corporate environmental responsibilities and targets in line with contemporary civil society expectations. Some more recent instruments devised by industry include the International Cyanide Management Code which links producers, transporters and users of cyanide in the mining sector (www.cyanidecode.org), the Equator Principles in the banking sector (www.equator-principles.com), and the sustainability vision of the palm oil

TABLE 14.1
Common Environmental Management Tools

- Environmental Assessment
- Environmental Monitoring and Audits
- Environmental Management Systems
- Supply-Chain Management
- Extended Producer Responsibility
- Product Service Systems
- Green Design
- Green Procurement
- Environmental/Sustainability Reports

Source: OEF Online www.oilandgasforum.net/oefonline and others.

industry in Southeast Asia (www.sustainable-palmoil.org). Voluntary codes are still weak in the upstream business sectors such as advertising, marketing, and purchasing. It should be noted that such codes can only be considered as (collective) management tools when they are accompanied by a formal action program to ensure application. The cyanide code for example has established an institute to administer the implementation, overseen by an international multi-stakeholder board of directors. With a more extensive and rigorous follow-up implementation of voluntary codes they may at some future time take their place among the formal groups of management instruments accepted by society at large. This may however depend on a closer linking with regulatory instruments, as shown by the experience in the Netherlands where the system of *industry covenants* has been accepted only because of the clear threat of government standards if industry did not itself make a binding commitment to "voluntary" targets (www.greenplans.rri.org).

Regulations and Codes

From the early beginnings of **command and control standards** and regulations, (still necessary in order to maintain a minimum acceptable level of performance), regulatory policy has added increasingly sophisticated mechanisms.

In the first instance "smart regulations" (Gunningham 2002) try to deal with those circumstances where traditional standards have reached the limit of their effectiveness, and where conventional enforcement approaches cannot cope. This approach leads to for example environmental quality standards rather than emission standards, negotiated agreements (either individual or on a sector basis), creating a market for emissions trading schemes, and so on. An interesting and powerful example is the obligatory reporting of emissions or of non-compliance. This often leads to major efforts by industry to improve environmental performance in order to avoid adverse publicity or even litigation.

There has also been a gradual trend to require the use of certain environmental management tools as a condition of permits, or of an operating license. The mandatory use of EIA for major projects was an early example of this. The use of EMS as a way of easing the burden of government inspection and enforcement is also increasing, as for example in the offshore industry in Norway (http://www.oilandgasforum.net/management/regula/nationalprofiles.htm#norway). In many places there are legal requirement to undertake certain types of site or risk audits prior to property transactions. Companies and financing institutions already routinely require environmental liability audits prior to major contracts or corporate mergers.

Economic instruments have been advocated as overcoming many of the limitations of conventional standards. Table 14.2 lists a number of

TABLE 14.2
Economic Instruments in Environmental Policy

- Emissions Trading/Tradable Permits
  — Acid Rain program, lead reduction credits, in US. Fisheries ITQs in New Zealand, etc. UK emissions trading for CO2.
- Emissions taxes, Pollution levies
  — Water pollution charges, NOx tax (e.g., Sweden), Landfill Taxes
- Eco-taxes on products
  — Motor fuels, Plastic bags, retail pesticides, NiCad batteries
- Deposit-Refund systems
  — bottles for soft drinks and beer, motor vehicles
- Abatement subsidies

From: S. Smith, University College, London.

instruments that have come into use in various parts of the world. All of them are implemented by regulation.

As environmental policy looks further upstream and regulations try harder to address the drivers of environmental change, they focus more on influencing values, management practices, fiscal regimes and consumption patterns than environmental parameters per se. Thus product legislation is often invoked for environmental reasons through for example product specifications (e.g., CFC-free, dolphin-friendly tuna), product bans (e.g., DDT pesticides, lead glazes, or asbestos), product take-back requirements (or its variant, product deposit schemes) on, for example, batteries, tires, or packaging, and so forth. Legally mandated labeling on certain chemical products now may include mention of procedures for waste disposal.

The environmental impact of other legislation is often overlooked in discussions of environmental laws, and yet is increasingly important if an upstream approach to action is to be effective. The legal foundations of economic and physical planning, natural resource development, energy, agriculture, transport and urban development, public health and indeed almost all areas of human endeavor strongly influence our environmental futures, and cannot be easily rectified through an add-on layer of environmental legislation after the fact. In theory Strategic Environmental Assessment (SIA) (http://www.strath.ac.uk/Departments/GSES/research/initiative2.htm) can be used to reveal the relationships, and it is unfortunate that it is not more used at the highest levels.

Conventional environmental impact assessment is more commonly applied, most usually to a specified group of projects comprising either certain listed activities, or a minimum size of fiscal investment (or both). Much variation exists in the way civil society is involved in the EIA process and the degree of transparency in the final decision making. Regrettably EIA is frequently seen and applied, by government and industry alike, as a procedure simply to obtain a permit rather than as a project optimization tool for the design and operation of the activity proposed (www.uneptie.org/pc/pc/tools/eia.htm).

There is now much talk about **international environmental law**. In fact, we should remind ourselves that international agreements are almost always activated through national legislation, thus giving countries the maximum flexibility in how to formulate the implementation programs.

The effective application of all of the above requires clear environmental targets and end-points. Agreed policy goals and environmental standards still remain a basic requirement.

Command and control approaches will not become obsolete anytime soon and will remain an important element of a more holistic approach to environmental legislation. We can thus end this section with a brief look at the **environmental standards** that provide the target for many compliance-based solutions. While emission standards may change as technologies improve, the ability of the environment to absorb wastes and pollutants is generally fixed. Much work has gone into developing ambient standards, and standards from one country tend to be adopted by others, but it has to be admitted that much depends on local factors and there is no substitute for doing one's own homework. National standards, let alone international, standards are best reserved for circumstances where there is little variation, as for example human health standards. In fact even the WHO human exposure standards are recommendations rather than legal limits. The box below gives references to some of the better-known international standards or criteria. National equivalents are even more numerous and can be found from appropriate regulatory bodies.

Chemical safety -IPCS—www.inchem.org/
WHO Drinking Water Standards—http://bookorders.who.int/bookorders/index.jsp
WHO Air quality standards—http://bookorders.who.int/bookorders/index.jsp
Workplace exposure and safety ILO—http://www.ilo.org/ilolex/english/subjlst.htm
Carcenogenic substances—IARC—www.iarc.fr/
Pollution prevention standards in industry—http://lnweb18.worldbank.org/ESSD/sdvext.nsf/05ByDocName/FrequentlyAskedQuestions FAQs
http://ifcln1.ifc.org/ifcext/enviro.nsf/Content/EnvironmentalGuide lines

International Frameworks for Environmental Action

The world is now a global machine of trade, services, finance, information, and even culture. Environmental action is shaped by the interplay of these forces as well as by its own dynamics. Nevertheless their greater preoccupation with environmental quality issues gives developed countries a natural leadership role in initiating, and in financing environmental action programs in the framework of global policies and priorities for development. Developed countries can use the existing frameworks and initiatives to catch up and learn from the often expensive environmental lessons learned by industrialized communities. International action programs also result in a more level playing field for trade and social exchange by standardizing the standards and approaches, and by providing assistance to countries in need.

This section will examine some of the international decisions and actions which define such programs and which shape the solutions we all want to see applied.

Major International Policy Decisions

A number of major international meetings over the past two decades has provided the broad policy direction and set some priorities for international action. It is important to emphasize that these events were gatherings of world leaders who arrived at the decisions after extensive discussions and negotiations. The conclusions were shaped by the need to reach a consensus between often radically different political and cultural viewpoints. Vague to some, inadequate to others, these conclusions nevertheless represent the only common global platform for moving forward. It should be noted that international assistance agencies, including financing institutions, base many of their program priorities on such outcomes.

The most influential meeting in recent times was undoubtedly the Earth Summit in Rio de Janeiro in 1992, under the formal title of World Conference on Environment and Development (WCED) (http://www.ecouncil.ac.cr/about/ftp/riodoc.htm). The summit ended with a declaration, and the enunciation of principles on forests and other subjects. But the main outcome was the adoption of a forward action plan, the so-called Agenda 21 that described a series of issues to be addressed, the major partners and their responsibilities, and gave indications of targets and objectives. Agenda 21 is still frequently the reference point in environmental debate.

The intervening years saw an evolution in vision where environment and development became more strongly coupled. Progress since Rio was reviewed in 2002 at the World Summit on Sustainable Development (WSSD) in Johannesburg (www.johannesburgsummit.org). WSSD also produced a declaration by heads of governments, and further interpreted the priorities for Agenda 21 in its Plan of Implementation. This defined relevant principles for action, identified the relevant partners and actors, and listed a number of new instruments to be used to move ahead. Prominent among these was the section on "Sustainable Consumption and Production" which called for a more upstream approach based on prevention principles. Like the earlier documents, this is now providing guidance to the lending and technical assistance priorities of international organizations, setting a trend which soon also trickles down to private sector decision-making (especially in major corporations).

Summits on other sustainability issues have also taken place. From an environmental standpoint the recent international meetings to develop a Strategic Approach to International Chemicals Management (SAICM) (http://www.chem.unep.ch/saicm/) stand out. Building on recent success with international control programs of selected substances, SAICM has the ambition of developing a global approach to chemicals management generally, based heavily on the system elements already developed by the OECD, UNEP, WHO, ILO, and FAO.

Some of the mechanisms being used to implement the policy decisions made at these events are described below.

Treaties and Conventions

Increasingly, the idea of a level playing field in a global economy requires collective, agreed action across national boundaries. **Multilateral Environmental Agreements** (MEAs, or often just referred to as international "Conventions" due to their basis in legal obligations agreed by nation states) have defined action in a number of areas where transfrontier problems demand also transfrontier solutions. The Basel Convention on hazardous waste (www.basel.int) and the UN Framework Convention on Climate Change (http://unfccc.int) are two instruments that are well known; but other agreements also cover toxic chemicals, ozone depleting substances, marine disposal of waste, coastal discharges, air emissions, biotechnology and in the ecological arena, trade in endangered species, wetland protection, world heritage sites and monuments, among others. A common approach in the pollution-related instruments is a phase-out of defined substances, restrictions in trade in wastes and other substances, encouragement of alternative technologies. Prior notification, exchange of information and a standardized set of definitions and classifications is a common element to ensure that global action can be efficiently harmonized. Signatory governments are expected to implement the agreement at national level through appropriate regulations.

In a small number of cases international jurisdictions impose binding regulations and directives concerning environmental behavior, but only where a supranational executive body has been created by member states. This is most visible in the European Union, and to some extent also the OECD with its Council regulations. Of course federal states such as the

U.S., Canada, Germany, India, and many others already make national requirements that apply at state level.

Globally agreed approaches to environmental action are also appearing in the corporate sector. Many major companies now require subsidiaries companies, suppliers and contractors to have ISO 14000 certification, to publish corporate environmental performance reports in accordance with GRI guidelines, to undertake site and product assessments, and so on.

Some International Action Programs

A number of international environmental programs have been built on the concepts, frameworks and agreements discussed so far. While these are still young, the recognition of pro-active prevention-based solutions is gradually taking hold in many regions.

WSSD called for a "**10 year framework of programs** to assist countries and regionsto adopt sustainable patterns of consumption and production." (See Appendix.) UNEP and UND DESA were designated to take a lead in bringing relevant individuals and institutions together to define further action in their sphere of endeavor. This program began a series of regional consultations to define priorities and actions, while simultaneously promoting increased use of environmental tools such as eco-design, green procurement, life-cycle management, cleaner and safer production, and energy-efficient technologies. A number of developing countries and regions are now adopting polices to promote more sustainability in production and consumption.

The **Cleaner Production** program came into its 17th year in 2005, with continued high-level acceptance at both national and international levels. Under this program UNEP and UNIDO joined forces in *1994* to create a worldwide network of National Cleaner Production Centres (NCPC), with almost 100 centers around the world, including 25 directly administered by these agencies (http://www.uneptie.org/pc/cp/ncpc/home.htm). The NCPCs undertake outreach, assessment, advisory, and consulting work for national clients on preventive approaches to pollution control. The work of the Centres is now broadening to include life-cycle management and sustainable consumption issues. Cleaner Production policy in SE Asia is also being promoted by the Asian Development Bank in order to improve the uptake of preventive approaches to pollution reduction by regional member governments. A parallel initiative, the Green Productivity program (http://www.apo-tokyo.org/gp/), is being undertaken by the Asian Productivity Organisation (APO). The Asia-Pacific Roundtable for Sustainable Consumption and Production (http://www.aprscp.org) is a regional forum for exchange of experience among professionals and administrators in the region interested in preventive approaches. There are similar roundtables in Europe and in Africa, and a number of national roundtables have been in operation for some years including several pollution prevention roundtables in North America (http://www.p2.org/ and http://www.c2p2online.com). No access.National governments, including China, are attempting to incorporate cleaner production (http://www.chinacp.com/eng/cppolicystrategy/circular_economy.html) policies in national legislation and procedures. Japan has put much effort into the creation of a "recycling society" (http://www.mofa.go.jp/policy/economy/japan/pm0112.html) based on some of the same principles.

A number of action programs have been adopted to respond to the increasing dispersion of **potentially harmful chemicals**. These are commonly agreed to include toxic heavy metals, persistent organic pollutants (POPs), endocrine disruptors, ozone depleting substances, carcinogens such as asbestos, and other popular categories. (Waste chemicals are usually dealt with under a separate program.) Industrialized countries have mostly already created programs to address these substances, although admittedly there is still much to do. Conversely, developing countries are often ill-prepared to deal with chemical safety issues. Key problems include the lack of capacity at government level to handle chemicals issues, large existing stocks of obsolete pesticides, and generally insufficient information for decision-making. Global structures such as the International Forum on Chemical Safety (IFCS) (www.who.int/ifcs) have been created to encourage a harmonized approach but the lack of capacity by many counties remains a serious obstacle.

While the combined information and scientific programs of the OECD and several UN agencies such as WHO, ILO, FAO and UNEP (http://www.who.int/ipcs/en) are bringing forward much useful information and a coordinated approach to testing and assessment procedures, there are also several programs to limit the use (and the trade) in undesirable chemicals. The Basel Convention (www.basel.int) which restricts trade in toxic waste is already well known, as is the Montreal Protocol (http://www.unep.org/ozone/Treaties_and_Ratification/2B_montreal_protocol.asp) through which the manufacture, trade, and use of ozone-depleting substances are being phased out. Following the entry into force in 2004 of the Rotterdam Convention (www.pic.int) trade in toxic chemicals is required to be more transparent. The Stockholm Convention (www.pops.int) seeks to phase out the manufacture and use of a set of listed chemicals, most of them polluting and persistent organic chemicals including a number of well-known pesticides. In order to help countries to introduce a more systematic approach to chemicals management, the Strategic Approach to International Chemicals Management (SAICM) has been launched by UNEP (http://www.chem.unep.ch/saicm). This is coming at a time when the European Union is undertaking regional consultations on a more comprehensive regulatory approach to chemicals assessment and registration, the REACH program (http://europa.eu.int/comm/enterprise/

reach). The European Union has already adopted a regional approach to hazardous industrial installations through its SEVESO II Directive *SEVESO II* (http://europa.eu.int/comm/environment/seveso/).

It has often been observed that major **industrial accidents** also affect the public, but that the civil community is rarely consulted on emergency response plans, and is generally not well prepared in case of industrial accidents or natural disasters. Natural disasters often lead to chemical contamination when they destroy pesticide depots or factories. UNEP joined in 1988 with a number of national agencies and with the chemicals industry to develop the program APELL. This approach of involving local communities in disaster planning is now becoming more common. Other sectors such as mining, transport, oil and gas are also strengthening their community outreach. The APELL process, and the programs to promote it, can be found at (www.uneptie.org/apell).

Integrated product policy (IPP) was developed as a policy instrument for member states by the European Commission. (http://europa.eu.int/comm/environment/ipp/home.htm). IPP is based on the realization that environmental impact can only be reduced if there is also an emphasis on evaluating and regulating impacts from products in addition to those from their manufacture. As many products are now manufactured offshore, control of products is often the only remaining choice. A variety of techniques is now in use to influence the environmental impact of products and their use. Eco-design (or design for environment) can reduce the impact of products or assist in their recycling. Product service systems evaluate the possibilities of substituting a service for a physical product (e.g., rental instead of purchase), while green procurement influences the purchasing policies in favor of ecologically sound products. In addition to ordinary consumers, governments and major companies have enormous potential leverage in many markets to influence the supply of better products. A variety of take-back schemes can be linked to the issues of product design (we can call this "reverse distribution"), and for some products a decision has been made simply ban their manufacture or use (e.g., CFCs, products from certain endangered species, and so on). Extended producer responsibility and supply-chain management were mentioned earlier.

A large number of initiatives have been undertaken in the **waste and pollution management** area. Much of this is concerned with introducing pollution management regulations and standards, and the introduction of treatment technologies. The World Bank Pollution Abatement Guidelines (World Bank 1998) are gradually incorporating a wider approach, while still trying to give concrete guidelines for project planners and assessors on environmental and pollution standards. Training in management of hazardous waste was first commenced at the global level by UNEP in 1992 together with the International Solid Waste Association (ISWA). This program included the publication of a series of manuals to encourage training by suitable national institutes and has a strong emphasis on waste reduction and cleaner production. The Basel Convention, through its Regional Centres also sponsors training, with strong emphasis on implementation of the convention. There are extensive publications on waste management including minimization (http://www.uneptie.org/pc/pc/waste/waste.htm) and (www.iswa.org). Waste management generally, but especially waste-water, has also become one of the priority areas of the Global Program of Action (GPA) of Land-based Sources of Marine Pollution (http://www.gpa.unep.org). Through GPA and the global Regional Seas Programs, extensive assistance has been given to coastal communities in how to manage all sorts of waste.

Financial institutions have also become more coordinated around the environmental agenda. The Financial Institutions (FI) Initiative launched by UNEP in 1992 now consists of more than 200 institutions who have signed a common environmental policy statement (www.unepfi.net). In June 2003 a series of institutions adopted the Equator Principles (www.equator-principles.com). A number of stock exchanges now require certain environmental actions before companies can be listed. Rating agencies have found a thriving business in evaluating the environmental and sustainability performance of listed companies to allow investors with environmental criteria to profile the companies in which they are considering investing. Very large investors such as pension funds are increasingly discriminating on the basis of such ratings. Two examples can be seen on (www.sustainability-index.com) and (www.sam-group.com).

In addition to the greening of individual institutions several **international environmental financing mechanisms** have been established to help developing countries meet the ever-growing need for green investment and technology transfer. The Global Environment Facility (GEF) (www.gefweb.org) provides funds for the marginal extra costs required for compliance with multilateral agreements in areas of biodiversity, climate change, ozone layer depletion, international waters and chemicals control. The Multilateral Ozone Fund was established to assist counties to comply with the requirements of the Montreal Protocol. The Kyoto Protocol was only agreed by developing countries after a financial transfer mechanism of the clean development mechanism (CDM) was added to the text (http://cdm.unfccc.int).

Much of the **global standardization of environmental management tools** has been achieved through the work of the International Organization for Standardization (ISO) (www.iso.org) based in Geneva. Tools such as ISO 14000, Environmental Audits, LCA, and others now have agreed procedures and methodologies that ensure that their results are recognized by all countries, thus replacing the earlier multitude of nontransportable national standards. (See http://www.iso.org/iso/en/iso9000-14000/iso14000/iso14000index.html.) One initiative outside the ambit of ISO is the Global Reporting Initiative (GRI) which has developed its own governance and procedures (www.globalreporting.org).

International business groups have also developed programs and actions to assist their members in improving environmental performance and compliance. The Responsible Care program developed by the U.S. chemicals industry in the 1980s and subsequently expanded to have a global reach requires companies to achieve a higher level of environmental and safety performance than is required under legislation. The International Chamber of Commerce (ICC) (www.iccwbo.org) produced a set of environmental guidelines which advocated a "beyond compliance" approach that included also waste minimization, public engagement, and improved environmental technologies. This led to collaborative work with UNEP to produce training manuals on environmental management systems and management guidance materials also for small size companies. The World Business Council for Sustainable Development (WBCSD) (www.wbcsd.org) is strongly promoting its "eco-efficiency" approach, and has also spearheaded a program to look further ahead at future products and services. In the U.S., the Global Environmental Management Initiative (GEMI) (www.gemi.org) promotes more widespread use of systematic management tools by large companies. The GEMI materials and manuals are available for use beyond its membership. Numerous industry associations and federations in sectors such as chemicals, mining, oil and gas, iron and steel, construction and manufacturing of various sorts are also conducting environmental programs. In 2002, twenty-one of these associations agreed to prepare sustainability outlook reports for their sectors, to be eventually published by UNEP for WSSD. The reports can be seen on (www.uneptie.org/outreach/wssd/contributions/sector_reports/reports.htm). The Global Compact is an industry initiative led by the Office of the Secretary General of the UN who invited business to join with the UN to adopt and promote a more sustainable approach to business activities (www.unglobalcompact.org/Portal/). Many major companies have now joined to share their experience and reach out to nonmember companies. Member companies are asked to demonstrate their respect for a series of principles covering human rights, labor rights, anti-corruption and environment.

Environmental Principles in the Global Compact

Principle 7: Businesses should support a precautionary approach to environmental challenges;
Principle 8: undertake initiatives to promote greater environmental responsibility; and
Principle 9: encourage the development and diffusion of environmentally friendly technologies.

Solutions to Nature Conservation

Much of the above discussion is necessarily focused on pollution type issues. There are many initiatives also in nature conservation. One of the best known is the World Heritage Sites initiative sponsored by UNESCO (http://whc.unesco.org). Nature conservation programs can best be accessed through UNESCO, UNEP, IUCN, WWF, and other international groups.

Conclusions

The international community is reacting to environmental issues at two levels: (a) assisting national initiatives on an individual basis using established concepts and approaches, and (b) putting in place ideas, systems, and programs that deal in a more strategic and holistic fashion with environmental stress.

Through various mechanisms of dialogue and cooperation many policies and programs have been established to provide solutions that are accessible to developing countries. All stakeholders benefit from the international standardization of techniques and tools.

The complexity of the environmental agenda requires a range of different tools and approaches. Traditional pollution control approaches continue to be needed everywhere, however they need now to be supplemented also by newer innovative solutions that are better able to deal with the underlying factors of environmental degradation.

Many of the holistic programs look at upstream action in prevention, using more integrated, life-cycle approaches that address the driving forces of environmental change rather than addressing the problems (and the wastes) after they have been created.

Several concepts and ideas such as industrial ecology are concerned with optimizing our industrial system overall, with the understanding that the individual components (e.g., production plants) will then themselves have to be addressed as a second step.

Relatively few established programs and concepts have an explicit social or ecological endpoint as a target, rather they are based on improving the functioning of the present system by reducing wastes. TNS and the function-based approach stand out as having a clear sustainability endpoint, however they have still to be operationalized on a broad scale.

A number of international programs based on the holistic concepts have already been put in place. While there has been good uptake, the impact of these programs still needs to be operationalized more widely—old ideas of end-of-pipe intervention die hard. The most visible programs to date are the initiatives favoring the installation of cleaner technologies and the more widespread use of environmental management systems in industry. While use of LCA and EIA is becoming more common, they still lack an effective international management framework for their use.

Product-cycle and consumption programs are now ready to move from the stage of concept to broader application. There is a need for stronger "champions" at the political level as these systematic life-cycle approaches require a more collective change in behavior and thinking than is necessary with established pollution control approaches.

Given the magnitude of the task ahead, new mechanisms of environmental finance are indispensable. Scarce public funds will not have the ability to fund the entire range of necessary actions. Future funds will have to come from the various activities and products that result in environmental damage in the first place.

Traditional solutions of technology, science-based standards and rigorous assessment methodologies transposed into regulations are still applicable as a platform for moving on to more management-based approaches and holistic solutions.

Further progress in developing treatment technologies can help to make these simpler, more affordable and easier to apply and operate. In the longer term the environmental technologies that will have the most impact are those that make production less wasteful and products less resource-intensive.

Finally, it should be stressed that while action on the ground is what is needed, if this proceeds outside a clear holistic conceptual framework, its impact will inevitably be limited, and in particular, be relatively expensive than if newer ideas were to underpin the actions.

Review Questions

1. The evolution of "smart regulation" has been slower than advances in technologies and in environmental management tools. It is important to build better bridges between the environmental profession and the regulators. What might some of these bridges be that bring the various "solution sectors" more into concordance?
2. The impact of consumption patterns and end-of-the-life products is gradually becoming more obvious in all societies, yet environmental solutions for this area have been weak, and are often not taking place in a coherent policy framework. How can the upstream solutions be better integrated into a holistic model for action? Where does the responsibility for action lie? Who is the "first mover"?
3. Reliable data underlying environmental action is often difficult to find. There is a tendency by some stakeholders not to collect or to dispute data that might show the problem is becoming worse, thus delaying action on urgent problems. How can global data and information systems improve this situation, and what is an appropriate framework for such global information?
4. The mainstreaming of environmental considerations into traditional business management has been slow. There are many organizational and attitudinal obstacles, yet mainstreaming it is essential if holistic solutions are to take root. How can the environmental professional assist in this process? What are some things that environmentalists can do to speed up the process?

5. Transfer of environmental solutions between countries faces many obstacles because the context and conditions in which such solutions operate vary greatly. What processes can be devised to adapt western solutions to the needs of other countries, and how can we monitor and validate the success of application?

References

Agenda 21 http://www.un.org/esa/sustdev/agenda21.htm.

Frosch Robert A., Gallopoulos, N. E. 1989. Strategies for manufacturing. *Scientific American* September 261(3):94–102.

Füssler C., Cramer A., van der Vegt S., *Raising the Bar*, Greenleaf Publishing 2004 ISBN 1874719829.

Gunningham N., Grabosky P. 1998. *Smart Regulation—designing environmental policy*. Clarendon Press.

Gunningham N., Sinclair D. 2002. *Leaders and Laggards—next generation environmental regulation*. Greenleaf Publishing.

Hawken P., Lovins A., Lovins H. 2003. *Natural Capitalism*. Rocky Mountain Institute.

Hawken, P., Lovins, A., Lovins, H. 2000. *Natural Capitalism*. Boston: Little Brown & Co.

Inami, H. 2001. Developing mechanisms for a 3R society: Focusing on reducing pollution, reusing resource and recycling wastes. NRI Papers No. 21. (Available at http://www.nri.co.jp/english/opinion/papers/2001/pdf/np200121.pdf.)

Industry and Environment, Vol.18 No.2–3, Environmental Management Tools.

Tellus Institute. 2002. *Halfway to the Future: Reflections on the Global Condition*.

UNEP. 2003. Global Environment Outlook, GEO 3.

von Weizsäcker E., Lovins A., and Lovins H. 1997. Factor Four Doubling Wealth—Halving Resource Use.

World Bank. 1998. Pollution Prevention and Abatement Handbook.

World Summit on Sustainable Development "Plan of Action" http://www.un.org/esa/sustdev/documents/WSSD_POI_PD/English/POIToc.htm.

Worldwatch Institute. 2004. State of the World—Special Focus: The Consumer Society. January.

Appendix

WSSD Plan of Implementation

III—Changing Unsustainable Consumption and Production Patterns

13. Fundamental changes in the way societies produce and consume are indispensable for achieving global sustainable development. All countries should promote sustainable consumption and production

patterns, with the developed countries taking the lead and with all countries benefiting from the process, taking into account the Rio principles, including, inter alia, the principle of common but differentiated responsibilities as set out in principle 7 of the Rio Declaration on Environment and Development. Governments, relevant international organizations, the private sector and all major groups should play an active role in changing unsustainable consumption and production patterns. This would include the actions at all levels set out below.

14. Encourage and promote the development of a 10-year framework of programs in support of regional and national initiatives to accelerate the shift towards sustainable consumption and production to promote social and economic development within the carrying capacity of ecosystems by addressing and, where appropriate, delinking economic growth and environmental degradation through improving efficiency and sustainability in the use of resources and production processes, and reducing resource degradation, pollution and waste. All countries should take action, with developed countries taking the lead, taking into account the development needs and capabilities of developing countries through mobilization, from all sources, of financial and technical assistance and capacity-building for developing countries. This would require actions at all levels to:
    (a) Identify specific activities, tools, policies, measures and monitoring and assessment mechanisms, including, where appropriate, life-cycle analysis and national indicators for measuring progress, bearing in mind that standards applied by some countries may be inappropriate and of unwarranted economic and social cost to other countries, in particular developing countries;
    (b) Adopt and implement policies and measures aimed at promoting sustainable patterns of production and consumption, applying, inter alia, the polluter-pays principle described in principle 16 of the Rio Declaration on Environment and Development;
    (c) Develop production and consumption policies to improve the products and services provided, while reducing environmental and health impacts, using, where appropriate, science-based approaches, such as life-cycle analysis;
    (d) Develop awareness-raising programs on the importance of sustainable production and consumption patterns, particularly among youth and the relevant segments in all countries, especially in developed countries, through, inter alia, education, public and consumer information, advertising and other media, taking into account local, national and regional cultural values;
    (e) Develop and adopt, where appropriate, on a voluntary basis, effective, transparent, verifiable, non-misleading and non-discriminatory consumer information tools to provide infor-

mation relating to sustainable consumption and production, including human health and safety aspects. These tools should not be used as disguised trade barriers;

(f) Increase eco-efficiency, with financial support from all sources, where mutually agreed, for capacity-building, technology transfer and exchange of technology with developing countries and countries with economies in transition, in cooperation with relevant international organizations.

15. Increase investment in cleaner production and eco-efficiency in all countries through, inter alia, incentives and support schemes and policies directed at establishing appropriate regulatory, financial and legal frameworks. This would include actions at all levels to:
    (a) Establish and support cleaner production programs and centres and more efficient production methods by providing, inter alia, incentives and capacity-building to assist enterprises, especially small and medium-sized enterprises and particularly in developing countries, in improving productivity and sustainable development;
    (b) Provide incentives for investment in cleaner production and eco-efficiency in all countries, such as state-financed loans, venture capital, technical assistance and training programs for small and medium-sized companies while avoiding trade-distorting measures inconsistent with WTO rules;
    (c) Collect and disseminate information on cost-effective examples in cleaner production, eco-efficiency and environmental management, and promote the exchange of best practices and know-how on environmentally sound technologies between public and private institutions;
    (d) Provide training programs to small and medium-sized enterprises on the use of information and communication technologies.
16. Integrate the issue of production and consumption patterns into sustainable development policies, programs and strategies, including, where applicable, into poverty reduction strategies.
17. Enhance corporate environmental and social responsibility and accountability. This would include actions at all levels to:
    (a) Encourage industry to improve social and environmental performance through voluntary initiatives, including environmental management systems, codes of conduct, certification and public reporting on environmental and social issues, taking into account such initiatives as the International Organization for Standardization (ISO) standards and Global Reporting Initiative guidelines on sustainability reporting, bearing in mind principle 11 of the Rio Declaration on Environment and Development;
    (b) Encourage dialogue between enterprises and the communities in which they operate and other stakeholders;

   (c) Encourage financial institutions to incorporate sustainable development considerations into their decision-making processes;
   (d) Develop workplace-based partnerships and programs, including training and education programs.

18. Encourage relevant authorities at all levels to take sustainable development considerations into account in decision-making, including on national and local development planning, investment in infrastructure, business development and public procurement. This would include actions at all levels to:
   (a) Provide support for the development of sustainable development strategies and programs, including in decision-making on investment in infrastructure and business development;
   (b) Continue to promote the internalization of environmental costs and the use of economic instruments, taking into account the approach that the polluter should, in principle, bear the costs of pollution, with due regard to the public interest and without distorting international trade and investment;
   (c) Promote public procurement policies that encourage development and diffusion of environmentally sound goods and services;
   (d) Provide capacity-building and training to assist relevant authorities with regard to the implementation of the initiatives listed in the present paragraph;
   (e) Use environmental impact assessment procedures.

(For entire document, see www.johannesburgsummit.org/html/documents/summit_docs/2309_planfinal.htm.)

Part IV

Summary

CHAPTER 15

The Ultimate Goal of Zero Pollution

Franklin J. Agardy

Introduction

In 1971 Paul Ehrlich wrote a book titled, *The Population Bomb* He expressed the opinion that, "Doubling the population normally much more than doubles environmental deterioration." We have before us in this book a very comprehensive and detailed summary of environmental solutions addressing a broad spectrum of global pollution problems. However, to those of us familiar with environmental solutions, and indeed this book does contain a combined knowledge easily measured in hundreds of years of experience, there is a decided lack of addressing the overall impact of population and population growth on our ability to approach zero pollution. Recognizing that world population now exceeds 6 billion with an additional 80 million added each year, it should come as no surprise that more than 1 billion people lack safe drinking water and that more than 2 billion people lack adequate sanitation. Furthermore, about 47% of the global community now reside in urban areas, thus concentrating many negative effects that result in greater and more concentrated pollution.

Therein lies the conflict; on the one hand we recognize that population drives pollution, on the other hand we attempt to define and come to grips with sustainable development. The real question then is can the implementation of advanced technology achieve the ultimate goal of zero pollution globally, or be limited only to the most advanced and wealthy of the world's nations?

As we march through this chapter, I have taken the liberty of either quoting or paraphrasing the words of other chapter contributors. I leave it to the reader to cross-reference the sources should the reader so desire.

Sustainability

While global population continues to increase there appears to be little concerted governmental efforts to control or reduce birth rates or close an almost open door policy regarding transmigration. Regardless of one's position on population control, it is inconceivable that population control policies would ever be proposed or implemented in the foreseeable future. Unfortunately population control appears to be in the hands of infectious diseases, uneven food distribution, limited or polluted water resources, nature, and war (including genocide). There are exceptions to this in that many of the more industrialized nations are exhibiting patterns of population decline (when one factors out immigration). On a global scale, population continues to increase. When this growth is coupled with an almost universal migration to urban population centers, pressures on the environment take an even greater toll on one's ability to approach the goal of zero pollution.

Increasing population accelerates consumption, thus it is inevitable that pollution will also increase. Furthermore, pollution does not simply keep pace with population but tends to expand at a greater rate than the population. At a time when increased environmental pollution controls are needed, there are escalating economic pressures which mitigate against implementing more comprehensive pollution control measures, and, in some cases, even result in the relaxation of existing pollution programs and requirements. Interestingly, as pointed out by several chapter contributors, the demand for manufactured goods of a more complex nature (e.g., electronics, fertilizers, pharmaceuticals, etc.) places further pressure on the environment. All of this adds an increasing chemical assault upon the human habitat, in addition to the biological (disease) assaults, both occurring without a viable means of protecting human health on a truly global scale.

World Population, Economics, and Environmental Insults

World Population

According to the Internal Energy Annual 2002 report, the world population increase, as measured by region, can be summarized as follows:

Region	***Population (Millions)***		***Percent***
	1980	*2002*	***Increase***
North America	319.14	412.00	+29.1
Central & South America	290.78	421.32	44.9
Western Europe	433.18	481.09	11.1
Eastern Europe	360.87	387.31	7.3
Middle East	91.76	170.67	86.0
Africa	464.81	795.49	71.1
Asia & Oceania	2,454.69	3,442.15	40.2
World Total	**4,415.20**	**6,090.00**	**37.9**

Clearly, a major impediment to achieving zero pollution in many parts of the world is the rapid increase in population without the necessary increase in economic growth.

World Economics

The Wikipedia Free Encyclopedia (www.wikipedia.org) has posted economic data in terms of gross domestic product (GDP) based on the U.S. dollar as of 2003, as follows (selected nations for illustrative purposes).

Country	*2003 GDP (nominal) Percent of World*	
	Millions of US$	*GDP*
United States	10,881,609	29.9
Japan	4,326,444	11.9
Germany	2,400,665	6.6
United Kingdom	1,794,868	4.9
France	1,747,973	4.8
Italy	1,465,895	4.0
China	1,409,852	3.9
Sub Total (the big 7)	**24,027,306**	**66.1**
The Next 18	**7,710,396**	**21.2**
The Remaining 158	**4,618,538**	**12.7**

Reviewing the above, one comes to the conclusion that when comparing population against gross domestic product, the ability of most nations to be able to adequately address pollution problems is severely limited. As an example, the "big 7" have a combined population (year 2000) of 1,575,300,000 and control 66.1% of the global GDP. Thus, 66.1% of the GDP is vested in nations containing 25.8% of the population. Conversely, the remaining 74.2% of the world population is limited to controlling only 33.9% of the GDP. Having said that, it should also be noted that other differences between developing nations and industrialized nations severely impact the environment and therefore limit the achievement of environmental objectives. These include infrastructure (lack thereof), population density, fresh water, labor (training, high unemployment), economy, social issues (education, health, corruption), lack of or slow (to function) legal systems, and lax law enforcement.

Environmental Insults

Environmental insults have been addressed by most of the chapter contributors but there is value in singling out chemicals and their impact on the environment, as this appears to be of major concern to the most industrialized nations today. Excellent data is available with regard to production within the United States and so this can serve as a model of what is occurring and will occur worldwide in the coming years.

Chemical Production

A significant portion of the U.S. GDP is represented by the chemical industry. Over the past 50 years, the chemical industrial has grown by 1,261%. This growth is compared to population and GDP growth shown above. These data clearly show that the pollution potential rate from the production and use of chemicals is significantly greater than population growth.

In addition to the growth in chemical production, the toxicity of the chemicals being produced is equally important. In this regard, the most significant category is the production of synthetic organic chemicals. These are the chemicals that are not found in nature, although, in the broad sense, even the production of naturally occurring chemicals, fall into the category of synthetic chemicals. As a consequence, they are usually more difficult to assimilate into the environment. This is particularly true for those compounds that have an inherent resistance to biodegradation, such as halogenated organic compounds,[1] which tend to be both toxic and persistent once introduced into the environment.

The results of a study reported in *Chemical and Engineering News* show an increase in organic chemical production of some 28% over the 10-year period from 1987 to 1996. Other sources estimate that the production of synthetic organic chemicals is increasing by a factor of 10 every 35 years. Although the United States produces tremendous quantities of these chemicals, not all of these chemicals remain in domestic use, as the United States is a major world supplier. However, although many chemicals are exported, much of the pollution associated with their domestic production remains in the vicinity of the manufacturing facilities. For example, it has been estimated that over seven billion pounds of toxic chemicals are released into the environment each year and of this amount over 115 million pounds are recognized as being carcinogenic.

Pharmaceuticals

There are approximately 1,200 corporations in the United States that produce what would appear to be environmentally benign pharmaceuticals, which are presumed to be receptor specific. Although pharmaceuticals have limited access to water resources (i.e., because of their apparent point of application to either humans or animals), this does not necessarily mean that they are not a significant source of pollution. First of all, the production of pharmaceuticals produces wastes which are discharged to the environment at the manufacturing locations (i.e., as a point source). This is no different from the production of any chemical, except that pharmaceuticals are not regulated as toxic compounds. Pharmaceuticals also occur in human

1. These are organic compounds that contain halogens. The common halogens are chlorine, bromine, fluorine, and iodine.

wastes that are received at a publicly owned treatment works (POTW). Wastewater from the POTW, which may still contain pharmaceuticals, is discharged to a body of water (i.e., another point source).

In addition to point sources of pollution, pharmaceuticals also occur as non-point sources of pollution. When antibiotics are administered to livestock, feedlot wastes containing these antibiotics are discharged as non-point pollution in surface water runoff. Even the U.S. Geological Service has reported the presence of antibiotics in surface and ground waters in the vicinity of animal feeding operations. Is it no wonder that Sweden banned all nontherapeutic use of antibiotics in agriculture in 1986 and that the World Health Organization recommended ending the use of all antibiotics (used in human medicine) in animal feed in a 1997 report.

As a consequence of both point source and nonpoint sources of pollution, pharmaceuticals have a very wide distribution throughout the environment. Once these pharmaceuticals enter the environment, it is not unrealistic to assume that what is one person's health improving drug can become another person's poison.

Pesticides

Pesticides by their very nature are meant to be toxic and therefore are a hazard to the environment (a poison by any other name is still a poison). In the United States, pesticides are used on 900,000 farms and reportedly found in 70 million households. Pesticide usage in the United States stood at 647 million pounds in 1964. Overall usage peaked in 1979 at 1,144 million pounds and seems to have leveled to 950 million pounds. However, the past accumulation of pesticides in the environment, continues to be released into water resources today. Therefore, past practices (the use of pesticides which may be currently banned and out of production[2]) and continued use of pesticides tend to provide an increasing source of pollution to water resources.

Because pesticides are toxic, it might be expected that the government would prohibit the occurrence of any pesticide in drinking water. Sadly, the government allows small amounts of selected pesticides in drinking water while the vast majority of pesticides that are regulated in food go unregulated in drinking water.[3]

Pesticide manufacturing facilities and POTWs do discharge pesticide polluted wastewater to the environment (i.e., as a point source of pollution). However, this is not the major source of pesticide contamination in the

2. For example, 1,1-(2,2,2-Trichoroethylidene)bis(4-chlorobenzene), or DDT, was banned over 30 years ago but still pollutes surface and groundwaters today.
3. This is an excellent example of the insanity of standard based policies that allow the consumption of different toxic chemicals, at low levels, in drinking water and food.

United States. Most pollution of surface water and ground water resources results from pesticide applications to agricultural lands (i.e., a nonpoint source of pollution). As a consequence, pesticides leach to groundwater, are washed off the land by precipitation and irrigation into surface water, or are transported long distances by the wind. As long as pesticides are used in agriculture, they will always be a source of drinking water pollution.

Given the obvious hazards of pesticide use, it is reasonable that the government could ban all pesticide use. In fact, there are very good arguments for, at least, the ban of all chlorinated pesticides (i.e., organic compounds containing chlorine). Such a ban, however, could have a significant effect on food production. As a result, an immediate and total ban on pesticides, even chlorinated pesticides, is not very probable. More likely than not, the EPA will just continue to ban specific pesticides on a case by case basis. Given this scenario, the elimination of pesticide use in the United States is highly unlikely. Thus, it is also equally infeasible to keep pesticides out of the environment.

Chemical Dependence

In the United States, population growth begets economic growth which begets chemical industry growth which begets pollution. Modern society is dependent on chemicals and their products. This dependence manifests itself globally today. This dependence will continue to cause an increase in the magnitude and diversity of pollution that is beyond current government policies to control; regardless of the national wealth, control is a matter of degree.

To put this in further perspective, in 2001, the American Chemical Society registry included over 30 million chemical substances. One can only wonder at the number of new chemical compounds our advanced science and technology will produce over the next 50 years, particularly when we are presently developing approximately 2,000 new chemicals each year. This is a very serious condition, particularly since the vast majority of the existing chemical compounds remain untested, unregulated and untreated. In 1997, the United Nations concluded that:

> From a global perspective the environment has continued to degrade during the past decade, and significant environmental problems remain deeply embedded in the socio-economic fabric of nations in all regions.

Our demand for chemicals is embedded in the socioeconomic fabric of the United States as well as most other countries. It is hard to conceive of a future condition that will reverse this dependence. Thus, governmental policies that were established to protect the purity of drinking water are in a never-ending race to realistically keep pace with pollution.

The March Toward Zero Pollution

As one progresses thorough the text, chapter by chapter, a pattern develops wherein one can, under the proper circumstances, approach, if not totally reach the goal of zero pollution. Having addressed the issue of population, we now turn to solutions. It is obvious from the text that although there are many proposed and demonstrated approaches by which to achieve zero pollution, there are several necessary platforms or foundations upon which technical solutions rest. These include, but are not limited to: (1) education, (2) economics, (3) legislation, and (4) regulation. An added dimension is the fact that today pollution does not respect frontiers, and emissions released in one county often impact its neighbors or, in some cases the whole planet.

The tools described in this text address the wide spectrum of proposed and proven approaches available today by which to deal with pollution. We certainly recognize that zero pollution can only be achieved by addressing all aspects of resource conservation coupled with reduction, to the maximum, of residuals even before treatment technologies are employed. Examples such as extensive recycling of materials, environmentally balanced industrial complexes, and alternate energy sources, all help to reduce the need to treat remaining residuals. Treatment technologies available today include optimized biological systems, and advanced physical and chemical treatment, which, when applied in sequence, can, and has been demonstrated to result in zero pollution.

Summary

As one contributing author put it, there is no free lunch when it comes to waste treatment. This is particularly true with developing countries. Based on the population dynamic/GDP relationships, it appears that only a fraction of today's global community has the means to achieve zero pollution. However, the fact that technology exists to achieve this goal, although currently limited by economics, should not lead one to the conclusion that on a global scale we are akin to Sisyphus and that our task is Sisyphean. Each step taken to control population, to reduce the pressure towards urbanization (reduction of mass), to educate the population, to reduce the stress on natural resources, to maximize the reuse of residuals and implement levels of technology, reduces the overall burden on the planet, thus making our Sisyphean task less hopeless—there is light at the end of the tunnel.

Recommended Resources

American Chemical Society, CAS Registry, March 31, 2001.

Ehrlich, P. R. 1971. *The Population Bomb*. Cutchogue: Buccaneer Books, Inc.

Executive Office of the President, Counsel on Environmental Quality, 17th Annual Report, 1986.

Environmental Defense, Environmental Defense Fund, New York, 2001.

Karliner, J. 1997. *The Corporate Planet*. San Francisco: Sierra Club Books.

Meyer, M. T., et al. 1999. Occurrence of antibiotics in liquid waste at confined animal feeding operations and in surface and ground water. U.S. Geological Survey, Dec. 21.

Sullivan, P., F. J. Agardy, J. J. Clarke. 2005. *The Environmental Science of Drinking Water*. Amsterdam/New York: Elsevier.

Thornton, J. 2000. *Pandora's Poison, Chlorine, Health, and a New Environmental Strategy.* Cambridge, MA: MIT Press.

Index